MICROBIOLOGY FOR SUSTAINABLE AGRICULTURE, SOIL HEALTH, AND ENVIRONMENTAL PROTECTION

MICROBIOLOGY FOR SUSTAINABLE AGRICULTURE, SOIL HEALTH, AND ENVIRONMENTAL PROTECTION

Edited by

Deepak Kumar Verma

Apple Academic Press Inc.	Apple Academic Press Inc.
3333 Mistwell Crescent	1265 Goldenrod Circle NE
Oakville, ON L6L 0A2	Palm Bay, Florida 32905
Canada USA	USA

© 2019 by Apple Academic Press, Inc.

First issued in paperback 2021

Exclusive worldwide distribution by CRC Press, a member of Taylor & Francis Group

No claim to original U.S. Government works

ISBN-13: 978-1-77463-535-3 (pbk)

ISBN-13: 978-1-77188-669-7 (hbk)

Library and Archives Canada Cataloguing in Publication

Microbiology for sustainable agriculture, soil health, and environmental protection / edited by Deepak Kumar Verma.

Includes bibliographical references and index.
Issued in print and electronic formats.
ISBN 978-1-77188-669-7 (hardcover).--ISBN 978-1-351-24706-1 (PDF)

1. Soil microbiology. 2. Sustainable agriculture. 3. Environmental protection.
I. Verma, Deepak Kumar, 1986-, editor

| QR111.M53 2018 | 579'.1757 | C2018-906784-5 | C2018-906785-3 |

Library of Congress Cataloging-in-Publication Data

Names: Verma, Deepak Kumar, 1986- editor.

Title: Microbiology for sustainable agriculture, soil health, and environmental protection / [edited by] Deepak Kumar Verma.

Description: Toronto; New Jersey : Apple Academic Press, 2019. | Includes bibliographical references and index.

Identifiers: LCCN 2018057885 (print) | LCCN 2018058641 (ebook) | ISBN 9781351247061 (ebook) | ISBN 9781771886697 (hardcover : alk. paper)

Subjects: LCSH: Agricultural microbiology. | Sustainable agriculture. | Environmental engineering.

Classification: LCC QR51 (ebook) | LCC QR51 .M454 2019 (print) | DDC 630.276--dc23

LC record available at https://lccn.loc.gov/2018057885

Apple Academic Press also publishes its books in a variety of electronic formats. Some content that appears in print may not be available in electronic format. For information about Apple Academic Press products, visit our website at **www.appleacademicpress.com** and the CRC Press website at **www.crcpress.com**

CONTENTS

ABOUT THE EDITOR

 **Deepak Kumar Verma** is an agricultural science professional and is currently a PhD Research Scholar in the specialization of food processing engineering in the Agricultural and Food Engineering Department at the Indian Institute of Technology, Kharagpur (WB), India. In 2012, he received a DST-INSPIRE Fellowship for PhD study from the Department of Science & Technology (DST), Ministry of Science and Technology, Government of India. Mr. Verma is currently working on the research project "Isolation and Characterization of Aroma Volatile and Flavoring Compounds from Aromatic and Non-Aromatic Rice Cultivars of India." His previous research work included "Physico-Chemical and Cooking Characteristics of Azad Basmati (CSAR 839-3): A Newly Evolved Variety of Basmati Rice (*Oryza sativa* L.)." He earned his BSc degree in agricultural science from the Faculty of Agriculture at Gorakhpur University, Gorakhpur, and his MSc in Agricultural Biochemistry in 2011. He also received an award from the Department of Agricultural Biochemistry, Chandra Shekhar Azad University of Agriculture and Technology, Kanpur, India. Apart from his area of specialization in plant biochemistry, he has also built a sound background in *plant physiology, microbiology, plant pathology, genetics and plant breeding, plant biotechnology and genetic engineering, seed science and technology,* and *food science and technology*. In addition, he is member of several professional bodies, and his activities and accomplishments include conferences, seminars, workshops, training, and the publication of research articles, books, and book chapters.

CONTRIBUTORS

Bavita Asthir
Senior Biochemist-cum-Head, Department of Biochemistry, College of Basic Sciences and Humanities, Punjab Agriculture University, Ludhiana 141004, Punjab, India

Sunil Banskar
Microbial Culture Collection, National Centre for Cell Science, Savitribai Phule University of Pune Campus, Ganeshkhind, Pune 411007, Maharashtra, India

Anil Kumar Dwivedi
Associate Professor, Pollution and Environmental Assay Research Laboratory (PEARL), Department of Botany, Deen Dayal Upadhyay Gorakhpur University, Gorakhpur 273009, Uttar Pradesh, India

Huey-Min Hwang
Professor of Biology, Department of Biology, Jackson State University, Jackson, MS 39217, USA

Arpana H. Jobanputra
Assistant Professor, Department of Microbiology, PSGVPM's, SIP Arts, GBP Science & STSKVS Commerce College, District Nandurbar, Shahada 425409, Maharashtra, India

Madhu Kamle
Department of Forestry, North Eastern Regional Institute of Science and Technology, Nirjuli-791109, Arunachal Pradesh, India

Balraj Kaur
Department of Biochemistry, College of Basic Sciences and Humanities, Punjab Agriculture University, Ludhiana 141004, Punjab, India

Adesh Kumar
Department of Plant Pathology, Punjab Agriculture University, Ludhiana 141004, Punjab, India

Ajay Kumar
Lecturer, Plant Protection, Department of Plant Protection, Chaudhary Charan Singh University, Meerut 200005, Uttar Pradesh, India

Nirdesh Kumar
M.Sc. Agriculture (Plant Protection) Student, Department of Plant Protection, Chaudhary Charan Singh University, Meerut 200005, Uttar Pradesh, India

Pradeep Kumar
Department of Forestry, North Eastern Regional Institute of Science and Technology, Nirjuli-791109, Arunachal Pradesh, India

Satish Kumar
Scientist, ICAR-National Institute of Abiotic Stress Management, Baramati, Pune 411007, Maharashtra, India

Vipul Kumar
Assistant Professor (Plant Protection), School of Agriculture, Lovely Professional University,
Phagwara 144411, Punjab, India

Ezhil Malar S.
Assistant Professor, Department of Biotechnology, Valliammal College for Women,
Chennai 600102, Tamil Nadu, India

Pawan Kumar Maurya
Department of Biochemistry, Central University of Haryana, Jant-Pali,
Mahendergarh (Haryana)-123031, India

Balaram Mohapatra
Department of Biotechnology, Indian Institute of Technology, Kharagpur 721302, West Bengal, India

Abhay K. Pandey
Research Associate, PHM-Division, National Institute of Plant Health Management,
Ministry of Agriculture and Farmers Welfare, Government of India, Rajendranagar 500030
Hyderabad, Telangana, India

Manjula P. Patil
Department of Microbiology, PSGVPM's, SIP Arts, GBP Science & STSKVS Commerce College,
District Nandurbar, Shahada 425409, Maharashtra, India

Dhiraj Paul
DST Young Scientist, Microbial Culture Collection, National Centre for Cell Science,
Savitribai Phule University of Pune Campus, Ganeshkhind, Pune 411007, Maharashtra, India

Reshma Prakash
M.Sc. Agriculture (Mycology and Plant Pathology) Student, Department of Mycology
and Plant Pathology, Institute of Agriculture Science, Banaras Hindu University,
Varanasi 221005, Uttar Pradesh, India

Roomi Rawal
Department of Entomology, Chaudhary Charan Singh Haryana Agricultural University,
Hisar 125004, Haryana, India

Mamta Sahu
Assistant Professor, Department of Biotechnology and Microbiology, Saaii College of
Medical Science & Technology, Kanpur 209203, Uttar Pradesh, India

Yogesh S. Shouche
Scientist G, Microbial Culture Collection, National Centre for Cell Science, Savitribai Phule
University of Pune Campus, Ganeshkhind, Pune 411007, Maharashtra, India

Ravi Kant Singh
Amity Institute of Biotechnology, Amity University Chhattisgarh, Raipur, C.G.-493225, India

Shikha Srivastava
Department of Botany, Deen Dayal Upadhyay Gorakhpur University, Gorakhpur 273009,
Uttar Pradesh, India

Diganggana Talukdar
Teaching Associate, Plant Pathology and Microbiology, College of Horticulture Under College of Agricultural Engineering and Post-harvest Technology, Central Agricultural University, Ranipool 737135, East Sikkim, India

Deepak Kumar Verma
Agricultural and Food Engineering Department, Indian Institute of Technology, Kharagpur 721302, West Bengal, India

Roni Yulianto
Department of Grassland Ecology, Development of Technology Science, International Development Education and Cooperation (IDEC), Hiroshima University, Hiroshima, Japan

Ali Tan Kee Zuan
Senior Lecturer, Department of Land Management, Universiti Putra Malaysia, 43400 UPM Serdang, Selangor Darul Ehsan, Malaysia

ABBREVIATIONS

ACC	1-aminocyclopropane-1-carboxylate
ACCD	1-aminocyclopropane-1-carboxylate deaminase
ACCO	1-aminocyclopropane-1-carboxylic acid oxidase
AChE	acetylcholinesterase
ADP-ribose	adenosine diphosphate ribose
AMO	ammonia monooxygenase
Anf	alternative nitrogen fixation
AOA	ammonia-oxidizing archaea
AOB	ammonia-oxidizing bacteria
ATP	adenosine triphosphate
B-rAs	biorational approaches
BCAs	biocontrol agents
BDP	Bengal Delta Plain
BGA	blue green algae
BOD	biological oxygen demand
bp	base pair
Bt	*Bacillus thuringiensis*
C-iSM	culture-independent shotgun metagenomics
CAHs	chlorinated aliphatic hydrocarbons
CFB	Cytophaga-Flavobacterium-Bacteroides
CO	carbon monoxide
CO_2	carbon dioxide
COD	chemical oxygen demand
CTAB	cetrimidetetradecyltrimethyl ammonium bromide
$CuSO_4$	copper sulfate
DAPG	2,4-diacetylphloroglucinol
DBPs	disinfection by-products
DDT	dichloro-diphenyl-trichloroethane
DGGE	denaturing gradient gel electrophoresis
DNA	deoxyribonucleic acid
DO_2	dissolved oxygen
DOC	dissolved organic carbon
DRAT–DRAG	dinitrogenase reductase ADP-ribosyltransferase–dinitrogenase reductase-activating glycohydrolase
DSF	diffusible signal factor
EBPR	enhanced biological phosphorus removal
ePGPR	extracellular plant growth–promoting rhizobacteria
EPS	exo-polysaccharides
ET	ethylene
FACS	fluorescence activated cell sorting

FAO	Food and Agricultural Organization
FIAM	free ion activity model
FISH	fluorescence in situ hybridization
FW	Fusarium wilt
GABA	gamma-aminobutyric acid
H_2SO_4	sulfuric acid
HA	harzianic acid
HAO	hydroxylamine oxidoreductase
HCl	hydrochloric acid
HCN	hydrogen cyanide
HMWCs	high molecular weight compounds
HYVs	high-yielding varieties
IAA	indole-3-acetic acid
IARI	Indian Agricultural Research Institute
iPGPR	intracellular plant growth–promoting rhizobacteria
IR	induced resistance
ISR	induced systemic resistance
ITS	internal transcribed spacer
JA	jasmonate
K_2SO_4	potassium sulfate
LMWCs	low-molecular-weight compounds
LPS	lipopolysaccharides
MAs	microbial agents
MCLGs	maximum contaminant level goals
MLSA	multilocus sequence analysis
Mo	molybdenum
MoFe	protein molybdenum–iron protein
N_2	nitrogen
NaOH	sodium hydroxide
NGS	next-generation sequencing
NH_3	ammonia
NH_4^+	ammonium ion
Nif	nitrogen fixation
NO_2^-	nitrite
NO_2	nitrogen dioxide
NO_3^-	nitrate
NOB	nitrite-oxidizing bacteria
NPR	nodule-promoting rhizobacteria
NXR	nitrite oxidoreductase
ORP	oxidation–reduction potential
PAHs	polycyclic aromatic hydrocarbons
PAL	phenylalanine ammonia-lyase
PAOs	polyphosphate-accumulating organisms
PCA	phenazine-1-carboxylic acid
PCBs	polychlorinated biphenyls
PCR	polymerase chain reaction
PDM	plant disease management

PGPB	plant growth–promoting bacteria
PGPR	plant growth–promoting rhizobacteria
PGPR–PSB	plant growth–promoting rhizobacteria–phosphate solubilizing bacteria
PGRs	plant growth regulators
PHPR	plant health–promoting rhizobacteria
POX	peroxidase
PPO	polyphenol oxidase
PR	proteins pathogenesis-related protein
PSB	phosphate solubilizing bacteria
PSMs	phosphate-solubilizing microorganisms
qPCR	quantitative PCR
QS	quorum sensing
RAPD	random amplified polymorphic DNA
rDNA	ribosomal DNA
RES-PCR	repetitive element sequence-based PCR
RFLP	restriction fragment length polymorphism
rRNA	ribosomal RNA
RT-PCR	reverse-transcriptase PCR
RTV	rice tungro virus
SAR	systemic acquired resistance
SIP	single isotope probing
SO$_2$	sulfur dioxide
SS	suspended solids
T1SS	type 1 secretion system
T2SS	type 2 secretion system
T3SS	type 3 secretion system
T4SS	type 4 secretion system
T5SS	type 5 secretion system
T-DNA	transfer DNA
T-IIA	tanshinone IIA
T-RFLP	terminal-restriction fragment length polymorphism
THM	trihalomethanes
TSM	*Trichoderma* selective media
UNs	United Nations
USA	United States of America
USDAAPHIS	US Department of Agriculture's Animal and Plant Health Inspection Service
USEPA	United State Environmental Protection Agency
USGS	United States Geological Survey
VFe protein	vanadium–iron protein
Vnf	vanadium nitrogen fixation
VOCs	volatile organic compounds
VSCs	volatile sulfide compounds
WHO	World Health Organization
WWTPs	wastewater treatment plants

PREFACE

Microbiology is the study of microscopic living organisms, which are recognized for their ubiquitous presence, diverse metabolic activity, and unique survival strategies under extreme conditions. These microscopic organisms include fungi, bacteria, viruses (infectious agents at the borderline of life) algae and protozoa. These microorganisms surrounding our environments with their diversity and abundance in universe are poorly explored. In microbiology, we are concerned with the organisms' form, structure, classification, reproduction, physiology, and metabolism. It includes the study of (1) their distribution in nature, (2) their relationship to each other and other living organisms, (3) their effects on human beings and on other animals and plants, (4) their abilities to make physical and chemical changes in our environment, and (5) their reactions to physical and chemical agents. Thus, microbiology, being a part of science, has given birth to the several branches, namely, mycology, bacteriology, virology, protozoology, parasitology, phycology or algology, microbial morphology, microbial physiology, microbial taxonomy, microbial genetics, molecular biology, microbial ecology, food microbiology, dairy microbiology, aquatic microbiology, industrial microbiology, etc., each being pursued as a specialty in itself.

Agricultural microbiology is one among them; it is concerned with the relationships between microbes and agricultural crops, with an emphasis on improving yields and combating plant diseases. This branch presents as synthetic research field that is responsible for the transfer of knowledge from general microbiology and microbial ecology to agricultural biotechnology. Agricultural microbiology has a major goal as comprehensive analysis of symbiotic microorganisms, namely, bacteria and fungi, interacting with agriculture. In agriculture, the entire food and agricultural process is permeated by these organisms. While the most visible and important role of agriculture is probably producing and delivering food, microbiology is critical to other agricultural sectors as well, for example, for bioremediation of agricultural wastes and for production of energy. Some microorganisms are an integral part of successful food production, whereas others are a constant source of trouble for agricultural endeavors. Microbial influences on food and agriculture have produced both advancements and disasters that have punctuated human history. Some examples of microbe-driven outcomes set the stage for describing how important it is to seize research opportunities in food and agriculture microbiology.

In agricultural education and research, the study of microbiology has undergone tremendous changes in the past few decades in all of the abovementioned areas. This covers all human endeavors in the broad sense, such as the transmission, absorption, and acquisition of knowledge for the better means of understanding the processes that lead to scientific farming, and as we all know, agriculture is a backbone of economy all over the globe. This book is a general consensus on the need for a comprehensive treatment of recent advances and innovations in microbiology.

The book, *Microbiology for Sustainable Agriculture, Soil Health, and Environmental Protection*, is divided into in four main parts. *Part 1: Microbiology for Sustainable Agriculture and Improved Production* contains two chapters devoted entirely to plant growth–promoting rhizobacteria (PGPR), which describe the recent trends and research toward the development of sustainable agriculture and its beneficial application to improved crop production.

Part 2: Microbiology for Crop Disease Management and Pathogenic Control in Sustainable Environment consists of three different chapters, of which one is based on disease management of agricultural and horticultural crop plants through microbial control and also elaborates how the microbial control may be a potential solution in the sustainable environment and agriculture. The second chapter is devoted to canker diseases, which are known as important diseases of citrus; it also describes their impact on worldwide production and the methodology to reduce the losses and to improve citrus production, whereas the third chapter explores the pathogenic control of various crop disease through *Trichoderma* spp. employed for its potential application in disease management to improve sustainable agricultural production.

Part 3: Microbiology for Soil Health and Crop Productivity Improvement contains one chapter, which describes the activity and mechanism of nitrogenase enzyme in soil. This enzyme is well known for its role in nitrogen fixation, and this nitrogen fixation is very important for soil health, which is necessary for improvement of crop production and productivity.

Part 4: Microbiology for Environmental Security and Pollution Control contains two chapters entirely devoted to environmental pollution and its control. One chapter describes the interaction of microbes in aqueous environments and its effects, whereas another chapter is based on an eco-friendly approach, that is, pesticides bioremediation. This approach is very important in exploration of a clean environment.

With contributions from a broad range of leading professors and scientists, this book focuses on areas of agricultural microbiology as we discussed above and will provide guidance to students, instructors, and researchers of agriculture. It covers the topics expected by students, instructors, and researchers taking agricultural microbiology courses as major. In addition, agricultural science professionals who are seeking recent advanced and innovative knowledge in agricultural microbiology will find this book helpful. It is envisaged that this book will also serve as a reference source for individuals engaged in research, processing, and product development in scientific areas of agricultural microbiology.

With great pleasure, I would like to extend my sincere thanks to all the learned contributors for the magnificent work they did their timely responses, their excellent devoted contributions to detail, and accuracy of information, and their consistent support and cooperation, which has made my task as an editor a pleasure.

I feel that I covered most of the topics expected for agricultural microbiology in this text. It is hoped that this edition will stimulate discussion and generate helpful comments to improve upon future editions. Efforts are made to cross reference the chapters as such.

Finally, I acknowledge Almighty God, who provided all the inspirations, insights, positive thoughts, and channels to complete this book project.

Deepak Kumar Verma
Editor

PART 1

Microbiology for Sustainable Agriculture and Improved Production

PLANT GROWTH-PROMOTING RHIZOBACTERIA: AN ECO-FRIENDLY APPROACH FOR SUSTAINABLE AGRICULTURE AND IMPROVED CROP PRODUCTION

DEEPAK KUMAR VERMA[1*], ABHAY K. PANDEY[2*],
BALARAM MOHAPATRA[3*], SHIKHA SRIVASTAVA[4], VIPUL KUMAR[5],
DIGANGGANA TALUKDAR[6], RONI YULIANTO[7], ALI TAN KEE ZUAN[8*],
ARPANA H. JOBANPUTRA[9], and BAVITA ASTHIR[10]

[1]Agricultural and Food Engineering Department, Indian Institute of Technology, Kharagpur 721302, West Bengal, India

[2]PHM-Division, National Institute of Plant Health Management, Ministry of Agriculture and Farmers Welfare, Government of India, Rajendranagar 500030 Hyderabad, Telangana, India

[3]Department of Biotechnology, Indian Institute of Technology, Kharagpur 721302, West Bengal, India

[4]Department of Botany, Deen Dayal Upadhyay Gorakhpur University, Gorakhpur 273009, Uttar Pradesh, India

[5]Department of Plant Protection, School of Agriculture, Lovely Professional University, Phagwara 144411, Punjab, India

[6]Department of Plant Pathology and Microbiology, College of Horticulture Under College of Agricultural Engineering and Post-harvest Technology, Central Agricultural University, Ranipool 737135, East Sikkim, India

[7]Department of Grassland Ecology, Development of Technology Science, International Development Education and Cooperation (IDEC), Hiroshima University, Hiroshima, Japan

[8]Department of Land Management, Universiti Putra Malaysia, 43400 UPM Serdang, Selangor Darul Ehsan, Malaysia

[9]Department of Microbiology, PSGVPM's, SIP Arts, GBP Science & STSKVS Commerce College, Shahada 425409, District Nandurbar, Maharashtra, India

[10]Department of Biochemistry, College of Basic Sciences and Humanities, Punjab Agriculture University, Ludhiana 141004, Punjab, India

*Corresponding authors. E-mail: deepak.verma@agfe.iitkgp.ernet.in/ rajadkv@rediffmail.com; abhaykumarpandey.ku@gmail.com; balarammohapatra09@gmail.com; tkz@upm.edu.my

1.1 INTRODUCTION

Since ancient time, microbial utilization for plant improvements has been carried out (Bhattacharyya and Jha, 2012). The application of microbes to improve the plant nutrients availability is a vital practice and needed for agriculture ecosystem (Freitas et al., 2007). In the recent years, the use of plant growth–promoting rhizobacteria (PGPR) for sustainable agriculture has increased globally since it has the abilities to release beneficial phytohormones (Table 1.1 and Fig. 1.1) to promote crop physiological growth and development. In addition to producing chemical substances, PGPRs are also known for their other important functional mechanisms in crop plants (Table 1.2). Foundation of plant growth–promoting bacteria application in agriculture has been started when Theophrastus (372–287 BC) recommended mixing of different soil samples to eliminate defects by which that soil life could sustain longer (Tisdale and Nelson, 1975). Certainly, after the invention of microscope, the technical approach behind it became clear. PGPRs are root bacteria colonizing plant's rhizosphere that could improve plant growth through various mechanisms such as solubilization of phosphate, quorum sensing, nitrogen fixation, etc. (Bhattacharyya and Jha, 2012). PGPRs are bacterial species that are associated with the plant rhizosphere and have beneficial effect on plant's growth and crop yield. The "PGPR" term was suggested by Kloepper and Schroth (1978) to assemble rhizospheric bacteria that have multiple beneficial effects on plant growth in which they act together with the plant host and enhance their nutrient uptake by different mode (Fig. 1.2) (Vessey, 2003). Such effects in various plants are increased in vigor, nutrients uptake, biomass, yield, early seedling emergence, enhancement of root proliferation, and so on (Kloepper, 1993; Tan et al., 2014). PGPR has a high potential to be used as a substitute of chemical fertilizers; thus, the market demand is high. PGPR is also used as biocontrol agents, and the ineffectiveness of PGPR in the field have often been accredited to their inability to colonize plant roots (Bloemberg and Lugtenberg, 2001). PGPRs are also referred as nodule-promoting rhizobacteria or plant health–promoting rhizobacteria and are associated with the rhizosphere which is a key soil ecological environment for plant–microbe interactions (Hayat et al., 2010). The PGPR includes both endophytic and free-living bacteria in plant root systems that cause imperceptible and asymptomatic infections (Sturz and Nowak, 2000). These significant evidences are promising for a paradigm move for a more microbial dominated or at least highly mutual view of the association between plant and microbiota. In this chapter, we describe the important PGPR species and their mode of action/application/role in improving sustainable agriculture.

TABLE 1.1 PGPR Produce Growth-promoting Chemical Substances for Agricultural Crops.

Chemical substances	PGPRs
ACCD	*Acinetobacter* spp., *Burkholderia* spp., *Enterobacter* spp., *Kluyvera ascorbate, Pseudomonas* spp., *P. aeruginosa, P. fluorescens, P. jessenii, Rahnella aquatilis, Ste. maltophilia*
Ammonia	*Azotobacter* spp., *Bacillus* spp., *Bacillus* species PSB10, *Bradyrhizobium* spp., *E. asburiae, Klebsiella* spp., *Mesorhizobium* spp., *Pseudomonas* spp., *P. aeruginosa, P. putida, Rhizobium* spp.
Cytokinin	*Rhi. Leguminosarum*
EPS	*Bradyrhizobium* spp., *E. asburiae, Klebsiella* spp., *Mesorhizobium* spp., *P. aeruginosa, Rhizobium* spp.
Gibberellin	*Azot. chroococcum*
HCN	*Azotobacter* spp., *Bacillus* spp., *Bacillus* species PSB10, *Bradyrhizobium* spp., *E. asburiae, Klebsiella* spp., *Mesorhizobium* spp., *Pseudomonas* spp., *P. aeruginosa, P. putida, Rhizobium* spp., *Serratia marcescens*
IAA	*Acinetobacter* spp., *Azospirillum* spp., *Azomonas* sp. RJ4, *Azos. amazonense, Azos. brasilense, Azotobacter* spp., *Azot. chroococcum, Bacillus* spp., *B. subtilis, Bacillus* sp. PSB10, *Bacillus* sp. RJ31, *Bradyrhizobium* spp., *Brad. japonicum, Brevibacillus* spp., *Burkholderia* spp., *Cellulomonas* spp., *Enterobacter* spp., *E. asburiae, Flavobacterium* spp., *Klebsiella* spp., *Kle. oxytoca, Mesorhizobium* spp., *Paenibacillus polymyxa, Pseudomonas* spp., *P. aeruginosa, P. fluorescens, P. fluorescens* GRS$_1$, *P. fluorescens* PRS$_9$, *P. jessenii, P. putida, Pseudomonas* sp. A3R3, *Pseudomonas* sp. RJ10, *Rah. Aquatilis, Rhizobium* spp., *Rhizobium phaseoli, Rhodococcus* spp., *Ser. marcescens, Sphingomonas* spp., *St. maltophilia, Variovorax paradoxus, Xanthomonas* sp. RJ3

TABLE 1.1   (Continued)

Chemical substances	PGPRs
Kinetin	*Azot. chroococcum*
Siderophores	*Acinetobacter* spp., *Azos. amazonense, Azotobacter* spp., *Azot. chroococcum, Bacillus* spp., *Bacillus* species PSB10, *Bradyrhizobium* spp., *Brad. japonicum, Bravibacterium* spp., *Burkholderia* spp., *Enterobacter* spp., *E. asburiae, Flavobacterium* spp., *Klebsiella* spp., *Klu. ascorbate, Mesorhizobium* spp., *Mesorhizobium ciceri, Pae. polymyxa, Proteus vulgaris, Pseudomonas* spp., *P. aeruginosa. P. aeruginosa* 4EA, *P. fluorescens, P. fluorescens* GRS$_1$, *P. fluorescens* PRS$_9$, *P. jessenii, P. putida, Pseudomonas* sp. A3R3, *Rah. aquatilis, Ralstonia metallidurans, Rhizobium* spp., *Rhi. ciceri, Rhi. meliloti, Rhodococcus* spp., *Ser. marcescens, V. paradoxus*

ACCD, 1-aminocyclopropane-1-carboxylate deaminase; EPS, exo-polysaccharides; HCN, hydrogen cyanide; IAA, indole-3-acetic acid; PGPR, plant growth–promoting rhizobacteria; *Azos.*, *Azospirillum*; *Azot.*, *Azotobacter*; *B.*, *Bacillus*; *Brad.*, *Bradyrhizobium*; *E.*, *Enterobacter*; *Kle.*, *Klebsiella*; *Klu.*, *Kluyvera*; *M.*, *Mesorhizobium*; *Pae.*, *Paenibacillus*; *P.*, *Pseudomonas*; *Rah.*, *Rahnella*; *Rhi.*, *Rhizobium*; *Ser.*, *Serratia*; *St.*, *Stenotrophomonas*; *V.*, *Variovorax*.

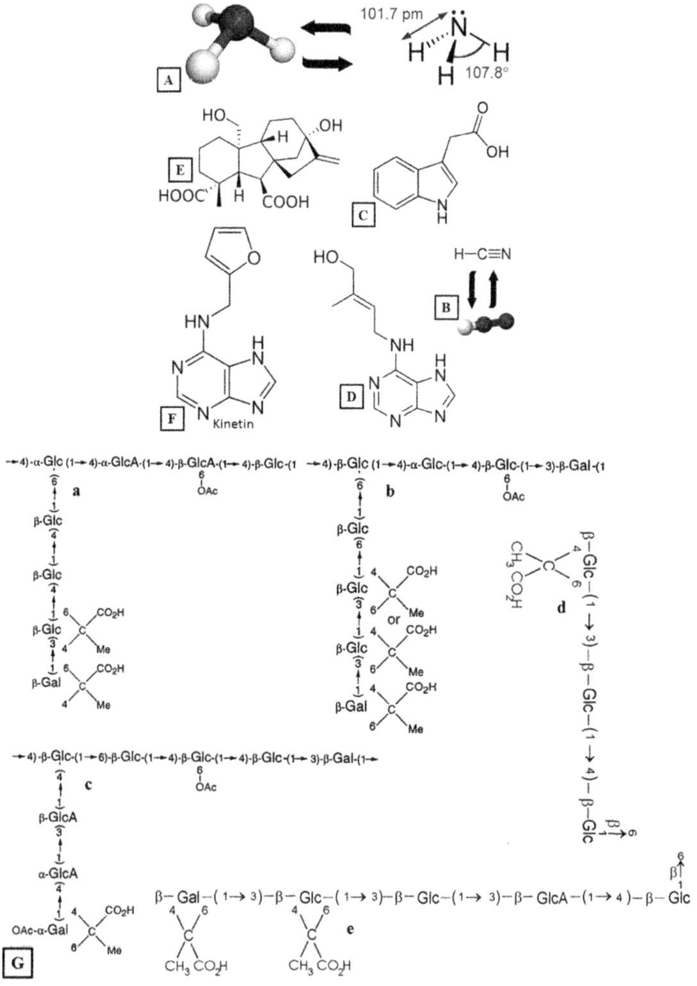

FIGURE 1.1 Structure of chemical substances produced by plant growth–promoting rhizobacteria (PGPR) for growth promotion of agricultural crops. (A) Ammonia, (B) hydrogen cyanide (HCN), (C) indole-3-acetic acid (IAA), (D) cytokinin, (E) gibberellin, (F) kinetin, and (G) exo-polysaccharides structure of *Rhizobium* spp. [(a) *R. leguminosarum* (*Source*: Adapted from Robertsen et al., 1981; McNeil et al., 1986; O'Neil et al., 1991; Dudman et al., 1983); (b) *R. tropici* CIAT899 (*Source*: Adapted from Gil-Serrano et al., 1990); (c) *Rhizobium* sp. strain NGR234 (*Source*: Adapted from Djordjevic et al., 1986); (d) *R. leguminosarum* bv. trifolii 4S (*Source*: Adapted from Amemura et al., 1983); and (e) *R. leguminosarum* bv. viciae 248 (*Source*: Adapted from Canter-Cremers et al., 1991) in which Glc: glucose, Gal: galactose, and GlcA: glucuronic acid].

TABLE 1.2 Important Direct and Indirect Functional Mechanisms of PGPR Strains in Crop Plants.

Functional mechanism	PGPR strains
Antibiotic resistance	*Azospirillum amazonense*
Antifungal activity	*Acinetobacter* spp., *Azotobacter* spp., *Bacillus* spp., *Bacillus subtilis*, *Mesorhizobium* spp., *Pseudomonas* spp., *Pseudomonas chlororaphis*, *Pseudomonas fluorescens*, *Pseudomonas putida*
Biocontrol potentials	*Pseudomonas* spp.
Biological nitrogen (N_2) fixation	*Acinetobacter* spp., *Pseudomonas* spp., *Lysinibacillus xylantilyticus*, *Bradyrhizobium japonicum*
Heavy metal mobilization	*Psychrobacter* sp. SRS8
Heavy metal solubilization	*Bradyrhizobium* sp. 750, *Burkholderia* spp., *Ochrobactrum cytisi*, *Pseudomonas* spp., *Pseudomonas jessenii*
Induced systemic resistance	*Pseudomonas fluorescens*
Metal resistance	*Kluyvera ascorbata*
Nitrogenase activity	*Azospirillum amazonense*, *Azospirillum brasilense*, *Klebsiella oxytoca*, *Stenotrophomonas maltophilia*
Pb and Cd resistance	*Pseudomonas putida*
Phosphate solubilization	*Acinetobacter* spp. *Azospirillum* spp., *Azospirillum amazonense*, *Azospirillum brasilense*, *Azotobacter* spp., *Azotobacter chroococcum*, *Bacillus* spp., *Bacillus subtilis*, *Bradyrhizobium* spp., *Burkholderia* spp., *Enterobacter* spp., *Enterobacter asburiae*, *Klebsiella* spp., *Klebsiella oxytoca*, *Pseudomonas* spp., *Pseudomonas aeruginosa*, *Pseudomonas fluorescens*, *Pseudomonas fluorescens* GRS_1, *Pseudomonas fluorescens* PRS_9, *Pseudomonas jessenii*, *Pseudomonas putida*, *Rahnella aquatilis*, *Rhizobium* spp. *Stenotrophomonas maltophilia*
Zinc (Zn) resistance	*Gluconacetobacter diazotrophicus*
Zinc (Zn) solubilization	*Brevibacillus* spp.

PGPR, plant growth–promoting rhizobacteria.

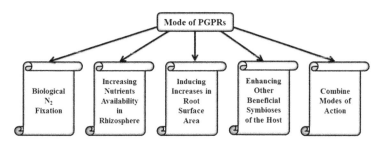

FIGURE 1.2   Mode of PGPR to enhance the nutrient uptake in host plants. (*Source*: Adapted from Vessey, 2003)

1.2   SOIL—AS DYNAMIC MEDIUM FOR MICROORGANISMS

Soil is a complex environment offering a variety of microhabitats for which microbial diversity in soil is much greater than that found in other environments. Each climate and soil type contains a microbial community precisely adapted to that specific habitat. Many microbes inhabit the pores between soil particles; others live in association with plants. The plant root surface (rhizoplane) and the region close to plant roots (rhizosphere) are important sites for microbial growth. It has been found that the microbial diversity in this rhizosphere region is much higher and has agricultural attention due to its beneficial way of making agriculture sustainable (Bonaterra et al., 2003). The rhizosphere is the soil–plant root interface and consists of the soil adhering to the root and its surrounding parts (Babalola, 2010) in which very significant and rigorous interactions are taking place between soil, microorganisms, and plant roots. The microbes found in this zone are known to be called plant root colonizing microorganisms. Many types of plant and microbe interaction occur in this zone (commensalism, mutualism, and parasitism), but in all cases, the microbe and the plant have established the capacity to communicate with each other. Plants secrete chemical compounds (high and low molecular weight) from their roots, termed as root exudates shown in Table 1.3. These compounds act as signal molecules/carbon sources for microbial nutrition (Antoun and Prevost, 2006) which in turn increases microbial biomass and activity (Bashan and de-Bashan, 2005). So overall it's a two way signaling cross talk that employ molecular lexicons.

TABLE 1.3   Compounds Excreted by Microorganisms Free Root of Cereal Crops.

Class of compounds	Compounds
Volatile	Acetoin, CO_2, ethanol, ethylene, iso-amyl alcohol, isobutanol, isobutyric acid
LMWCs	Amino acids, nucleotides, organic acids, sugars, vitamins
HMWCs	Enzymes, polysaccharides

LMWCs, low-molecular-weight compounds; HMWCs, high-molecular-weight compounds.
Source: Prescott and Klein (2007).

1.3   INTERACTION BETWEEN MICROBES AND PLANTS

Microbe–plant interactions can be broadly divided into two classes: an epiphyte is a microbe that lives on the surface of plants; and those colonize internal plant tissues are called endophytes. However, some researchers suggested that plants–microbes interaction take place at three different layers, namely, endosphere, phyllosphere, and rhizosphere. Rhizosphere is the contact region between soil and root and in the soils, the microbes surround the roots. In 1904, Lorenz Hiltner coined the term "Rhizosphere." The mechanism of interaction depends on plant and microbes type. Endophytic microbes sense the chemical signal and reach at the root surface and produces counter signal that makes the plant root vulnerable for microbial penetration into the root. Once inside, the microbe multiply within root hair cells and results in a mass of root cells containing many microbial cells. Soil bacteria usually use this mechanism due to their nutritional adaptability thereby preventing other bacteria, from reaching the target (Lavelle and Spain, 2001). Epiphytes cover the plant root surface and solely metabolize the excreted chemicals from the plant. Plant exudates like amino acids and sugars are rich sources of energy and nutrition. In the rhizosphere, plant–root interaction is a combine effect of root–microbe interaction, root–root interaction, and root–insect interaction. In the soil microbial population varied both qualitatively and quantitatively in the rhizosphere from that in the soil. From time to time, the impact of root exudates on the proliferation of soil microorganisms around and inside roots and interactions between soil microorganisms, rhizosphere colonists, and plant hosts has been widely studied (Hartmann et al., 2008; Dennis et al., 2010; Friesen et al., 2011; Berendsen et al., 2012).

1.4   INVOLVEMENT OF MICROORGANISMS IN PLANT GROWTH PROMOTION

There are several microorganisms colonize the root of the plant and they include algae, protozoa, bacteria and fungi but bacteria are the most abundant among them and found to be 10^{11}–10^{12} per gram of the soil. They have been classified according to their impact on promoting yield, plant growth and the way they interact with roots. Some of them, being pathogenic but most are beneficial (Trivedi and Pandey, 2008; Babalola, 2010). In the Eubacterial domain, Gram-positive *Bacilli*, Gram-negative *Pseudomonas* spp. and Actinobacterial group are dominant PGPR. Among them, *Acinetobacter, Aeromonas, Alcaligenes, Arthrobacter, Azoarcus, Azospirillum, Azotobacter, Bacillus, Burkholderia, Clostridium, Enterobacter, Gluconacetobacter, Klebsiella, Pseudomonas,* and *Serratia* are considered as most important PGPR species because they have beneficial effects on agricultural crop plants directly and indirectly by enhancing soil fertility shown in Tables 1.4A–E and Figure 1.3 (Gupta et al., 2015; Prasad et al., 2015). Apart from this other effective PGPR strains include *Azorhizobium, Allorhizobium, Mesorhizobium, Bradyrhizobium, Sinorhizobium,* and *Rhizobium* for their skill to work as biofertilizers (Vessey, 2003).

1.5   MECHANISM OF ACTION OF PGPR FOR PLANT GROWTH

PGPR can benefit plant root by both direct and indirect mechanism within the soil microhabitat. The different role of PGPR strains in direct as well as indirect in biological growth promotion is shown in Table 1.1 and Figure 1.4. They have also been found to be effective against several plant pathogens as shown by Figure 1.4. Some extensively used PGPRs in agriculture is summarized in Table 1.5.

1.5.1   DIRECT MECHANISMS FOR PLANT GROWTH

1.5.1.1   ROOT COLONIZATION

The growth factor and solubilized plant residues near root zone helps bacteria colonize and survive in the root surface efficiently. The nature of

TABLE 1.4A Beneficial Effects of PGPRs on Cereals Crop.

Crop	Scientific name	Year of the work	Employed PGPRs	Effects on agricultural plants	References
Rice	*Oryza sativa*	2014–2015	*Bacillus subtilis* Ljb-4 *Lysinibacillus xylanilyticus* GDLY1 *Alcaligenes faecalis* B17 *Bradyrhizobium japonicum* HHB-02 *Rhizobium etli bv. mimosae Mim-1*	Enhancement of seedling early growth, promotion of plant and root growth, tiller numbers, plant dry weight, nutrient accumulations, and biological N_2 fixation rate	Tan et al. (2014, 2015)
		2014	*Pseudomonas aeruginosa* BHUJY16, *P. aeruginosa* BHUJY20, *P. fluorescens* BHUJY29, *P. putida* BHUJY23, *P. putida* BHUJY13	More effective and boost-up growth attributes, yield, and nutrient uptake	Lavakush et al. (2014)
			Azospirillum spp.	Improved growth and productivity result in better yield	Sahoo et al. (2014)
		2010	*Azospirillum brasilense* CW903, *Burkholderia pyrrocinia* CBPB-HOD, *Methylobacterium oryzae* CBMB20	Increased the plant growth (like shoot and root length and nutrient uptake)	Madhaiyan et al. (2010)
		2009	PGB4, PGT1, PGT2, PGT3, PGG1 and PGG2 (not identified, only characterized)	Isolate inoculation resulted in a significant increase in germination rates, plant height, root length, and dry matter production in rice seedlings	Ashrafuzzaman et al. (2009)
			Pseudomonas spp.	Increased IAA levels	Karnwal (2009)
			Azospirillum spp.	Grain yield and N content was improved	Pedraza et al. (2009)
		2008	*Bacillus* spp., *Paenibacillus* spp.	Significantly increase in the root and shoot growth of the plant	Beneduzi et al. (2008)

TABLE 1.4A (Continued)

Crop	Scientific name	Year of the work	Employed PGPRs	Effects on agricultural plants	References
			Azospirillum lipoferum	Increased root length, root surface area, and root volume	Boyer et al. (2008)
			Pseudomonas species	Pseudomonad isolated from rice showed comparatively higher ability to control bacterial and fungal root pathogens	Lawongsa et al. (2008)
			Azospirillum amazonense	Grain dry matter accumulation (7–11.6%), the number of panicles (3–18.6%), and nitrogen accumulation at grain maturation (3.5–18.5%) increased	Rodrigues et al. (2008)
		2007	Burkholderia vietnamensis	Inoculation increases the grain yield of rice	Govindarajan et al. (2007)
			Herbaspirillum sp. strain B501 gfp1	Increased dry and fresh weight, increased N fixation, and phytohormone synthesis	Zakria et al. (2007)
		2006	Azospirillum spp., Pseudomonas spp.	Increased shoot biomass and grain yield, increased N fixation, and phytohormone synthesis	Mirza et al. (2006)
		2004	Azospirillum brasilense (strains A3, A4, A7, A10, CDJA), Bacillus circulans P2, Bacillus sp. P3, Bacillus magaterium P5, Bacillus sp. Psd7 Streptomyces anthocysnicus, Pseudomonas aeruginosa Psd5, Pseudomonas pieketti Psd6, Pseudomonas fluorescens MTCC103	All the bacterial strains increased rice grain yield maximum up to 76.9% over uninoculated control	Thakuria et al. (2004)

TABLE 1.4A (Continued)

Crop	Scientific name	Year of the work	Employed PGPRs	Effects on agricultural plants	References
		2002	*Herbaspirillum seropedicae* Z67	Higher plant colonization, increased nitrogenase activity in leaf and stem, plant dry weight increased significantly (30%) with higher N content	James et al. (2002)
		2001	*Herbaspirillum* sp. strain B501	GFP-tagged cells of *Herbaspirillum* sp. strain B501gfp1 were apparently localized in intercellular spaces of 7-day-old seedlings of *O. officinalis* W0012	Elbeltagy et al. (2001)
			Rhizobacterial spp.	Inoculation with rhizobacterial strains active in IAA production had significant growth-promoting effects on rice and also relatively more positive effects on inoculated seedlings	Khalid et al. (2001)
			Azospirillum, Aeromonas veronii, Enterobacter cloacae	Increase in root area, plant biomass, and N fixation	Mehnaz et al. (2001)
			Rhizobium leguminosarum bv. Trifolii	Promotion of root and shoot growth, significant improvement in seedling vigor tends to increases in grain yield at maturity	Yamni et al. (2001)
		2000	*Burkholderia* spp., *Herbaspirillum* spp.	Medium to high increase in fresh weight of rice plant in response to inoculation with 19 strains, net increase of N content in grains compared to uninoculated controls	Baldani et al. (2000)
			Rhizobium leguminosarum	Rice inoculation with *R. leguminosarum* had significant growth-promoting effects on rice seedlings	Biswas et al. (2000)

TABLE 1.4A (Continued)

Crop	Scientific name	Year of the work	Employed PGPRs	Effects on agricultural plants	References
			Rhizobium leguminosarum	Rice growth–promoting effects upon inoculation on axenically grown rice seedlings were observed	Dazzo et al. (2000)
			Burkholderia vietnamiensis Azoarcus	Increased shoot and root weight and leaf surface, increased N fixation and phytohormone production are involved	Van et al. (2000)
Wheat	*Triticum aestivum*	2011	*Pseudomonas* spp.	Significantly increased soil enzyme activities, total productivity, and nutrient uptake	Sharma et al. (2011)
			Bacillus spp., *Enterobacter* spp., *Paenibacillus* spp.	Inoculating wheat seedlings produced more biomass compared to control. Exopolysaccharides producing PGPR protect the plant from Na toxicity by decreasing its uptake	Upadhyay et al. (2011)
		2010	*Pseudomonas* spp., *Enterobacter cloacae, S. ficaria*	Inoculation not only reduced the negative impact of salinity stress on wheat but also dilute the impact of ethylene on etiolated pea seedlings	Nadeem et al. (2010)
		2009	PGP microbes characterized in terms of plant growth promotion but not identified	Maximum increase in plant height, number of tillers/m², number of spikelets/spike, grain and straw yield with PGPR+RDF+compost	Akhtar et al. (2009)
			Klebsiella pneumonia	Significantly increased the root length and shoot length	Sachdev et al. (2009)
			Pseudomonas putida, P. aeruginosa, S. proteamaculans	*Pseudomonas putida* was more effective and a significant increase in plant height, root length, and chlorophyll content was observed compared to control	Zahir et al. (2009)

TABLE 1.4A (Continued)

Crop	Scientific name	Year of the work	Employed PGPRs	Effects on agricultural plants	References
		2008	*Pseudomonas fluorescens* SBW25, *Paenibacillus brasilensis* PB177	*Paenibacillus brasilensis* strongly inhibited the growth of pathogenic fungus *M. nivale* in dual culture plate assays	Jaderlund et al. (2008)
			Pseudomonas fluorescens SBW25; *Paenibacillus brasilensis* PB177	In some combinations increased plant dry weight Protection from pathogenic fungus	Jaderlund et al. (2008)
			Pseudomonas fluorescens	Significant increase in yield Regulate production of ethylene and elongate roots by hydrolyzing 1-aminocyclopropane-1-carboxylic acid	Naveed et al. (2008)
			Pseudomonas fluorescens	Plant protected from soilborne fungal pathogens Suppress soilborne fungal pathogens	Okubara and Bonsall (2008)
			Burkholderia caryophylli; *Pseudomonas fluorescens*	Significant root elongation, root weight and grain and straw yields. Increased ACC-deaminase activity, chitinase activity, phytohormone production and P solubilization, significantly improved growth when increased NPK are also added to the soil. Appropriate fertilizers doses combined with *Pseudomonads* strains employed better for growth and yield of wheat plants and also effective in nutrient utilization in an order to save fertilizer chemicals	Shaharoona et al. (2008)

TABLE 1.4A (Continued)

Crop	Scientific name	Year of the work	Employed PGPRs	Effects on agricultural plants	References
		2007	Bacillus cereus RC 18, Bacillus licheniformis RC08, Bacillus megaterium RC07, Bacillus subtilis RC11, Bacillus OSU-142, Bacillus M-13, Pseudomonas putida RC06, Paenibacillus polymyxa RC05 and RC14	All bacterial strains were efficient in indole acetic acid (IAA) production and significantly increased growth of wheat	Çakmakçi et al. (2007b)
			Strains of Azospirillum spp.	Increased shoot and root biomass Siderophore or phytohormone production	Fischer et al. (2007)
		2006	Azospirillum brasilense	Increased the quantity of photosynthetic pigments resulting in greener plants. Enhanced photosynthetic pigment production	Bashan et al. (2006)
			Pseudomonas sp.	A combined bio-inoculation of diacetyl-phloroglucinol producing PGPR and AMF and improved the nutritional quality of wheat grain	Roesti et al. (2006)
		2005	Pseudomonas denitrificans, Pseudomonas rathonis	All the bacterial strains had been found to increase plant growth of wheat in pot experiments	Egamberdiyeva (2005)
			Azospirillum lipoferum strains 15	Promoted development of wheat root system even under crude oil contamination in pot experiment in growth chamber	Muratova et al. (2005)
		2004	30 Isolates taken from wheat rhizosphere but not identified	Increased growth and yield Increased phytohormone biosynthesis	Khalid et al. (2004)

TABLE 1.4A (Continued)

Crop	Scientific name	Year of the work	Employed PGPRs	Effects on agricultural plants	References
		2003	*Rhizobacteria* (Unidentified)	Strain produced highest amount of auxin in nonsterilized soil and caused maximum increase in growth yield	Khalid et al. (2003)
		2001	*Rhizobacterial*	Inoculation with rhizobacterial isolates had significant growth-promoting effects on wheat	Khalid et al. (2001)
		2001	*Rhizobacteria* (unidentified)	Rhizobacterial strains active in IAA production had relatively more positive effects on inoculated seedlings	Khalid et al. (2001)
		2000	*Azospirillum brasilense*	Increased nitrogen fixation up to 85% production of IAA and siderophore, increased colonization potential	Kaushik et al. (2000)
Maize	*Zea mays*	2010	*R. leguminosarum*	Enhanced plant growth and biomass	Hadi and Bano (2010)
		2010	*Bacillus megaterium*	Exhibition of higher root hydraulic conductance under salt-stress, higher plasma membrane (PIP2) aquaporin amount in roots under salt-stress	Marulanda et al. (2010)
		2009	*Pseudomonas aeruginosa, Pseudomonas fluorescens, Ralstonia metallidurans*	Promoted plant growth, facilitated soil metal mobilization, enhanced Cr and Pb uptake	Braud et al. (2009)
		2009	*Azospirillum brasilense* Az39; *Bradyrhizobium japonicum* E109	Seed germination and nodule formation were promoted Production of phytohormones	Cassan et al. (2009)

TABLE 1.4A (Continued)

Crop	Scientific name	Year of the work	Employed PGPRs	Effects on agricultural plants	References
			Pseudomnas putida strain R-168, *Pseudomonas fluorescens* strain R-93, *Pseudomonas fluorescens* DSM 50090, *Pseudomonas putida* DSM291, *Azospirillum lipoferum* DSM 1691, *Azospirillum brasilense* DSM 1690	Plant height, seed weight, number of seed per ear and leaf area, shoot dry weight significantly increased	Gholami et al. (2009)
		2008	Commercially available Plant Growth Activator (PGA)	Greater plant height More efficient uptake of N and P	Adesemoye et al. (2008)
			Pseudomonas fluorescens (MPp4), *Burkholderia* sp. (MBp1, MBf21 and MBf15)	*B. megaterium* (*B5*), *B. cereus* sensu lato (*B25*) and *Bacillus* spp. (*B35*) showed the highest antagonistic activity against *Fusarium verticillium* pathovar and less disease incidence on plants. Production of glucanases, proteases or chitinases, as well as siderophores and auxins combinedly caused significant antagonism against the phytopathogen and hence higher plant growth	Hernández-Rodriguez et al. (2008)
			Rhizobium spp.; *Sinorhizobium* spp.	Increased shoot and root dry biomass Production of phytohormones and siderophores	Hossain and Martensson (2008)
			Pseudomonas species	Pseudomonad isolated from rice maize showed comparatively lower ability to control bacterial and fungal root pathogens	Lawongsa et al. (2008)

TABLE 1.4A (Continued)

Crop	Scientific name	Year of the work	Employed PGPRs	Effects on agricultural plants	References
			Pseudomonas spp.	Increased grain yield and nutrient uptake hydrolyses ACC	Naveed et al. (2008)
		2007	*Bacillus* spp.	Increasing salt concentrations, biological N fixation may be competitive, becoming a more economic and sustainable alternative to chemical fertilization. The bacterial inoculants increased the total N, P, and K contents of the shoot and root of maize in calcisol soil from 16% to 85% significantly as compared to the control counterpart	Egamberdiyeva (2007)
			Bacillus megaterium, B subtilis, Pseudomonas corrugata	Increase in grain yield Increase in fixed nitrogen, production of phytohormones, phosphate solubilization, production of antibiotics and siderophores	Kumar et al. (2007)
			Pseudomonas spp., *Enterobacter aerogenes, Flavobacterium ferrugineum*	PGPR enhanced the growth of maize under salinity but with variable efficacy. Overall, high chlorophyll content, relative water content and K+/Na+ ratio was observed in inoculated plant than uninoculated control	Nadeem et al. (2007)
			Bacillus spp., *Ochrobactrum* spp.	Significantly increased shoot and root dry weight Suppressed fungal pathogens	Principe et al. (2007)
		2006	*Azospirillum lipoferum* CRT1	Root growth was enhanced No explanation given El	El Zemrany et al. (2006)

TABLE 1.4A (Continued)

Crop	Scientific name	Year of the work	Employed PGPRs	Effects on agricultural plants	References
			Pseudomonas putida Gluconacetobacter azotocaptans Azospirillum lipoferum	Isolates exhibited plant growth–promoting effects, increased root/shoot weight when compared to uninoculated plants, fixation of nitrogen and IAA production, solubilization of phosphate and production of metabolic intermediates promoting less pathogen attack on plants	Mehnaz and Lazarovits (2006)
			Pseudomonas spp.,	Bacterium caused root elongation in maize	Shaharoona et al. (2006)
		2005	*Pseudomonas denitrificans, Pseudomonas rathonis*	All the bacterial strains had been found to increase plant growth of maize in pot experiments	Egamberdiyeva (2005)
			Azotobacter	Amendment of L-tryptophan (L–TRP) with/ without *Azotobacter* significantly increased the grain yield (18.4%), fresh biomass (16.7%), 1000-grain weight (14.5%), and total nitrogen uptake (40%)	Zahir et al. (2005)
		2003	*Enterobacter sakazakii* 8MR5, *Pseudomonas* sp. 4MKS8, *Klebsiella oxytoca* 10MKR7	Inoculation increased agronomic parameters	Babalola et al. (2003)
		2000	*Azotobacter* spp.	Inoculation with strain efficient in IAA production had significant growth-promoting effects on maize seedlings	Zahir et al. (2000)

TABLE 1.4A (Continued)

Crop	Scientific name	Year of the work	Employed PGPRs	Effects on agricultural plants	References
Barley	*Hordeum vulgare*	2007	*Bacillus licheniformis* RCO2, *Rhodobacter capsulatus* RCO4, *Paenibacilla polymyxa* RCO5, *Pseudomonas putida* RCO6	Net increase in root and shoot Improved uptake of Fe, N, Mn, and Zn Increased N fixation and production of phytohormones	Çakmakçi et al. (2007a)
		2006	*Bacillus* spp.	Increased root weight and shoot weight up to 16.7% and 347%, respectively	Canbolat et al. (2006)
Oat	*Avena sativa*	2008	*Azospirillum* spp. (ChO6 and ChO8), *Azotobacter* spp.(ChO5), *Pseudomonas* spp. (ChO9)	Significant increase in root length, shoot length, seedling vigor when treated with *B. subtilis* G-1. Formulation-based treatment significantly decreased stem rot (80%) and made higher pod yield	Yao et al. (2008)
Sorghum	*Sorghum bicolor*	2009	*B. cereus* (KBE7-8), *B. cereus* (NAS4-3), *Stenotrophomonas maltophilia* (KBS9-B)	Significant increase in fresh as well as dry weights of shoot and root Improve in length of stem, diameter of stem, leaf area, chlorophyll content, macro nutrients content in the leaf, number of tillers, number of panicles per hill, percentage of pithy grain per panicles, dry weights of 1000 grains and weight of grain per hill	Idris et al. (2009)
Rye	*Secale cereale*	2003	*Pseudomonas fluorescens*	Significant increase in foliar dry mass, production of siderophore, and suppression of fungal pathogens	Kurek and Jaroszuk-Scisel (2003)

TABLE 1.4B Beneficial Effects of PGPRs on Legumes and Oil Seed Crop.

Crop	Scientific name	Year of the work	Employed PGPRs	Effects on agricultural plants	References
Black gram	*Vigna mungo*	2008	*Pseudomonas aeruginosa* strain MKRh3	Decrease in net Cd accumulation, increase in extensive rooting, and enhanced plant growth	Ganesan (2008)
Common bean	*Phaseolus vulgaris*	2008	*Azospirillum brasilense* Sp245	Increase growth of roots were observed	Remans et al. (2008)
Chickpea *Cicer arietinum*		2011	*Mesorhizobium* spp.	Dual inoculation enhanced nutrition and growth of chick pea	Tavasolee et al. (2011)
		2010	PGPR isolates	Application of PGB4, PGT1, PGT2, PGT3, PGG1, and PGG2 induced indole acetic acid production, PGT3 isolate showed higher solubilization of phosphorus. Overall, significant increase in shoot length, root length and dry matter and seed germination under saline conditions	Mishra et al. (2010)
			Bacillus species PSB10	Significantly improved growth, nodulation, chlorophyll, leghemoglobin, seed yield and grain protein, decrease in uptake of Cr in roots, shoots, and grains	Wani and Khan (2010)

TABLE 1.4B (Continued)

Crop	Scientific name	Year of the work	Employed PGPRs	Effects on agricultural plants	References
		2009	M. mediterraneum LILM10	Increased nodule number, shoot dry weight and grain yield and NaCl tolerant	Romdhane et al. (2009)
			Pseudomonas spp.	Enhanced fresh and dry weight of plants even at 2 mM Ni concentration	Tank and Saraf (2009)
			M. metallidurans sp. nov	Resistance against the heavy metal	Vidal et al. (2009)
		2008	Enterobacter spp.	Effective against Fusarium avenaceum	Hynes et al. (2008)
			M. ciceri, M. mediterraneum	Higher growth and N fixing activity Increased antioxidant activity	Mhadhbi et al. (2008)
			Mesorhizobium RC3	Increased growth, nodulation, chlorophyll, leghemoglobin, nitrogen content, seed protein, and seed yield. Increased the dry matter accumulation, number of nodules, seed yield and grain protein by 71%, 86%, 36%, and 16%. Nitrogen in roots and shoots increased by 46% and 40%, respectively, at 136 mg Cr/kg	Wani et al. (2008)

TABLE 1.4B (Continued)

Crop	Scientific name	Year of the work	Employed PGPRs	Effects on agricultural plants	References
		2007	*Bradyrhizobium* RM8	Enhanced growth performance. Significant increase in legume grain yield, NP uptake when coinoculated with *Mesorhizobium, Bacillus* and *Pseudomonas* spp., When inoculated with *A. chrococcum+Bacillus*, it tripled the seed yield with higher protein content at 145 DAS. Triple inoculation yielded higher N concentration and uptake	Wani et al. (2007c)
			Rhizobium spp. RP5	Increase in dry matter, nodule numbers, root N, shoot-N, leghemoglobin, seed yield, and grain protein by 19%, 23%, 26%, 47%, 112%, 26%, and 8%, respectively, at 290 mg Ni/kg	Wani et al. (2007b)
		2006	*Mesorhizobium* spp.	Growth at pH 7/9 and others at pH 5/7	Rodrigues et al. (2006)
			Pseudomonas jesseniiPS06, Mesorhizobium ciceri C-2/2	The coinoculation treatment increased the seed yield (52% greater than the uninoculated control treatment) and nodule fresh weight	Valverde et al. (2006)
		2004	*M. ciceri, M. mediterraneum S. medicae*	Enhanced nodulation and CAT activity Significant nodule protein and SOD activity	Mhadhbi et al. (2004)

TABLE 1.4B (Continued)

Crop	Scientific name	Year of the work	Employed PGPRs	Effects on agricultural plants	References
		2003	*Acacia rhizobia*	Intracellular accumulation of free glutamate for salt and drought tolerance	Gal and Choi (2003)
		2001	*M. ciceri* ch-191	Higher tolerance was noticed on 200 mmol/L. Altered protein and LPS levels. Higher proline accumulation than glutamate	Soussi et al. (2001)
Cowpea	*Vigna unguiculata*	2008	*Streptomyces acidiscabies* E13	Promotion in cowpea growth under nickel stress	Dimkpa et al. (2008)
Greengram or Mungbean	*Vigna radiate*	2014	*Pseudomonas aeruginosa* GGRK21	Higher accumulation of drought specific amphibolites in plant parts (leaf and root) when inoculated	Sarma et al. (2014)
		2013	*Pseudomonas syringae* strain Mk1, *Pseudomonas fluorescens* strain Mk20, *Pseudomonas fluorescens* strain Biotype G, Mk25, *Rhizobium phaseoli* strain M6, *Rhizobium phaseoli* strain M9	Coinoculation of *Pseudomonas* and *Rhizobium* strains found increased phosphorus and protein concentration in grain Improved growth, physiology and quality of mung bean under salt-affected conditions	Ahmad et al. (2013)
		2010	*Rhizobium phaseoli*	Reduced effects of salinity and increased the plant height, number of nodules per plant, plant biomass, grain yield, and grain N concentration	Zahir et al. (2010)

TABLE 1.4B (Continued)

Crop	Scientific name	Year of the work	Employed PGPRs	Effects on agricultural plants	References
		2007	*Bradyrhizobium* sp. (vigna) RM8	Increase in nodule numbers by 82%, leghemoglobin by 120%, seed yield by 34%, grain protein by 13%, root N by 41% and shoot N by 37% at 290 mg Ni/kg soil	Wani et al. (2007a)
		2006	*Ochrobactrum*, *Bacillus cereus*	Lowers the toxicity of chromium to seedlings by reducing Cr (VI) to Cr (III)	Faisal and Hasnain (2006)
			Bradyrhizobium spp., *Pseudomonas* spp., *Bradyrhizobium* spp.	Bacterium promoted nodulation in mung bean	Shaharoona et al. (2006)
		2005	*Azotobacter* sp., *Pseudomonas* sp.	Controlled production of Indole acetic acid with tryptophan added	Ahmad et al. (2005)
			Pseudomonas putida KNP9	Stimulated the plant growth, reduced Pb and Cd uptake	Tripathi et al. (2005)
		1999	*Pseudomonas putida*	The ethylene production inhibited in the inoculated plants. The ethylene production was inhibited in inoculated cuttings	Mayak et al. (1999)
Hyacinth bean	*Lablab purpures*	2007	*Rhizobium* spp.	Greater HM accumulation in nodules than in roots and shoots	Younis (2007)
Lentil	*Lens esculentus*	2013	*Rhizobia* strains	Increased plant biomass, nodule number and nodule dry weight	Islam et al. (2013)

TABLE 1.4B (Continued)

Crop	Scientific name	Year of the work	Employed PGPRs	Effects on agricultural plants	References
			Rhizobium RL9	The strain showed high tolerance toward multiple heavy metals like: cadmium, chromium, nickel, lead, zinc and copper. Showed release of significant amount of indole acetic acid, siderophore, hydrogen cyanide, and ammonia. The strain RL9 significantly increased growth, nodulation, chlorophyll, leghemoglobin, nitrogen content, seed protein, and seed yield	Wani and Khan (2013)
		2012	*Rhizobium* RL9	The strain showed tolerance to Pb up to 1600 µg/mL. Showed release of significant amount of indole acetic acid, siderophore, hydrogen cyanide, and ammonia. RL9 showed significant increase in growth of plants, nodulation, chlorophyll content, N and seed protein content, seed yield	Wani and Khan (2012)
Pea	*Pisum sativum*	2006	*P. brassicacearum* AM3, *P. marginalis* Dp1	Inoculating plants produced longer roots, greater root density, and improved nutrient uptake. Bacteria counteracted the Cd-induced inhibition of nutrient uptake by plants	Safronova et al. (2006)
		2003	*Rhizobium leguminosarum* bv. viciae 128C53 K	Nodulation found enhanced in plants due to bacterium	Ma et al. (2003)

TABLE 1.4B (Continued)

Crop	Scientific name	Year of the work	Employed PGPRs	Effects on agricultural plants	References
Pigeon pea	*Cajanus cajan*	2010	*Sinorhizobium fredii* KCC5, *Pseudomonas fluorescens* LPK2	*P. fluorescens* LPK2 and *S. fredii* KCC5 showed chitinase and *b*-1,3-glu-canase activity	Kumar et al. (2010)
		2006	Unidentified PGPR and *Rhizobium*	Synergistic effects of PGPR and *Rhizobium* on nodulation and nitrogen fixation by pigeonpea observed	Tilak and Ranganayaki (2006)
Soybean	*Glycine max*	2011	*Pseudomonas* spp.	Significantly increased soil enzyme activities, total productivity, and nutrient uptake	Sharma et al. (2011)
		2005	*Pseudomas fluorescens*	Plant growth increase	Gupta et al. (2005)
		2001	*Pseudomonas fluorescens*	Increased production of plant growth regulators: cytokinin dihydrozeatin riboside (DHZR)	Garcia de Salamone et al. (2001)
		1999	*Pseudomonas cepacia*	Early growth was seen in soybean due to Rhizobacterium	Cattelan et al. (1999)
Ground nut or peanut	*Arachis hypogaea*	2013	*Bacillus*	Biological N fixation Increased total N, P, and K contents of the shoot and root of maize in calcisol soil from 16 to 85% significantly as compared to the control counterpart	El-Akhal et al. (2013)
		2007	*P. fluorescens* TDK1, *P. fluorescens* PF2, *P. fluorescens* RMD1	Increasing salt tolerance of groundnut	Saravanakumar and Samiyappan (2007)

TABLE 1.4B (Continued)

Crop	Scientific name	Year of the work	Employed PGPRs	Effects on agricultural plants	References
		2004	*Pseudomonas fluorescens* PGPR1, PGPR2, PGPR4	Enhance in pod yield, haulm yield and nodule dry weight over the control. Involvement of ACC deaminase and siderophore production promoted nodulation and yield of groundnut	Dey et al. (2004)
Mustard/canola/ rape/Indian mustard	*Brassica campestris*	2011	*Pseudomonas* spp. strain A3R3	The bacterium has proved its ability in biomass increase	Ma et al. (2011a)
	Brassica juncea, *Brassica napus,*	2009	*Pseudomonas* spp.	Rate of seed germination and seedling growth was significantly higher. Tolerance against salinity stress	Jalili et al. (2009)
		2008	*Pseudomonas tolaasiiACC23,* *Pseudomonas fluorescens* ACC9, *Alcaligenes* sp. ZN4, *Mycobacterium* sp. ACC14	Protection against the inhibitory effects of cadmium	Dell'Amico et al. (2008)
			Bacillus edaphicus	Stimulated plant growth, facilitated soil Pb mobilization, enhanced Pb accumulation	Sheng et al. (2008)
		2007	*Mesorhizobium loti* MP6	*Mesorhizobium loti* MP6-coated seeds enhanced seed germination, early vegetative growth, and grain yield as compared to control	Chandra et al. (2007)

TABLE 1.4B (Continued)

Crop	Scientific name	Year of the work	Employed PGPRs	Effects on agricultural plants	References
		2006	*Sinorhizobium* sp. Pb002	Increased the efficiency of lead phyto-extraction by *B. juncea* plants	Di Gregorio et al. (2006)
			Methylobacterium fujisawaense	Bacterium promoted root elongation in canola	Madhaiyam et al. (2006)
			Pseudomonas sp., *Bacillus* sp.	Stimulated plant growth and decreased Cr (VI) content	Rajkumar et al. (2006)
			Xanthomonas sp. RJ3, *Azomonas* sp. RJ4, *Pseudomonas* sp. RJ10, *Bacillus* sp. RJ31	Stimulated plant growth and increased cadmium accumulation	Sheng and Xia (2006)
			Azotobacter chroococcum HKN-5, *Bacillus megateri-um* HKP-1, *Bacillus mucil-laginosus* HKK-1	Protected plant from metal toxicity, stimulated plant growth	Wu et al. (2006)
			Bacillus subtilis SJ-101	Facilitated Ni accumulation	Zaidi et al. (2006)
		2005	*Rhodococcus* spp., *Variovorax paradoxus* spp., *Flavobacterium*	Stimulating root elongation, cadmium (Cd) tolerant PGPR strain protected the plant metal from toxicity. A signifi-cant improvement in plant growth was observed at toxic Cd concentration	Belimov et al. (2005)

TABLE 1.4B (Continued)

Crop	Scientific name	Year of the work	Employed PGPRs	Effects on agricultural plants	References
			Rhizobacterial strains A3 and S32	Promoted the plant growth under chromium stress	Rajkumar et al. (2005)
		2003	Bacillus circulans DUC1, B. firmus DUC2, B. globisporus DUC3	Root and shoot elongation were enhanced by bacterial inoculation	Ghosh et al. (2003)
		2002	Rhizobacteria (unidentified)	Significant correlation between auxin production by PGPR in vitro and growth promotion of inoculated rapeseed seedlings in the modified jar experiments were observed	Asghar et al. (2002)
		2001	Alcaligenes spp., Bacillus pumilus, Pseudomonas spp., Variovorax paradoxus	Inoculated plant demonstrated more vigorous growth than the uninoculated (control)	Belimov et al. (2001)
			Enterobacter cloacae, Enterobacter cloacae CAL2, Pseudomonas putida UW4	Both root and shoot length significantly increased. Significant increases in root and shoot lengths were observed	Saleh and Glick (2001)
		1996	Rhizobium leguminosarum	Production of IAA and cytokinin with higher seedling growth	Noel et al. (1996)
Sunflower	Helianthus annuus	2011	Psychrobacter sp. SRS8	Stimulated plant growth and Ni accumulation in both plant species with increased plant biomass, chlorophyll, and protein content	Ma et al. (2011b)

TABLE 1.4B (Continued)

Crop	Scientific name	Year of the work	Employed PGPRs	Effects on agricultural plants	References
		2008	*Bacillus weihenstephanensis* strain SM3	Increased plant biomass and the accumulation of Cu and Zn in the root and shoot systems, also augmented the concentrations of water soluble Ni, Cu, and Zn in soil with their metal mobilizing potential. The bacterium can increase plant biomass and accumulation of trace elements like Cu, Ni, and Zn in the roots and shoot	Rajkumar et al. (2008)
		2005	*Ochrobactrum intermedium*	Increased plant growth and decreased Cr(VI) uptake	Faisal and Hasnain (2005)
Sesame		2006	*Paenibacillus polymyxa* E681	Strain E681 showed reduced disease incidence in greenhouse trial, significant reduction in pre- and postemergence damping-off compared to the nontreated	Ryu et al. (2006)

TABLE 1.4C   Beneficial Effects of PGPRs on Forage and Economic Crops.

Crop	Scientific name	Year of the work	Employed PGPRs	Effects on agricultural plants	References
Forage crops					
Lucerne (also called Alfalfa)	*Medicago sativa*	2003	*Pseudomonas fluorescens Avm, Rhizobium leguminosarum* bv *phaseoli* CPMex46	Improved Cu and Fe translocation from root to shoot	Carrillo-Castaneda et al. (2003)
			Bacillus pumillus, B. lichenformis	The effect was variable with respect to different combination of bacteria	Medina et al. (2003)
		1994	*Bacillus cereus* UW85	Filtrates of cultures suppressed alfalfa disease caused by *P. medicaginis* and inhibited the growth of the pathogen in an agar plate assay.	Silo-Suh et al. (1994)
Red clover	*Trifolium pratense*	2003	*Brevibacillus* spp.	Single as well as dual inoculation caused positive effect under lead contamination	Vivas et al. (2003)
Subterranean clover	Trifolium *subterraneum*	1986	*Pseudomonas putida*	A significant increase in shoot and root dry weight, was observed when PGPR was present	Meyer and Linderman (1986)

TABLE 1.4C (Continued)

Crop	Scientific name	Year of the work	Employed PGPRs	Effects on agricultural plants	References
Economic crops					
Cotton	Gossypium hirsutum	2016	Brevibacillus brevis	Have beneficial effects on cotton crop through combined modes of actions	Nehra et al. (2016)
		2010	Pseudomonas putida Rs-198	Protection against salt stress and promoted cotton seedling growth with 30% higher germination rate	Yao et al. (2010)
				Plant height, fresh weight and dry weight of cotton seedling were increased by 12.8%, 30.7% and 10.0%, compared to the control. Increased absorption of Mg^{2+}, K^+, and Ca^{2+} and decreased the uptake of the Na^+ from the soil. improved the production of endogenous indole acetic acid (IAA) content and reduced the ABA content of cotton seedling under salt stress	
		2007	Azotobacter chroococcum, Azospirillum lipoferum	Seed yield (21%), plant height (5%) and microbial population in soil (41%) increased over their respective controls while boll weight and staple length remained statistically unaffected	Anjum et al. (2007)
			Klebsiella oxytoca	In addition to significant increase in height and dry weight of cotton plants, inoculation with PGPR uptake of major nutrients like N, P, K, and Ca increased while Na deceased	Yue et al. (2007)
		2006	Bacillus subtilis FZB 24®	Higher production of IAA and siderophore and hence plant growth	Yao et al. (2006)

TABLE 1.4C (Continued)

Crop	Scientific name	Year of the work	Employed PGPRs	Effects on agricultural plants	References
Sugar cane	*Saccharum officinarum*	2005	*Azotobacter* spp., *Azospirillum* spp.,	Higher endophytic colonization of sugarcane	Tejera et al. (2005)
		2003	*Gluconacetobacter diazotrophicus* strains	The endophytic establishment of *G. diazotrophicus* within stems of sugarcane was confirmed by the scanning electron microscopy	Muñoz-Rojas et al. (2003)
		2001	*Gluconacetobacter diazotrophicus*	Localized host defense	James et al. (2001)
		1961	*Beijerinckia*	Higher plant colonization, higher sugar content in all parts of plant	Dobereiner (1961)
Saffron	*Crocus sativus*	2008	*Bacillus subtilis*	*B. subtilis* FZB24 significantly increased leaf length, flowers per corm, weight of the first flower stigma, total stigma biomass. Significant increase in production of picrocrocin, crocetin, and safranal as well as corm growth	Sharaf-Eldin et al. (2008)
Tobacco		2003	*Bacillus pumilus* SE 34, *Streptomyces marcescens* 90–116	Tobacco blue mold (TBM)	Zhang et al. (2003)
		2000	*Pseudomonas fluorescens*	Tobacco necrosis virus (TNV)	Park and Kloepper (2000)

ABA, abscisic acid.

TABLE 1.4D Beneficial Effects of PGPRs on Vegetables Crop.

Crop	Scientific name	Year of the work	Employed PGPRs	Effects on agricultural plants	References
Artichoke	*Cynara scolymus*	2012	*Pseudomonas putida*, *Azospirillum* spp., *Azotobacter* spp.,	Have significant affect in germination. Known to increase radical and shoot length, shoot weight Significant increase in radicle and shoot length, shoot weight, coefficient of velocity of germination, seedling vigority index, and significant decrease in mean time of germination	Jahanian et al. (2012)
Cucumber	*Cucumis sativus*	2011	*Bacillus*	Development in lateral roots	Sokolova et al. (2011)
		2006	*Bacillus*	Relatively higher availability of P and K in soils planted in cucumber	Han et al. (2006)
		2005	*P. putida* KD	Increased plant protection against oomycete *Pythium ultimum*	Rezzonoco et al. (2005)
		2002	*Bacillus* spp.	Cotton aphids	Stout et al. (2002)
Carrot	*Daucus carota*	2013	*Rhizobium leguminosarum* spp.	The dry weight of the inoculated seedlings (shoots and roots) increased Concentrations of N, P, and Ca were significantly higher in inoculated plants, indicating that they had higher potential for nutrient uptake than control plants	Flores-Felix et al. (2013)
Eggplant	*Solanum melongena*	2005	*Bacillus megaterium* *Bacillus mucilaginosus*	Higher P and K solubility in vitro. Inoculation of these strains with amendment of its respective rock P or K materials increased the availability of P and K in soil, enhanced N, P, and K uptake, and promoted growth	Han and Lee (2005)

TABLE 1.4D (Continued)

Crop	Scientific name	Year of the work	Employed PGPRs	Effects on agricultural plants	References
Kidney bean	*Phaseolus vulgaris*	2013	*Bradyrhizobium* spp.	Enhanced drought tolerance, IAA and EPS production; nodulation, nodule ARA, nodule N	Uma et al. (2013)
		2008	*R. tropici* coinoculated with *Paenibacillus polymyxa*	Enhanced plant height, shoot dry weight, and nodule number	Figueiredo et al. (2008)
			R. elti	Enhanced nodules, nitrogenase activity and biomass production. Higher tolerance than wild type strains	Suarez et al. (2008)
		2006	*Mesorhizobium* spp.	Overproduction of 60 kDa protein by all the isolates. All the isolates revealed more tolerance to 20 °C than 37 °C	Rodrigues et al. (2006)
		2001	*Rhizobium* sp. DDSS69	Induction of 135 and 119 kDa proteins. Variation in the protein profile of stressed and nonstressed cells	Sardesai and Babu (2001)
Lettuce	*Lectuca sativa*	2015	*Azospirillum* inoculation	Promoted higher biomass and improved lettuce quality when grown under salt-stress. Higher ascorbic acid content with lower oxidation rate in inoculated plants. Higher chlorophyll content, hue, Chroma, L and lower browning intensity as well as extended storage life	Gabriela et al. (2015)
		2013	*Rhizobium leguminosarum* strain PEPV16	The dry weight of the inoculated seedlings (shoots and roots) was more than twice with respect to the un-inoculated seedlings. Concentrations of N, P, and Ca were significantly higher in inoculated plants, indicating that they had higher potential for nutrient uptake than control plants	Flores-Felix et al. (2013)
		2010	*Pseudomonas mendocina*	Enhanced plant biomass. However, aggregate stability decreased under salinity even with inoculation	Kohler et al. (2010)

TABLE 1.4D (Continued)

Crop	Scientific name	Year of the work	Employed PGPRs	Effects on agricultural plants	References
		2007	*Pseudomonas putida*CC-R2-4, *Bacillus subtilis* CC-pg104	Significant increase in shoot length and root length achieved through encapsulated inoculant	Rekha et al. (2007)
		1999	*Agrobacterium* spp., *Alcaligenes piechaudii Comamonas acidovorans*	Production of high level of IAA (76 mM) and other growth-promoting substances	Barazani and Friedman (1999)
		1996	*Rhizobium leguminosarum*	Increased IAA and cytokinin production and higher seedling growth rate	Noel et al. (1996)
Pepper, Black	*Piper nigrum*	2012	*Paenibacillus glucanolyticus* strain IISRBK2	Inoculation with strain observed to increase tissue dry mass of black pepper ranged between 37.0–68.3%	Sangeeth et al. (2012)
Pepper, Bell/Red	*Capsicum annuum*	2010 2008	*Azospirillum brasilense* CW903, *Burkholderia pyrrocinia* CBPB-HOD, *Methylobacterium oryzae* CBMB20	Increased the plant growth (like shoot and root length and nutrient uptake	Madhaiyan et al. (2010)
			Paenibacillus polymyxa	Significantly increased the biomass of plants and elicited induced systemic resistance against bacterial spot pathogen *Xanthomonas axonopodis* pv. *vesicatoria* untreated plants	Phi et al. (2010)

TABLE 1.4D (Continued)

Crop	Scientific name	Year of the work	Employed PGPRs	Effects on agricultural plants	References
		2006	Bacillus megaterium var. phosphaticum, Bacillus mucilaginosus	The results showed that there was a relatively higher availability of P and K in soils planted in pepper	Han et al. (2006)
		2005	Bacillus cereus	A 1.38-fold increase in fw and .28-fold fw gain in roots with gibberellic acid production	Joo et al. (2005)
		2004	P. putida GR12-2, Achromobacter piechaudii ARV8	Inoculating plants were able to maintain their growth under water limited conditions. The PGPR dilute the negative impact of stress induced ethylene on root growth by the activity of their ACC-deaminase enzyme	Mayak et al. (2004b)
Potatoes (Stored)	Solanum tuberosum	2010	Bacillus spp.	Both the strains enhanced the auxin content of inoculated plants up to 71.4% and 433%, respectively, as compared to noninoculated plants	Ahmed and Hasnain (2010)
		2009	Bacillus cepacia strain OSU-7	All PGPRs showed antagonism against the pathogen	Recep et al. (2009)
Radish	Raphanus sativus	1998	Bradyrhizobium spp., Rhizobium leguminosarum	Prospectively used as PGPR in nonlegumes	Antoun et al. (1998)

TABLE 1.4D (Continued)

Crop	Scientific name	Year of the work	Employed PGPRs	Effects on agricultural plants	References
Spinach	*Amaranthus* spp.	2007	*Bacillus cereus* RC 18, *Bacillus licheniformis* RC08, *Bacillus megaterium* RC07, *Bacillus subtilis* RC11, *Bacillus* OSU-142, *Bacillus* M-13, *Pseudomonas putida* RC06, *Paenibacillus polymyxa* RC05 and RC14	Efficient in IAA production and significant increase in spinach growth observed	Çakmakçi et al. (2007b)
Tomato	*Solanum lycopersicum*	2014	*Sphingomonas* sp. LK11	Tomato plants inoculated with endophytic *Sphingomonas* sp. LK11 showed significantly increased growth attributes (shoot length, chlorophyll contents, shoot, and root dry weights) compared to the control. This indicated that such phyto-hormones-producing strains could help in increasing crop growth	Khan et al. (2014)
		2010	*Pseudomonas fluorescens*, *P. aeruginosa*, *P. stutzeri*	All PGPR strains enhanced the root and shoot growth of tomato. Sodium contents (Na) were low in plants inoculated with *P. stutzeri* and showed relatively better growth compared to other two strains	Tank and Saraf (2010)

TABLE 1.4D (Continued)

Crop	Scientific name	Year of the work	Employed PGPRs	Effects on agricultural plants	References
		2008	*Collimonas fungivorans, Pseudomonas fluorescens WCS365*	Increase in total amount of organic acid and also a strong increase of the amount of citric acid, less succinic acid as a mode of inhibiting the pathogens	Kamilova et al. (2008)
		2007	*Pseudomonas brassicacaerum, P. marginali, P. oryzihabitans, P. putida, Alcaligenes xylosoxidans*	Higher ACC deaminase activity, Increased root elongation and root biomass as well as colonization ability	Belimov et al. (2007)
		2006	*P. fluorescens* WCS 365	Increase in total amount of organic acid. A strong increase of the amount of citric acid, less succinic acid as a mode of inhibiting the pathogens	Kamilova et al. (2006)
		2005	*P. putida* KD	Increased plant protection against oomycete *Pythium ultimum*	Rezzonoco et al. (2005)
		2004	*Achromobacter piechaudii* ARV8	Inoculation increased fresh and dry weight as well as water use efficiency of tomato by decreasing the ethylene production under stress	Mayak et al. (2004a)
			P. putida GR12-2, *Achromobacter piechaudii* ARV8	Inoculating plants were able to maintain their growth under water limited conditions. The PGPR dilute the negative impact of stress induced ethylene on root growth by the activity of their ACC-deaminase enzyme	Mayak et al. (2004b)
		2000	*Bacillus amyloliquefaciens*	Used against Tomato mottle virus, reduced ToMoV incidence and disease severity	Murphy et al. (2000)

TABLE 1.4D (Continued)

Crop	Scientific name	Year of the work	Employed PGPRs	Effects on agricultural plants	References
	Lycopersicon esculentom	2014	*Bacillus megaterium*	Growth, stomatal conductance, shoot hormone concentration, competition were observed	Porcel et al. (2014)
		2013	*Chryseobacterium* spp. C138, *Pseudomonas fluorescens* N21.4	Siderophore production increased as bacterial biomass increased after 16 h of culture	Radzki et al. (2013)
		2012	*Bacillus* spp.	Dual inoculation of PGPR provide great control of root-knot nematode on tomato than single inoculation	Liu et al. (2012)
		2010	*Azospirillum brasiiense* CW903, *Burkholderia pyrrocinia* CBPB-HOD, *Methylobacterium oryzae* CBMB20	Increased the plant growth (like shoot and root length and nutrient uptake	Madhaiyan et al. (2010)
			Pseudomonas putida, *Azotobacter chroococcum*, *Azosprillum lipoferum*	Inoculation enhanced lycopene and antioxidant activity, shoot and fruit potassium content. Dual inoculation caused significant effect on lycopene, antioxidant activity and potassium content of tomato	Ordookhani et al. (2010)
		2007	*Methylobacterium oryzae*, *Berkholderia*sp.	Reduction in the level of toxicity of Ni and Cd in tomato and promote plant growth	Madhaiyan et al. (2007)

TABLE 1.4D (Continued)

Crop	Scientific name	Year of the work	Employed PGPRs	Effects on agricultural plants	References
			Bacillus subtilis BEBISbs (BS13).	Higher shoot dry wt, fruit/plant and plant dry wt observed	Mena-Violante and Olalde-Portugal (2007)
		2004	Pseudomonas spp.	Synergistic effect on root weight and root architecture and improved mineral nutrition by increasing P content	Gamalero et al. (2004)
White clover	Trifolium repens	2006	Brevibacillus	Enhanced plant growth and nutrition of plants and decreased zinc concentration in plant tissues	Vivas et al. (2006)
		2002	Pseudomonas spp.	Acyrthosiphon kondoi	Kempster et al. (2002)

fw, fresh weight.

TABLE 1.4E Beneficial Effects of PGPRs on Fruits Crops.

Crop	Scientific name	Year of the work	Employed PGPRs	Effects on agricultural plants	References
Apple	Malus domestica	2012	Bacillus spp., Pseudomonas spp.	The severity of root rot diseases of fungus decreased by bacteria	Dohroo and Sharma (2012)
		2007	Bacillus M3, Bacillus OSU-142, Microbacterium FS01	Increased fruit yield, plant growth and leaf P and Zn contents. Number of fruits per plant significantly increased	Karlidag et al. (2007)
Avocado		2006	Pseudomonas fluorescens PCL1606	Antibiotic, viz., phenazine-1-carboxylic acid and phenazine-1-carboxamide production found with biocontrol potential	Cazorla et al. (2006)

TABLE 1.4E (Continued)

Crop	Scientific name	Year of the work	Employed PGPRs	Effects on agricultural plants	References
Kiwi		2010	*Comamonas acidovorans*	Production of IAA and increase in rooting and root length of plants	Erturk et al. (2010)
Sweet cherry	*Prunus avium*	2006	*Pseudomonas BA-8, Bacillus OSU-142*	Increased mineral contents of tomato and cucumber fruit as compared to control treatment. Improved N, P, Mg, Ca, Na, K, Cu, Mn, Fe, and Zn contents of the fruit	Esitken et al. (2006)
Strawberry	*Fragaria × ananassa*	2015	*Phyllobacterium endophyticum* strain PEPV15	Strain PEPV15 was able to solubilize moderate amounts of phosphate (5 mm radius around the colonies). The strain grew on the CAS indicator medium where the colonies were surrounded by a yellow-orange halo (3.5 mm radius around colonies) indicative of the siderophore production	Flores-Felix et al. (2015)
		2009	*Pseudomonas BA-8, Bacillus OSU-142, Bacillus M3*	Root and foliar application of PGPR strains were significantly increased yield per plant, total soluble solids, total sugar and reduced sugar, but decreased titratable acidity as compared with the control	Pirlak and Kose (2009)
Raspberry cv Heritage	*Rubus* spp.	2006	*Bacillus M3*	Stimulated plant growth and resulted in significant yield increase. A significant increase in yield (33.9% and 74.9%), cane length (13.6% and 15.0%), number of cluster per cane (25.4% and 28.7%), and number of berries per cane (25.1% and 36.0%) with N, P and Ca contents of raspberry leaves obtained	Orhan et al. (2006)
Jojoba	*Azospirillum brasilense*	2015		Higher plant growth and increased stability toward high salt 160 mM	Gonzalez et al. (2015)

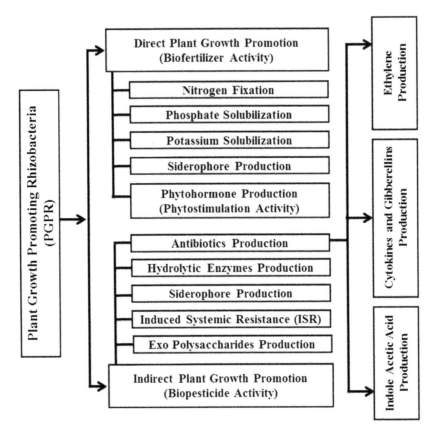

FIGURE 1.3   Diagram for direct and indirect beneficial effects of PGPR on agricultural crop.

bacterial flagella, pili, lipopolysaccharides, biofilm activity, and exopolysaccharides is the most important factors determining the roots colonization by PGPR. Plants secrete substances into the soil referred to as root exudates, which contain amino acids, proteins, carbohydrates, organic acids, vitamins, and other nutrients, and these substances act as substrate for symbiosis with the root (Benizri et al., 2001).

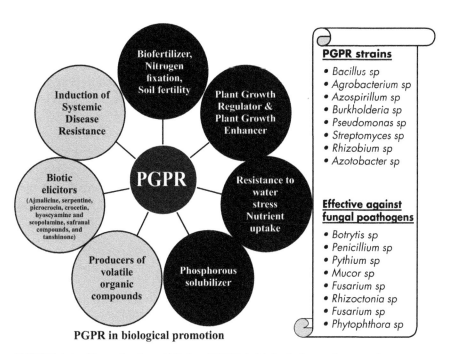

FIGURE 1.4 **(See color insert.)** Role of PGPR in biological growth promotion.

TABLE 1.5   Selection of Widely Used Major PGPR and Their Role.

PGPR	Beneficial traits
Bacillus spp.	• N_2 fixer
	• Phosphate solubilizer
	• Potassium solubilizer
Pseudomonas spp.	• Associative N_2 fixer
	• Phosphate solubilizer
	• Produces siderophore
	• Produces auxin and cytokinin
	• Produces ACC deaminase

Source: De Bruijn (2013).

1.5.1.2   PRODUCTION OF GROWTH REGULATORS

Phytohormones production by PGPR is an important mechanism due to which the bacteria promote plant growth, from germination to senescence through the phytohormones production such as indole-3-acetic acid (IAA), auxins, ethylene, cytokinin and gibberellin within the root zone (Gnanamanickam, 2006; Kloepper et al., 2007). In the rhizospheric soil, some pathogenic symbiotic free-living rhizobacterial species along with PGPR are found to be produced IAA and gibberellic acid and, thus, perform a key role in many plants in the root surface area and number of root tips enhancement (Han et al., 2005). Some plant growth regulators produced by PGPR strains in different crops has been summarized in Table 1.6.

1.5.1.3   NITROGEN FIXATION

As nitrogen (N_2) forms the principal constituent in terms of amino acids, nucleotides, other cellular metabolites in plant metabolism and cannot be fixed from atmosphere, many free living bacteria (*Azotobacter, Azospirilum*) and symbiotically associated (*Rhizobia*) fix atmospheric N_2 and make available ammonia to plants and are called diazotrophs (Babalola, 2010). Nitrogen is a major plant nutrient. Due to its loss by rainfall and mineral leaching it causes a limiting factor in the agricultural ecosystem. In the soil, some PGPR strains, namely, *Rhizobium* spp., *Pantoea agglomerans*, and *Klebsiella pneumoniae* fixed atmospheric N_2 and profit it to the plants (Riggs et al., 2001). In chickpea and tomatoes *Pseudomonas fluorescens* is reported to promote the nodulation (Parmar and Dadarwal, 1999; Minorsky, 2008), thereby enhancing plant height and flowering and fruiting capability. In the soil symbiotic and nonsymbiotic fixation of N_2 by microorganisms and increase crop yield could substitute to the use of nitrogenous fertilizers (Vessey, 2003). In leguminous crops this studies has been proved by several investigators (Dobereiner, 1997; Esitken et al., 2006). Species of *Azotobacter, Bacillus, Beijerinckia* are mostly involved in symbiotic N_2 fixation, while free-living diazotrophs like *Azospirillum, Burkholderia, Azoarcus, Gluconacetobacter*, and *Pseudomonas* are play important role in nonsymbiotic nitrogen fixation (Reinhold-Hurek et al., 1993; Mirza et al., 2006; Bashan and de-Bashan, 2010). Investigation has also been carried out on the combined effect of symbiotic and

TABLE 1.6   PGRs Produced by PGPR.

Plant	PGRs	PGPR	References
Brassica	Chrome-azurol, siderophore, hydrocyanic acid, IAA	*Mesorhizobium loti* MP6	Chandra et al. (2007)
	Siderophores, IAA	*Pseudomonas tolaasii* ACC23	Dell'Amico et al. (2008)
Groundnut	Siderophores, IAA	*Pseudomonas fluorescens*	Dey et al. (2004)
Lettuce	IAA	*Agrobacterium sp.*	Barazani and Friedman (1999)
		Alcaligenes piechaudii	
		Comamonas acidovorans	
	Cytokinin	*Rhizobium leguminosarum*	Noel et al. (1996)
Maize	IAA	*Azotobacter* spp.	Zahir et al. (2000)
	Auxin	*Pseudomonas denitrificans*	Egamberdieva (2005)
Radish	IAA	*Bradyrhizobium* spp.	Antoun et al. (1998)
		Rhizobium leguminosarum	Antoun et al. (1998)
Rape	Cytokinin	*Rhizobium leguminosarum*	Noel et al. (1996)
Rice	IAA	*Aeromonas veronii*	Mehnaz et al. (2001)
		Azospirillum brasilense A3, A4, A7, A10, CDJA	Thakuria et al. (2004)
		Bacillus spp.	Beneduzi et al. (2008)
		Enterobacter cloacae	Mehnaz et al. (2001)
		Paenibacillus spp.	Beneduzi et al. (2008)
		Rhizobium leguminosarum	Biswas et al. (2000)
Sesbania	IAA	*Azotobacter* spp.	Ahmad et al. (2005)
Spinach	IAA	*Bacillus cereus* RC 18	Çakmakçi et al. (2007b)
Soya bean	Cytokinin	*Pseudomonas fluorescens*	Garcia de Salamone et al. (2001)
Wheat	IAA	*Azospirillum lipoferum* strains 15	Muratova et al. (2005)
		Azospirillum brasilense	Kaushik et al. (2000)
		Pseudomonas spp.	Roesti et al. (2006)
		Bacillus cereus RC 18	Çakmakçi et al. (2007b)
	Auxin	*Pseudomonas denitrificans*	Egamberdieva (2005)
	Cytokinin	*Paenibacillus polymyxa*	Timmusk et al. (1999)

PGRs, plant growth regulators.

nonsymbiotic microbes on plant growth promotion and in this regard, Dubey (1996) found that inoculation of *Bradyrhizobium* spp. in combination with *Pseudomonas striata* cause enhancement of nodule occupancy in soya bean resulted in the more biological N_2 fixation.

1.5.1.4   INCREASED UPTAKE OF MINERALS AND SOIL FERTILITY

Physical, physiological, and nutritional properties of the rhizospheric soil are found to be altered by PGPR strains. Apart from N_2, phosphorus is considered to be the second most major nutrient in the nutrition and development of plants including, metabolic processes of signal transduction, energy transfer, photosynthesis, macromolecular biosynthesis, and respiration chain (Chang and Yang, 2009), and most of the agricultural soil become deficient due to high phosphorus fixation ability that cannot be used by plants. So, the PGPRs are well known for their potential P solubilizing ability (mainly rock phosphate) by producing phosphatase and organic acids and mineralize the unavailable fixed phosphate and make it available to plant root. Soils contains large reservoir of total phosphorus (P) but tiny proportion of this total is supplied to the plants. The plants are only absorb phosphorus in two forms, that is, the monobasic ($H_2PO_4^{-1}$) and the diabasic ($H_2PO_4^{-2}$) (Glass, 1989). In this regards many phosphate-solubilizing microorganisms through secretion of organic acids or protons or acidification or and chelation and exchange reactions has been reported to alter the insoluble form of phosphorus to soluble form (Richardson et al., 2009; Hameeda et al., 2008) and are supplied to the plants. Some rhizobacteria are reported to raise uptake of other essential nutrient elements such as Ca, Mn, Cu, K, Fe, and Zn through the process of proton pump ATPase (Mantelin and Touraine, 2004). In this regard, Karlidag et al. (2007) reported that use of *Microbacterium* and *Bacillus* inoculants in crop plants is significantly improved the uptake of mineral elements. These rhizobacterial inoculants are also play a key role in soil fertility maintenance (Glass et al., 2002) and also help in the nutrients (unavailable forms) solubilization and facilitating its convey in crop plants (Glick, 1995).

1.5.1.5 PLANT GROWTH ENHANCEMENT

In agriculture system, many PGPRs are identified to promote the growth of plants, crop yield, seeds emergence. Improving environmental quality along with increasing demand for food has focused on the PGPR importance in agriculture. When PGPRs are sprayed to the mulberry plants, it has been found that they increase the leaf area, total chlorophyll content and biomass (Esitken et al., 2003). Another PGPR species, that is, *Azospirillum* spp. has been found to be very effective in the promotion of flowerings and fruiting in the agriculturally important crops (Dobbelaere et al., 2001). Sgroy et al. (2009) reported that some PGPR strains such as *Achromobacter xylosoxidans, Brevibacterium halotolerans, B. subtilis, Bacillus pumilus, Bacillus licheniformis,* and *Pseudomonas putida* having very important roles in escalating 1-aminocyclopropane-1-carboxylate (ACC) deaminase activity, cell elongation, and growth promotion. Additionally, before planting when some seeds of agriculture commodities were bacterized with PGPR mixture and rhizobia, it is found that the application enhanced the growth of the plants and disease resistance (Zehnder et al., 2001). Similarly, root inoculation of *Microbacterium* FS01 and *Bacillus* M3 in apple tree is also promoted the growth and yield of the apple by increasing the plant growth regulators production and mobilizing the nutrients uptake by the PGPR (Karlidag et al., 2007).

1.5.2 INDIRECT MECHANISMS FOR PLANT GROWTH

1.5.2.1 PRODUCTION OF ANTIBIOTICS

PGPR produces many different types of antibiotics which combat against phytopathogens and includes zwittermycin A, streptomycin, oligomycin A, butyrolactones, oomycin A, kanosamine, phenazine-1-carboxylic acid, pyrrolnitrin, pyoluteorin, xanthobaccin, viscosinamide, and 2,4-diacetyl phloroglucinol (2,4-DAPG) have been shown to be effective against phytopathogenic agents (Whipps, 2001). Among them, 2,4-DAPG is one of the most efficient antibiotics in the control of plant pathogens and can be produced by various strains of *Pseudomonas* (Fernando et al., 2006) have a broad-spectrum activity. This antibiotic has also revealed activity against a wide range of fungal species from the group of *Basidiomycetes, Deuteromycetes,* and *Ascomycetes,* including *Botrytis cinerea, Rhizoctonia solani, Sclerotinia sclerotiorum,* and *Verticillium dahliae* (Raaijmakers et al., 2002).

1.5.2.2 INDUCED SYSTEMIC RESISTANCE (ISR)

Induced systemic resistance (ISR) may define as ability of the plant defense to exclude or overcome completely or in some degree. The ISR mechanisms steps are (1) linked to growth promotion, that is, developmental escape, (2) reduced symptom expression, that is, physiological tolerance, (3) associated with microbial antagonisms in the rhizosphere, that is, environmental, and (4) induction of cell wall reinforcement, phytoalexins, and pathogenesis-related proteins induction, and "priming" of defense responses (resistance), that is, biochemical-resistance. Siderophores, antibiotics, N-acyl-homoserine lactones, volatile organic compounds (VOCs) [e.g., 3-hydroxy-2-butanone (acetoin), and 2,3-butandiol] are the substances involved in ISR. Once infection occurs, defense-related gene expression is activated by some PGPR species through effective resistance mechanisms priming, as reflected by earlier and stronger defense reaction. The pathway involved is jasmonic acid pathway. The mechanism of ISR has been described in Figure 1.5. PGPR promotes plant growth by taming growth-restricting conditions in indirect mode. To reduce the harmful microbes from the surrounding areas of roots and ISR, PGPR produces antagonistic substances which give protection against pathogens thus enhancing growth-promoting conditions (Weller et al., 2002). Some strains of PGPR have been known to be effective as inducing agents in different plants by producing salicylic acid which is responsible for the induction of ISR in plants. Against *Fusarium* wilt of carnation, ISR was observed first with *Pseudomonas* spp. (WCS417r) and also in in cucumber by some rhizobacteria against the *Colletotrichum orbiculare*. Seed bacterization of faba beans with *P. fluorescens* and *Rhizobium leguminosarum* caused significant diseases reduction of yellow mosaic potyvirus (Elbadry et al., 2006). The beans seeds treated with *P. fluorescens* also possessed significant reduction in halo blight disease occur due to *Pseudomonas syringae* pv. *phaseolicola* (Alstroem, 1991). Similarly, significant reduction in bacterial wilt and angular leaf spot diseases affected by *P. syringae* pv. *lachrymans* in cucumber was observed when seeds were bioprimed with rhizobacterial strains, namely, *P. putida* 89 B-27 and *Serratia marcescens* 90–166 and *B. pumilus* INR-7 (Kloepper et al., 1993; Wei et al., 1996).

At present, on diversity of hosts a number of strains of *Bacillus* such as *B. amyloliquefaciens, B. cereus, B. pumilus, B. mycoides, B. sphaericus,*

B. pasteurii, and *B. subtilis* are reported to cause significant reduction in disease incidence (Ryu et al., 2004) like bell pepper (*Capsicum annuum*), cucumber (*Cucumis sativus*), muskmelon (*Cucumis melo*), tobacco (*Nicotiana tabacum*), sugar beet (*Beta vulgaris*), tomato (*Solanum lycopersicum*), and watermelon (*Citrullus lanatus*) during greenhouse and field

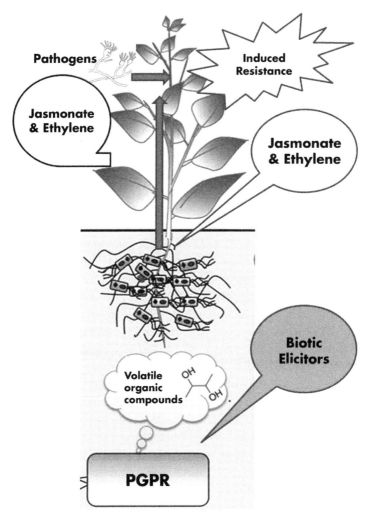

FIGURE 1.5 A pictorial presentation of mechanism of induced systemic resistance by PGPR.

conditions experiments. It is reported that in plants these strains of PGPR induce this type of systemic resistance through the activation of diverse defense-related enzymes, namely, β-1,3-glucanase, chitinases, peroxidase (POX), polyphenol oxidase (PPO), phenylalanine ammonia-lyase (PAL) (Dutta et al., 2008). Defense mechanism triggering in plants may be one of the key factors of ISR in plants treated with PGPR grown in pathogen-infested soil. The plants treated with PGPR are showed increase in POX, PAL, and PPO activity (Dutta et al., 2008). In this regard, Sailaja et al. (1997) reported that PAL, total phenol, POX and lipoxygenase level in plants is found to be increased when the seedlings are bacterized with *Bacillus* species, thereby, signifying the role of ISR in PGPR-mediated disease control. In another study, PPO, PAL and POX level are also raised in crops inoculated by pathogens when seeds were treated with *P. fluorescens* suspension (Sivakumar and Sharma, 2003).

1.5.2.3 PRODUCTION OF LYTIC ENZYMES

Several studies have been carried out which demonstrate that lytic enzyme production is by several PGPR strains which have a major role in the management of plant pathogens associated with root system specially *Fusarium oxysporium* and *R. solani*. These enzymes are classified as glucanases, proteases, cellulases, and chitinases (Dunne et al., 1997). Vivekananthan et al. (2004) found many PGPR strains excreting lytic enzymes killing pathogens or making their growth static. Lytic enzymes can reduce different polymeric substances such as cellulose, chitin, deoxyribonucleic acid (DNA) hemicellulose and proteins (Vivekananthan et al., 2004) and explained that laminarinase and extracellular chitinase produced by *P. stutzeri* lyse mycelia of *F. solani*. In a study among 26 *Pseudomonas* strains, strains AN-1-UHF, AN-5-UHF, PN-7-UHF and PN-13-UHF were reported to produce lytic enzymes especially proteolytic enzymes which have a very important role in plant growth promotion of apple and pear (Ruchi et al., 2012). *Bacillus* species isolated from different tomato rhizospheric soil are also found to be secreted several hydrolytic enzymes such as β-1,3-glucanase, protease, chitinase, and cellulose which have a vital role in plant growth promotion and plant diseases management (Kumar et al., 2012). Chitinolytic *Pseudomonas* isolate has also been showed a pronounced antifungal activity (Velazhahan et al., 1999). Additionally, when lytic producing rhizobacteria were combined with other biocontrol agents and were used in the field, a potent synergistic inhibitory effect against pathogens and in the promotion of plant growth has been observed (Dunne et al., 1998; Someya et al., 2007).

1.5.2.4 SIDEROPHORE PRODUCTION

Some living-organisms (fungi, bacteria, grasses, etc.) are involve in secretion of high-affinity iron chelating compounds which refers as siderophores and known as the strongest soluble Fe^{3+} binding agents. Iron is an essential element and plays an important role in different physiological processes such as respiration, photosynthesis, DNA synthesis, key factors of many enzymes, terminal oxido-reduction reactions, Fe–S cluster for reductive pool of cellular activities (Dellagi et al., 2009) but its availability of soluble iron is limited by its low solubility in neutral pH. In such Fe-limiting environments, PGPRs are known to be excellent producers of siderophores which are selective ferric ion chelators. Siderophores are produced by either plants or plant associated microorganisms (Loper and Buyer, 1991). In this context, *Pseudomonads* have received much attention over the past years, because of their role in the biological control of soilborne pathogens and in disease suppressive soil. Loper and Buyer (1991) reported production of siderophore by different bacterial genera, like pyoverdines by *Pseudomonas* spp.; hydroxamates by *Erwinia carotovora* and *Enterobacter cloacae*; catechols by *Agrobacterium tumefaciens* and *Erwinia chrysanthemi*; and rhizobactin by *Rhizobium meliloti*. Siderophores produced by PGPR could contribute to enhanced growth (Kloepper et al., 1980). When crops are grown on soil supplemented by pyoverdine or ferriopyoveridne, they exhibited enhancement of total iron content in the roots and ferric reductase activity and total chlorophyll content of the plants (Duijff et al., 1994). The microorganisms producing siderophores have also major role in disease-suppression of soilborne disease especially toward fusarium wilts which was exposed to be related to siderophore-mediated iron competition. Application of pyoverdine produced by *Pseudomonas* to soils advantageous to fusarium wilts (Kloepper et al., 1980). Similar results were reported in PGPR strain of *P. putida* WCS358 against *F. oxysporum* f. sp. *raphani* causing *Fusarium* wilt in radish (De Boer et al., 2003). Additionally, siderophores, namely, pyoveridine and pyocyanin are important in inducing systemic resistance in plants (Leeman et al., 1996; Meziane et al., 2005).

1.5.2.5 PRODUCTION OF VOLATILE ORGANIC COMPOUNDS

Some specific PGPR strains are found to release some mixed chemicals also known as VOCs which have a noteworthy role in plant growth promotion. These volatile compounds have also an important role in the mechanism for the stimulation of growth of plants by rhizobacteria. These compounds have also a major task in ISR mechanisms (Ryu et al., 2004). There are some volatile organic compounds, namely, 2,3-butanediol and acetoin which has been found to be released by certain PGPR strains like *B. subtilis* GB03, *B. amyloliquefaciens* IN937a, and *E. cloacae* JM22 have major role in plant growth promotion of *Arabidopsis thaliana* (Ryu et al., 2003). In *Arabidopsis* against *E. carotovora*, the compounds secreted by these *Bacillus* species have also been able to induce ISR (Ryan et al., 2001). VOCs produced by the rhizobacterial strains can act as signaling molecule in the mediation of plant–microbe interactions as volatiles produced by PGPR colonizing roots are generated at adequate dose to activate the plant responses (Ryu et al., 2003). Some plant volatiles having low molecular weight namely, jasmonates, terpenes, and green leaf components as an effective signal molecules for living organisms in different trophic levels has also been recognized (Farmer, 2001) which have several role in plant-defense mechanisms.

1.5.2.6 PRODUCTION OF BIOTIC ELICITORS

Biotic elicitors are chemicals or biofactors of diverse sources that activate phytoalexin accumulation, morphological and physiological responses in plants. The elicitors are abiotic, namely, inorganic compounds or metal ions or biotic elicitors which are essentially resultants of microorganisms like viruses, bacteria, and fungi as well as chemicals and plant cell wall components that are accumulated due to aggressive plants reaction in opposition to attack of phytopathogens. Biotic elicitors' treatment causes several defense reactions in plant. In crop plants PGPRs act as biotic elicitors, which can induce secondary products synthesis (Sekar and Kandavel, 2010). The main metabolites PGPR species produces in crops in stimulating the morphological and physiological responses are serpentine, ajmalicine, bocrocetin, picrocrocin, scopolamine, hyoscyamine, tanshinone, and safranal compounds.

1.6 MAJOR BENEFITS OF PGPR TO PLANTS

Recently, it has been studied that the enzyme ACC deaminase found in some PGPR species are responsible for stress resistance in plants. Considering all beneficial activity and owing to its nontoxic effects, the concept of PGPR-mediated plant growth promotion is gaining worldwide importance and acceptance (Babalola, 2002; Albino et al., 2006; Gnanamanickam, 2006; Wang et al., 2006) and applied on a wide range of crops including cereals, pulses, vegetables, oilseeds and plantation crops. Apart from the mentioned role, plant disease management through these microbes is a major advantage to the agriculture.

Bacteria belonging to *Pseudomonas* as PGPR were reported by many researchers (Burr et al., 1978; Kloepper and Schroth, 1978; Suslow and Schroth, 1982) in different crops like potato (*Solanum tuberosum*), radish (*Raphanus sativus*), sugar-beet (*B. vulgaris*), etc. Many *Pseudomonas* species suppress to the growth of plant pathogen, namely, *Fusarium* spp. Among Gram-positive bacteria, *Bacillus* and *Paenibacillus* are the most important PGPR. In vitro and field trial study suggested that they can suppress onion white rot diseases by inhibiting *S. cepivorum* by producing antibiotics. It has been reported that, PGPR mixture of *B. amyloliquefaciens* strain IN937a and *B. pumilus* strain IN937b protected plants by inducing systemic resistance against southern blight of tomato caused by *Sclerotium rolfsii*. Among bacterial domain, actinomycetes are the most abundant soil bacteria with high G+C content and can grow both aerobically and anaerobically with fungus like growth.

Recent work suggested that plant growth can be promoted by actinomycetes due to their ability to solubilize phosphate and production of phytohormone. In the green house experiment, it is confirmed that *Micromonospora endolithica* increases the available P content in soil. The beneficial traits are due to production of ACC deaminase and IAA. Currently, these bacteria are used to sustain agriculture as biofertilizers and biocontrol agents. So, a detailed understanding about the types of beneficial microbes, their in-depth mechanism of action, application, and beneficial behavior is necessary before soil microbial technology can be applied in the rhizosphere. Some beneficial effects of PGPR are summarized in Table 1.7.

TABLE 1.7   PGPR and Their Beneficial Effects on Various Plants.

PGPR	Plant species	Effect on plant species
Bacillus subtilis	*Arabidopsis thaliana*	Increase in foliar fresh weight
Citrobacter freundii	*Oryza sativa*	Biofertilizers, produced a significant increase in rice yield
Enterobacter sakazakii	*Zea mays*	Inoculation increases agronomic parameters of maize
Pseudomonas aeruginosa	*Abelmoschus esculentus, Lycopersicon esculentum*	Increase in growth, early fruiting and increased dry biomass
Pseudomonas spp.	*Zea mays*	Causes root elongation in maize
Rhizobium leguminosarum	*Pisum sativum*	Enhanced nodulation in plants

PGPR commercialization is also an important task for its advantageous application and it requires a proper amendment between industries and scientific organization. PGPR strains isolation, screening by pot and field trial, mass production, formulation, toxicology, field efficacy, viability, industrial linkages, and quality control are the some stages of commercialization of PGPRs (Bhattacharyya and Jha, 2012). Some important commercialized products are used in various agricultural/horticultural crops, developed by using different strains of PGPR (Table 1.8).

TABLE 1.8   Important Commercialized Products of PGPR Strains.

Commercial products	Employed PGPR strains	Intended crop
Azo-Green	*Azospirillum brasilense*	Turf and forage crops
Bio-save10	*Pseudomonas syringae*	Citrus and pome fruit
BlightBan A506	*Pseudomonas fluorescens*	Almond, apple, cherry, mushroom, peach, pear, potato, strawberry, and tomato
Blue Circle	*Burlkholderia cepacia*	Alfalfa, barley, beans, clover, cotton, maize, peas, sorghum, vegetables and wheat
Conquer	*Pseudomonas fluorescens*	Almond, apple, cherry, mushroom, peach, pear, potato, strawberry, and tomato
Deny	*Burlkholderia cepacia*	Alfalfa, barley, beans, clover, cotton, maize, peas, sorghum, vegetables, and wheat

TABLE 1.8 (Continued)

Commercial products	Employed PGPR strains	Intended crop
Diegall	*Agrobacterium radiobacter*	Fruit, nut, ornamental nursery stock and trees
Epic	*Bacillus subtilis*	Barley, beans, cotton, legumes peanut, pea, rice, and soybean
Galltrol-A	*Agrobacterium radiobacter*	Fruit, nut, ornamental nursery stock, and trees
HiStick N/T	*Bacillus subtilis*	Barley, beans, cotton, legumes peanut, pea, rice, and soybean
Intercept	*Burlkholderia cepacia*	Alfalfa, barley, beans, clover, cotton, maize, peas, sorghum, vegetables, and wheat
Kodiak	*Bacillus subtilis*	Barley, beans, cotton, legumes peanut, pea, rice, and soybean
Mycostop	*Streptomyces griseovirdis* K61	Field, ornamental, and vegetable crops
Nogall	*Agrobacterium radiobacter*	Fruit, nut, ornamental nursery stock and trees
Norbac 84 C	*Agrobacterium radiobacter*	
Quantum 4000	*Bacillus amyloliquefaciens* GB99	Broccoli, cabbage, cantaloupe, cauliflower, celery, cucumber, lettuce, ornamentals, peppers, tomato ,and watermelon
Rhizo-Plus	*Bacillus subtilis*	Barley, beans, cotton, legumes peanut, pea,
Serenade	*Bacillus subtilis*	Rice, and soybean
Subtilex	*Bacillus subtilis*	
Victus	*Pseudomonas fluorescens*	Almond, apple, cherry, mushroom, peach, pear, potato, strawberry, and tomato

Source: Modified from Bhattacharyya and Jha (2012).

1.7 CONCLUSION

Nitrogen, phosphorous, and potassium are the major chemical fertilizers applied in farm lands to enhance crop productivity. This activity leads to groundwater contamination due to extensive mineralization and leaching that ultimately leads to eutrophication and health risks. Excess use of these chemicals causes soil degradation and loss of biodiversity. Nowadays, the cost of fertilizer is steeply increasing as a consequence of the increasing energy prices. For these reasons, the interest in sustainable agriculture using balanced fertilization in organic way, with the help of microbes is of prime focus. The current need is to develop selected low cost effective and eco-friendly technology keeping an eye on farmer's productivity and profit as well as making agriculture sustainable for long term. The action of biofertilization and biocontrol are the promising and key roles of PGPR making agriculture productive in more organic and way, replacing synthetic fertilizers. Exploration of genes and traits involved in this process, their functional characterization, and metabolic engineering of their cell processes will result in detailed awareness of bacterial rhizospheric ecology, physiology, and its interaction with plant roots, which will assist more effective information and strategies for risk management and infection control.

1.8 SUMMARY

World population is increasing at maximal rate and food production is at risk. For getting higher food productivity use of chemical fertilizers has become a swift alternative. But the synthetic fertilizers are not eco-friendly due to their chemical toxicity that can cause pollution of soil, water and air, and can cause several side effects to the human health. They also destroy the soil fertility in a long run. As an ecoalternative of this, people tried of using organic fertilizer. Among all, conventional fertilizers which includes compost; household wastes and green manure can be used in farm fields but not as effective as chemical fertilizers. So, in the recent times researchers have established the way of organic farming through use of various microorganisms that are indigenous, beneficial and resides in the vicinity of the plant using plant-derived metabolites and help in the plant growth and development. By use of these beneficial organisms chemical pollution in farm lands can be prevented. These microorganisms are regarded as PGPR and promote the adequate nutrients supply to the host plants and warrant their suitable growth development and physiological regulation. The present chapter describes the different role of PGPR in agriculture ecosystem in direct as well as indirect way of mechanisms. The direct way of mechanisms includes, root colonization, production of growth regulators, nitrogen fixation, increased uptake of minerals and soil fertility, plant growth enhancement while indirect mechanisms are antibiotics, lytic enzymes, siderophores, volatile organic compounds and biotic elicitors productions and also ISR. Apart from these major benefits of PGPR to plants has also been taken into consideration in this chapter. Such microbes are used in the biofertilizers preparation, being a vital module of organic farming play key role in upholding long-term soil fertility and sustainability.

KEYWORDS

- agriculture ecosystem
- growth promotion
- plant health–promoting rhizobacteria
- plant pathogen
- plant-defense mechanisms
- plant–microbe interactions
- soil bacteria

REFERENCES

Adesemoye, A. O.; Torbert, H. A.; Kloepper, J. W. Enhanced Plant Nutrient Use Efficiency with PGPR and AMF in an Integrated Nutrient Management System. *Can. J. Microbiol.* **2008,** *54,* 876–886.

Ahmad, F.; Ahmad, I.; Khan, M. S. Indole Acetic Acid Production by the Indigenous Isolates of *Azotobacter* and *Fluorescent pseudomonas* in the Presence and Absence of Tryptophan. *Turk. J. Biol.* **2005,** *29,* 29–34.

Ahmad, M.; Zahir, Z. A.; Khalid, M. Efficacy of Rhizobium and Pseudomonas Strains to Improve Physiology, Ionic Balance and Quality of Mung Bean Under Salt-Affected Conditions on Farmer's Fields. *Plant Physiol. Biochem.* **2013,** *63,* 170–176.

Ahmed, A.; Hasnain, S. Auxin Producing *Bacillus* sp.: Auxin Quantification and Effect on the Growth *Solanum tuberosum. Pure Appl. Chem.* **2010,** *82,* 313–319.

Akhtar, M. J.; Asghar, H. N.; Shahzad, K.; Arshad, M. Role of Plant Growth Promoting Rhizobacteria Applied in Combination with Compost and Mineral Fertilizers to Improve Growth and Yield of Wheat (*Triticum aestivum* L.). *Pak. J. Bot.* **2009,** *41,* 381–390.

Albino, U.; Saridakis, D. P.; Ferreira, M. C.; Hungria, M.; Vinuesa, P.; Andrade, G. High Diversity of Diazotrophic Bacteria Associated with the Carnivorous Plant *Drosera villosa* var. Villosa Growing in Oligotrophic Habitats in Brazil. *Plant Soil* **2006,** *287,* 199–207.

Alstroem, S. Induction of Disease Resistance in Common Bean Susceptible to Halo Blight Bacterial Pathogen After Seed Bacterization with Rhizosphere Pseudomonads. *J. Gen. Appl. Microbiol.* **1991,** *37,* 495–501.

Amemura, A.; Harada, T.; Abe, M.; Higashi, S. Structural Studies on the Extracellular Acidic Polysaccharide from *Rhizobium trifolii* 4S. *Carbohydr. Res.* **1983,** *115,* 165–174.

Anjum, M. A.; Sajjad, M. R.; Akhtar, N.; Qureshi, M. A.; Iqbal, A.; Jami, A. R.; Hassan, M. Response of Cotton to Plant Growth Promoting Rhizobacteria (PGPR) Inoculation Under Different Levels of Nitrogen. *J. Agric. Res.* **2007**, *45*(2), 135–143.

Antoun, H.; Beauchamp, C. J.; Goussard, N.; Chabot, R.; Lalande, R. Potential of *Rhizobium* and *Bradyrhizobium* Species as Plant Growth Promoting Rhizobacteria on Non-Legumes: Effect on Radishes (*Raphanus sativus* L.). *Plant Soil* **1998**, *204*, 57–67.

Antoun H.; Prevost D. Ecology of Plant Growth Promoting Rhizobacteria. PGPR: *Biocontrol Biofertil.* **2006**, *15*, 1–38.

Asghar H. N.; Zahir Z. A.; Arshad M.; Khaliq A. Relationship Between In Vitro Production of Auxins by Rhizobacteria and Their Growth-Promoting Activities in *Brassica Junceal*. *Biol. Fertil. Soils* **2002**, *35*(23), 1–237.

Ashrafuzzaman M.; Hossen F. A.; Ismail M. R.; Hoque M. A.; Islam M. Z.; Shahidullah S. M.; Meon S. Efficiency of Plant Growth-Promoting Rhizobacteria (PGPR) for the Enhancement of Rice Growth. *Afr. J. Biotechnol.* **2009**, 8:1247–1252.

Babalola, O. O. *Interactions Between Striga hermonthica (Del.) Benth. and Fluorescent Rhizosphere Bacteria of Zea mays, L. and Sorghum bicolor L. Moench for Striga Suicidal Germination in Vigna unguiculata*; Department of Botany and Microbiology, University of Ibadan: Ibadan, 2002.

Babalola, O. O. Ethylene Quantification in Three Rhizobacterial Isolates from *Striga hermonthica*-Infested Maize and Sorghum. *Egypt. J. Biol.* **2010**, *12*, 1–5.

Babalola, O. O.; Osir, E. O.; Sanni, A. I.; Odhaimbo, G. D.; Bulimo, W. D. Amplification of 1-Aminocyclopropane-1-Carboxylic (ACC) Deaminase from Plant Growth Promoting Rhizobacteria in Striga-infested Soils. *Afr. J. Biotechnol.* **2003**, *2*(6), 157–160.

Baldani, V. L. D.; Baldani, J. I.; Dobereiner, J. Inoculation of Rice Plants with the Endophytic Diazotrophs *Herbaspirillums seropidicae*. *Biol. Fertil. Soils* **2000**, *30*, 485–491.

Barazani, O.; Friedman, J. Is IAA the Major Root Growth Factor Secreted from Plant-Growth-Mediating Bacteria? *J. Chem. Ecol.* **1999**, *25*(10), 2397–2406.

Bashan, Y.; Bustillos, J. J.; Leyva, L. A.; Hernandez, J. P.; Bacilio, M. Increase in Auxiliary Photoprotective Photosynthetic Pigments in Wheat Seedlings Induced by *Azospirillum brasilense*. *Biol. Fertil. Soil* **2006**, *42*, 279–285.

Bashan, Y.; de-Bashan, L. E. How the Plant Growth-Promoting Bacterium *Azospirillum* Promotes Plant Growth a Critical Assessment. *Adv. Agron.* **2010**, *108*, 77–136.

Bashan, Y.; de-Bashan, L. E. Fresh-Weight Measurements of Roots Provide Inaccurate Estimates of the Effects of Plant Growth-Promoting Bacteria on Root Growth: A Critical Examination. *Soil Biol. Biochem.* **2005**, *37*, 1795–1804.

Belimov, A. A.; Dodd, I. C.; Safronova, V. I.; Hontzeas, N.; Davies, W. J. *Pseudomonas brassicacearum* Strain Am3 Containing 1-Aminocyclopropane-1-Carboxylate Deaminase Can Show Both Pathogenic and Growth-Promoting Properties in its Interaction with Tomato. *J. Exp. Bot.* **2007**, *58*, 1485–1495.

Belimov, A. A.; Safronova, V.; Sergeyeva, T. A.; Egorova, T. N.; Matveyeva, V. A.; Tsyganov, V. E.; Borisov, A. Y.; Tikhonovich, I. A.; Kluge, C.; Preisfeld, A.; Dietz, K. J.; Stepanok, V. V. Characterization of Plant Growth Promoting Rhizobacteria Isolated from Polluted Soils and Containing 1-Aminocyclopropane-1-Carboxylate Deaminase. *Can. J. Microbiol.* **2001**, *47*, 642–652.

Belimov, A. A.; Hontzeas, N.; Safronova, V. I.; Demchinskaya, S. V.; Piluzza, G.; Bullitta, S.; Glick, B. R. Cadmium-Tolerant Plant Growth-Promoting Bacteria Associated with the Roots of Indian Mustard (*Brassica juncea* L. Czern.). *Soil Biol. Biochem.* **2005**, *7*, 241–250.

Beneduzi, A.; Peres, D.; Vargas, L. K.; Bodanese-Zanettini, M. H.; Passaglia, L. M. P. Evaluation of Genetic Diversity and Plant Growth Promoting Activities of Nitrogen-Fixing Bacilli Isolated from Rice Fields in South Brazil. *Appl. Soil Ecol.* **2008**, *39*, 311–320.

Benizri, E.; Baudoin, E.; Guckert, A. Root Colonization by Inoculated Plant Growth Promoting Rhizobacteria. *Biocontrol Sci. Technol.* **2001**, *11*, 557–574.

Berendsen, R. L.; Pieterse, C. M. J.; Bakker, P. A. H. M. The Rhizosphere Microbiome and Plant Health. *Trends Plant Sci.* **2012**, *17*(8), 478–486.

Bhattacharyya, P. N.; Jha D. K. Plant Growth-Promoting Rhizobacteria (PGPR): Emergence in Agriculture. *World J. Microbiol. Biotechnol.* **2012**, *28*, 1327–1350.

Biswas, J. C.; Ladha, J. K.; Dazzo, F. B.; Yanni, Y. G.; Rolf, B. G. Rhizobial Inoculation Influences Seedling Vigor and Yield of Rice. *Agron. J.* **2000**, *90*, 880–886.

Bloemberg, G. V.; Lugtenberg, B. J. J. Molecular Basis of Plant Growth Promotion and Biocontrol by Rhizobacteria. *Curr. Opin. Plant Biol.* **2001**, *4*, 343–350.

Bonaterra, A.; Ruz, L.; Badosa, E.; Pinochet, J.; Montesinos, E. Growth Promotion of Prunus Rootstocks by Root Treatment with Specific Bacterial Strains. *Plant Soil* **2003**, *255*, 555–569.

Boyer, M.; Bally, R.; Perrotto, S.; Chaintreuil, C.; Wisniewski-Dye, F. A Quorum-Quenching Approach to Identify Quorum-Sensing Regulated Functions in *Azospirillum lipoferum*. *Res. Microbiol.* **2008**, *159*, 699–708.

Braud, A.; Jezequel, K.; Bazot, S.; Lebeau, T. Enhanced Phytoextraction of an Agricultural Cr-, Hg- and Pb-Contaminated Soil by Bioaugmentation with Siderophore Producing Bacteria. *Chemosphere* **2009**, *74*, 280–286.

Burr, T. J.; Schroth, M. N.; Suslow, T. Increased Potato Yields by Treatment of Seed Pieces with Specific Strains of *Pseudomonas fluorescens* and *P. putida*. *Phytopathology* **1978**, *68*, 1377–1383.

Çakmakçi, R.; Donmez, M. F.; Erdogan, U. The Effect of Plant Growth Promoting Rhizobacteria on Barley Seedling Growth, Nutrient Uptake, Some Soil Properties, and Bacterial Counts. *Turk. J. Agric. For.* **2007a**, 31:189–199.

Çakmakçi, R.; Erat, M.; Erdoğan, Ü. G.; Dönmez, M. F. The Influence of PGPR on Growth Parameters, Antioxidant and Pentose Phosphate Oxidative Cycle Enzymes in Wheat and Spinach Plants. *J. Plant Nutr. Soil Sci.* **2007b**, 170:288–295.

Canbolat, M. Y.; Bilen, S.; Çakmakc, R.; Sahin, F.; Aydın, A. Effect of Plant Growth-Promoting Bacteria and Soil Compaction on Barley Seedling Growth, Nutrient Uptake, Soil Properties and Rhizosphere Microflora. *Biol. Fertil. Soils* **2006**, *42*, 350–357.

Canter-Cremers, H. C. J.; Stevens, K.; Lugtenberg, B. J. J.; Wijffelman, C. A.; Batley, M.; Redmond, J. W.; Breedveld, M.; Zevenhuizen, L. P. T. M. Unusual Structure of the Exopolysaccharide of *Rhizobium leguminosarum* bv. Viciae Strain 248. *Carbohydr. Res.* **1991**, *218*, 185–200.

Carrillo-Castaneda, G.; Munoz, J. J.; Peralta-Videa, J. R.; Gomez, E.; Gardea-Torresdey, J. L. Plant Growth-Promoting Bacteria Promote Copper and Iron Translocation from Root to Shoot in Alfalfa Seedlings. *J. Plant Nutr.* **2003**, *26*, 1801–1814.

Cassan, F.; Perrig, D.; Sgroy, V.; Masciarelli, O.; Penna, C.; Luna, V. *Azospirillum brasilense* Az39 and *Bradyrhizobium japonicum* E109, Inoculated Singly or in Combination,

Promote Seed Germination and Early Seedling Growth in Corn (*Zea mays* L.) and Soybean (*Glycine max* L.). *Eur. J. Soil Biol.* **2009**, *45*, 28–35.

Cattelan, A. J.; Hartel, P. G.; Fuhrmann, J. J. Screening for Plant Growth-Promoting Rhizobacteria to Promote Early Soybean Growth. *Soil Sci. Soc. Am. J.* **1999**, *63*, 1670–1680.

Cazorla, F. M.; Duckett, S. B.; Bergstrom, F. T.; Noreen, S.; Odik, R. et al. Biocontrol of Avocado Dematophora Root Rot by the Antagonistic *Pseudomonas fluorescens* PCL 1606 Correlates with the Production 2-Hexyl-5-Propyl Resorcinol. *Mol. Plant Microbe Interact.* **2006**, *19*, 418–428.

Chandra, S.; Choure, K.; Dubey, R. C.; Maheshwari, D. K. Rhizosphere Competent *Mesorhizobium loti* MP6 Induces Root Hair Curling, Inhibits *Sclerotinia sclerotiorum* and Enhances Growth of Indian Mustard (*Brassica campestris*). *Braz. J. Microbiol.* **2007**, *38*, 124–130.

Chang, C. H.; Yang, S. S. Thermo-Tolerant Phosphate-Solubilizing Microbes for Multifunctional Biofertilizer Preparation. *Bioresour. Technol.* **2009**, *100*, 1648–1658.

Dazzo, F. B.; Yanni, Y. G.; Rizk, R.; De Bruijn, F. J.; Rademaker, J.; Squartini, A.; Corich, V.; Mateos, P.; Martinez-Molina, E.; Velázquez, E.; Biswas, J. C; Hernandez, R. J.; Ladha, J. K.; Hill, J.; Weinman, J.; Rolfe, B. G.; Vega-Hernández, M.; Bradford, J. J.; Hollingsworth, R. I.; Ostrom, P.; Marshall, E.; Jain, T.; Orgambide, G.; Philip-Hollingsworth, S.; Triplett, E.; Malik, K. A.; Maya-Flores, J.; Hartmann, A.; Umali-Garcia, M.; Izaguirre-Mayoral, M. L. Progress in Multinational Collaborative Studies on the Beneficial Association Between Rhizobium leguminosarum bv. Trifolii and Rice. In *The Quest for Nitrogen Fixation in Rice*; Ladha J. K., Reddy P. M., Eds.; IRR1: Los Banos, Philippines, 2000; pp 167–189.

De Boer, M.; Bom, P.; Kindt, F.; Keurentjes, J. J. B.; van der Sluis, I.; van Loon, L. C.; Bakker, P. A. H. M. Control of *Fusarium* Wilt of Radish by Combining *Pseudomonas putida* Strains That Have Different Disease-Suppressive Mechanisms. *Phytopathology* **2003**, *93*, 626–632.

De Bruijn, F. J. *Molecular Microbial Ecology of the Rhizosphere*; de Bruijn, F. J., Ed.; John Wiley & Sons, Inc.; Hoboken, New Jersey, 2013; Vols. 1 and 2.

Dell'Amico, E.; Cavalca, L.; Andreoni, V. Improvement of *Brassica napus* Growth Under Cadmium Stress by Cadmium Resistant Rhizobacteria. *Soil Biol. Biochem.* **2008**, *40*, 74–84.

Dellagi, A.; Segond, D.; Rigault, M.; Fagard, M.; Simon, C.; Saindrenan, P.; Expert, D. Microbial Siderophores Exert a Subtle Role in *Arabidopsis* During Infection by Manipulating the Immune Response and the Iron Status. *Plant Physiol.* **2009**, *150*, 1687–1696.

Dennis P. G.; Miller A. J.; Hirsch P. R. Are Root Exudates More Important Than Other Sources of Rhizodeposits in Structuring Rhizosphere Bacterial Communities? *FEMS Microbiol. Ecol.* **2010**, *72*, 313–327.

Dey, R.; Pal, K. K.; Bhatt, D. M.; Chauhan, S. M. Growth Promotion and Yield Enhancement of Peanut (*Arachis hypogaea* L.) by Application of Plant Growth-Promoting Rhizobacteria. *Microbiol. Res.* **2004**, *159*, 371–394.

Di Gregorio, S.; Barbafieri, M.; Lampis, S.; Sanangelantoni, A. M.; Tassi, E.; Vallini, G. Combined Application of Triton X-100 and *Sinorhizobium* sp. Pb002 Inoculum for the Improvement of Lead Phytoextraction by *Brassica juncea* in EDTA Amended Soil. *Chemosphere* **2006**, *63*, 293–299.

Dimkpa, C.; Aleš, S.; Dirk, M.; Georg, B.; Erika, K. Hydroxamate Siderophores Produced by *Streptomyces acidiscabies* E13 Bind Nickel and Promote Growth in Cowpea (*Vigna unguiculata* L.) Under Nickel Stress. *Can. J. Microb.* **2008**, *54*, 163–172.

Djordjevic, S. P.; Batley, M.; Redmond, J. W.; Rolfe, B. G. The Structure of the Exopolysaccharide from *Rhizobium* sp. Strain ANU280 (NGR234). *Carbohydr. Res.* **1986**, *148*, 87–99.

Dobbelaere, S.; Croonenborghs, A.; Thys, A.; Ptacek, D. et al. Responses of Agronomically Important Crops to Inoculation with *Azospirillum*. *Aust. J. Plant Physiol.* **2001**, *28*, 871–879.

Dobereiner, J. Biological Nitrogen Fixation in the Tropics: Social and Economic Contributions. *Soil Biol. Biochem.* **1997**, *29*, 771–774.

Dobereiner, J. Nitrogen-Fixing Bacteria of the Genus Beijerinckia Derx in the Rhizosphere of Sugar Cane. *Plant Soil* **1961**, *15*, 211–216.

Dohroo, A.; Sharma, D. R. Role of Plant Growth Promoting Rhizobacteria, Arbuscular Mycorrhizal Fungi and Their Helper Bacteria on Growth Parameters and Root Rot of Apple. *World J. Sci. Technol.* **2012**, *2*, 35–8.

Dubey, S. K. Combined Effect of *Bradyrhizobium japonicum* and Phosphate-Solubilizing *Pseudomonas striata* on Nodulation, Yield Attributes and Yield of Rainfed Soybean (*Glycine max*) Under Different Sources of Phosphorus in Vertisols. *Ind. J. Microbiol.* **1996**, *33*, 61–65.

Dudman, W. F.; Franzén, L.-E.; Darvill, J. E.; McNeil, M.; Darvill, A. G.; Albersheim, P. The Structure of the Acidic Polysaccharide Secreted by *Rhizobium phaseoli* Strain 127 K36. *Carbohydr. Res.* **1983**, *117*, 141–156.

Duijff, B. J.; Bakker, P. A. H. M.; Schippers, B. Suppression of *Fusarium* Wilt of Carnation by *Pseudomonas putida* WCS358 at Different Levels of Disease Incidence and Iron Availability. *Biocontrol Sci. Technol.* **1994**, *4*, 279–288.

Dunne, C.; Crowley, J. J.; Moënne-Loccoz, Y.; Dowling, D. N.; de Bruijn, F. J.; O'Gara, F. Biological Control of *Pythium ultimum* by *Stenotrophomonas maltophilia* W81 Is Mediated by an Extracellular Proteolytic Activity. *Microbiology* **1997**, *143*, 3921–3391.

Dunne, C.; Moënne-Loccoz, Y.; McCarthy, J.; Higgins, P.; Powell, J.; Dowling, D. N.; O'Gara, F. Combining Proteolytic and Phloroglucinol-Producing Bacteria for Improved Biocontrol of Pythium-Mediated Damping-Off of Sugar Beet. *Pathology* **1998**, *47*, 299–307.

Dutta, S.; Mishra, A. K.; Dileep Kumar, B. S. Induction of Systemic Resistance Against Fusarial Wilt in Pigeon Pea Through Interaction of Plant Growth Promoting Rhizobacteria and Rhizobia. *Soil Biol. Biochem.* **2008**, *40*, 452–461.

Egamberdieva, D. Plant Growth Promoting Rhizobacteria Isolated from a Calsisol in Semi Arid Region of Uzbekistan: Biochemical Characterization and Effectiveness. *J. Plant Nutr. Soil Sci.* **2005**, *168*, 94–99.

Egamberdiyeva, D. Characterization of *Pseudomonas* Species Isolated from the Rhizosphere of Plants Grown in Serozem Soil, Semi-Arid Region of Uzbekistan. *Sci. World J.* **2005**, *5*, 501–509.

Egamberdiyeva, D. The Effect of Plant Growth Promoting Bacteria on Growth and Nutrient Uptake of Maize in Two Different Soils. *Appl. Soil Ecol.* **2007**, *36*, 184–189.

El Zemrany, H.; Cortet, J.; Lutz, M. P.; Chabert, A.; Baudoin, E.; Haurat, J.; Maughan, N.; Felix, D.; Defago, G.; Bally, R.; Moenne-Loccoz, Y. Field Survival of the Phytostimulator *Azospirillum lipoferum* CRT1 and Functional Impact on Maize Crop, Biodegradation of

Crop Residues, and Soil Faunal Indicators in a Context of Decreasing Nitrogen Fertilisation. *Soil Biol. Biochem.* **2006,** *38,* 1712–1726.

El-Akhal, M. R.; Rincon, A.; Coba de la Peña, T.; Lucas, M. M.; El Mourabit, N.; Barrijal, S.; Pueyo, J. J. Effects of Salt Stress and Rhizobial Inoculation on Growth and Nitrogen Fixation of three Peanut Cultivars. *Plant Biol.* **2013,** *15,* 415–421.

Elbadry, M.; Taha, R. M.; Eldougdoug, K. A.; Gamal-Eldin, H. Induction of Systemic Resistance in Faba Bean (*Vicia faba* L.) to Bean Yellow Mosaic Potyvirus (BYMV) via Seed Bacterization with Plant Growth Promoting Rhizobacteria. *J. Plant Dis. Prot.* **2006,** *113,* 247–251.

Elbeltagy, A.; Nishioka, K.; Sato, T. Endophytic Colonization and in Planta Nitrogen Fixation by a *Herbaspirillum* sp. Isolated from Wild Rice Species. *Appl. Environ. Microbiol.* **2001,** *67,* 5285–5293.

Erturk, Y.; Ercisli, S.; Haznedar, A.; Cakmakci, R. Effects of Plant Growth Promoting Rhizobacteria (PGPR) on Rooting and Root Growth of Kiwifruit (*Actinidia deliciosa*) Stem Cuttings. *Biol. Res.* **2010,** *43,* 91–98.

Esitken, A.; Karlidag, H.; Ercisli, S.; Turan, M.; Sahin, F. The Effect of Spraying a Growth Promoting Bacterium on the Yield, Growth and Nutrient Element Composition of Leaves of Apricot (*Prunus armeniaca* L. cv. *hacihaliloglu*). *Aust. J. Agric. Res.* **2003,** *54,* 377–380.

Esitken, A.; Pirlak, L.; Turan, M.; Sahin, F. Effects of Floral and Foliar Application of Plant Growth Promoting Rhizobacteria (PGPR) on Yield, Growth and Nutrition of Sweet Cherry. *Sci. Hortic.* **2006,** *110,* 324–327.

Faisal, M.; Hasnain, S. Bacterial Cr (VI) Reduction Concurrently Improves Sunflower (*Helianthus annuus* L.) Growth. *Biotechnol. Lett.* **2005,** *27,* 943–947.

Faisal, M.; Hasnain, S. Growth Stimulatory Effect of *Ochrobactrum intermedium* and *Bacillus cereus* on *Vigna radiata* Plants. *Lett. Appl. Microbiol.* **2006,** *43,* 461–466.

Farmer, E. E. Surface-to-Air Signals. *Nature* **2001,** *411,* 854–856.

Fernando, W. G. D.; Nakkeeran, S.; Zhang, Y. Biosynthesis of Antibiotics by PGPR and Its Relation in Biocontrol of Plant Diseases. In *PGPR: Biocontrol Biofertilization*; Siddiqui, Z. A. Ed.; Springer: Netherlands, 2006; pp. 67–109.

Figueiredo, M. V. B.; Burity, H. A.; Martinez, C. R.; Chanway, C. P. Alleviation of Drought Stress in Common Bean (*Phaseolus vulgaris* L.) by Co-Inoculation with *Paenibacillus polymyxa* and *Rhizobium tropici. Appl. Soil Ecol.* **2008,** *40,* 182–188.

Fischer, S. E.; Fischer, S. I.; Magris, S.; Mori, G. B. Isolation and Characterization of Bacteria from the Rhizosphere of Wheat. *World J. Microbiol. Biotechnol.* **2007,** *23,* 895–903.

Flores-Felix, J. D.; Menendez, E.; Rivera, L. P. Use of *Rhizobium leguminosarum* as a Potential Biofertilizer for *Lactuca sativa* and *Daucus carota* Crops. *J. Plant Nutr. Soil Sci.* **2013,** *176,* 876–882.

Flores-Felix, J. D.; Silva, L. R.; Rivera, L. P. Plants Probiotics as a Tool to Produce Highly Functional Fruits: The Case of Phyllobacterium and Vitamin C in Strawberries. *PLoS ONE* **2015,** *10,* e0122281.

Freitas, A. D. S.; Vieira, C. L.; Santos, C. E. R. S.; Stamford, N. P.; Lyra, M. C. C. P. Caracterização de rizóbios isolados de jacatupé cultivado em solo salino no Estado de Pernanbuco, Brasil. *Bragantia* **2007,** *66,* 497–504 (Article in Portuguese).

Friesen, M. L.; Porter, S. S.; Stark, S. C.; Wettberg, E. J.; Sachs, J. L.; Martinez-Romero, E. Microbially Mediated Plant Functional Traits. *Annu. Rev. Ecol. Evol. Syst.* **2011**, *42*, 23–46.

Gabriela, F.; Casanovas, E. M.; Quillehauquy, V.; Yommi, A. K.; Goni, M. G.; Roura, S. I.; Barassi, C. A. Azospirillum Inoculation Effects on Growth, Product Quality and Storage Life of Lettuce Plants Grown Under Salt Stress. *Sci. Hortic.* **2015**, *195*, 154–162.

Gal, S. W.; Choi, Y. J. Isolation and Characterization of Salt Tolerance Rhizobia from *Acacia* Root Nodules. *Agric. Chem. Biotechnol.* **2003**, *46*, 58–62.

Gamalero, E.; Trotta, A.; Massa, N.; Copetta, A.; Martinotti, M. G.; Berta, G. Impact of Two Fluorescent Pseudomonads and an Arbuscular Mycorrhizal Fungus on Tomato Plant Growth, Root Architecture and P Acquisition. *Mycorrhiza* **2004**, 14:185–192.

Ganesan, V. Rhizoremediation of Cadmium Soil Using a Cadmium-Resistant Plant Growth-Promoting Rhizopseudomonad. *Curr. Microbiol.* **2008**, *56*, 403–407.

Garcia de Salamone, I. E.; Hynes, R. K.; Nelson, L. M. Cytokinin Production by Plant Growth Promoting Rhizobacteria and Selected Mutants. *Can. J. Microbiol.* **2001**, *47*, 404–411.

Gholami, A.; Shahsavani, S.; Nezarat, S. The Effect of Plant Growth Promoting Rhizobacteria (PGPR) on Germination, Seedling Growth and Yield of Maize. *Int. J. Biol. Life Sci.* **2009**, *1*, 35–40.

Ghosh, S.; Penterman, J. N.; Little, R. D.; Chavez, R.; Glick, B. R. Three Newly Isolated Plant Growth-Promoting Bacilli Facilitate the Seedling Growth Of Canola, *Brassica campestris*. *Plant Physiol. Biochem.* **2003**, *41*, 277–281.

Gil-Serrano, A.; Sanchez del Junco, A.; Tejero-Mateo, P. Structure of the Extracellular Acidic Polysaccharide Secreted by *Rhizobium leguminosarum* var. Phaseoli CIAT899. *Carbohydr. Res.* **1990**, *204*, 103–107.

Glass, A. D. M. *Plant Nutrition: An Introduction to Current Concepts;* Jones and Bartlett Publishers: Boston, 1989; p 234.

Glass, A. D. M.; Britto, D. T.; Kaiser, B. N. et al. The Regulation of Nitrate and Ammonium Transport Systems in plants. *J. Exp. Bot.* **2002**, *53*, 855–864.

Glick, B. R. The Enhancement of Plant Growth by Free Living Bacteria. *Can. J. Microbiol.* **1995**, *41*, 109–117.

Gnanamanickam, S. S. *Plant-Associated Bacteria*; Springer: Netherlands, 2006; pp. 195–218.

Gonzalez, A. J.; Larraburu, E. E.; Llorente, B. F. *Azospirillum brasilense* Increased Salt Tolerance of Jojoba During In Vitro Rooting. *Ind. Crop Prod.* **2015**, *76*, 41–48.

Govindarajan, M.; Balandreau, J.; Kwon, S. W. Effects of the inoculation of *Burkholderia vietnamensis* and Related Endophytic Diazotrophic Bacteria on Grain Yield of Rice. *Microb. Ecol.* **2007**, *55*, 21–37.

Gupta, A.; Rai, V.; Bagdwal, N.; Goel, R. *In-situ* Characterization of Mercury Resistant Growth Promoting Fluorescent Pseudomonads. *Microbiol. Res.* **2005**, *160*, 385–388.

Gupta, G.; Parihar, S. S.; Ahirwar, N. K.; Snehi, S. K.; Singh, V. Plant Growth Promoting Rhizobacteria (PGPR): Current and Future Prospects for Development of Sustainable Agriculture. *J. Microb. Biochem. Technol.* **2015**, *7*(2), 096–102.

Hadi, F.; Bano, A. Effect of Diazotrophs (*Rhizobium* and *Azatobactor*) on Growth of Maize (*Zea mays* L.) and Accumulation of Lead (PB) in Different Plant Parts. *P. J. Bot.* **2010**, *42*, 4363–4370.

Hameeda, B.; Harini, G.; Rupela, O. P.; Wani, S. P.; Reddy, G. Growth Promotion of Maize by Phosphate-Solubilizing Bacteria Isolated from Composts and Macrofauna. *Microbiol. Res.* **2008**, *163*, 234–242.

Han, J.; Sun, L.; Dong, X.; Cai, Z.; Sun, X.; Yang, H.; Wang, Y.; Song, W. Characterization of a Novel Plant Growth-Promoting Bacteria Strain *Delftia tsuruhatensis* HR4 both as a Diazotroph and a Potential Biocontrol Agent Against Various Plant Pathogens. *Syst. Appl. Microbiol.* **2005**, *28*, 66–76.

Han, H. S.; Lee, K. D. Phosphate and Potassium Solubilizing Bacteria Effect on Mineral Uptake, Soil Availability and Growth of Eggplant. *Res. J. Agric. Biol. Sci.* **2005**, *1*, 176–180.

Han, H. S.; Supanjani, S.; Lee, K. D. Effect of Co-Inoculation with Phosphate and Potassium Solubilizing Bacteria on Mineral Uptake and Growth of Pepper and Cucumber. *Plant Soil Environ.* **2006**, *52*, 130–136.

Hartmann, A.; Rothballer, M.; Schmid, M. Lorenz Hiltner, A Pioneer in Rhizosphere Microbial Ecology and Soil Bacteriology Research. *Plant Soil* **2008**, *312*, 7–14.

Hayat, R.; Ali, S.; Amara, U.; Khalid, R.; Ahmed, I. Soil Beneficial Bacteria and Their Role in Plant Growth Promotion: A Review. *Ann. Microbiol.* **2010**, *60*, 579–598.

Hernández-Rodríguez, A.; Heydrich-Pérez, M.; Acebo-Guerrero, Y.; Velázquez-del Valle, M. G.; Hernández-Lauzardo, A. N. Antagonistic Activity of Cuban Native Rhizobacteria Against *Fusarium verticillioides* (Sacc.) Nirenb. in Maize (*Zea mays* L.). *Appl. Soil. Ecol.* **2008**, *36*, 184–186.

Hossain, M. S.; Martensson, A. Potential Use of *Rhizobium* spp. to Improve Fitness of Nonnitrogen-Fixing Plants. *Acta Agric. Scand. B—Soil Plant Sci.* **2008**, *58*, 352–358.

Hynes, R. K.; Leung, G. C.; Hirkala, D. L.; Nelson, L. M. Isolation, Selection, and Characterization of Beneficial Rhizobacteria from Pea, Lentil and Chickpea Grown in Western Canada. *Can. J. Microbiol.* **2008**, *54*, 248–258.

Idris, A.; Labuschagne, N.; Korsten, L. Efficacy of Rhizobacteria for Growth Promotion in Sorghum Under Greenhouse Conditions and Selected Modes of Action Studies. *J. Agric. Sci.* **2009**, *147*, 17–30.

Islam, M. Z.; Sattar, M. A.; Ashrafuzzaman, M.; Berahim, Z.; Shamsuddoha, A. T. M. Evaluating Some Salinity Tolerant Rhizobacterial Strains to Lentil Production Under Salinity Stress. *Int. J. Agric. Biol.* **2013**, *15*, 499–504.

Jaderlund, L.; Arthurson, V.; Granhall, U.; Jansson, J. K. Specific Interactions Between Arbuscular Mycorrhizal Fungi and Plant Growth-Promoting Bacteria: As Revealed by Different Combinations. *FEMS Microbiol. Lett.* **2008**, *287*, 174–180.

Jahanian, A.; Chaichi, M. R.; Rezaei, K.; Rezayazdi, K.; Khavazi, K. The Effect of Plant Growth Promoting Rhizobacteria (PGPR) on Germination and Primary Growth of Artichoke (*Cynara scolymus*). *Int. J. Agric. Crop Sci.* **2012**, *4*, 923–929.

Jalili, F.; Khavazi, K.; Pazira, E. et al. Isolation and Characterization of ACC Deaminase-Producing Fluorescent Pseudomonads, to Alleviate Salinity Stress on Canola (*Brassica napus* L.) Growth. *J. Plant Physiol.* **2009**, *166*(6), 667–674.

James, E. K.; Gyaneshwar, P.; Mathan, N. et al. Infection and Colonization of Rice Seedlings by the Plant Growth-Promoting Bacterium *Herbaspirillum seropedicae* Z67. *Mol. Plant Microbe Interact.* **2002**, *15*, 894–906.

James, E. K.; Olivares, F. L.; de Oliveira, A. L. M.; dos Reis, F. B.; da Silva, L. G.; Reis, V. M. Further Observations on the Interaction Between Sugar Cane and *Gluconacetobacter diazotrophicus* Under Laboratory and Greenhouse Conditions. *J. Exp. Bot.* **2001,** *52*, 747–760.

Joo, G. J.; Kin, Y. M.; Kim, J. T.; Rhee, I. K.; Kim, J. H.; Lee, I. J. Gibberellins-Producing Rhizobacteria Increase Endogenous Gibberellins Content and Promote Growth of Red Peppers. *J. Microbiol.* **2005,** *43*(6), 510–515.

Kamilova, F.; Kravchenko, L. V.; Shaposhnikov, A. I.; Makarova, N.; Lugtenberg, B. Effects of the Tomato Pathogen *Fusarium oxysporum* f. sp. Radicis-Lycopersici and of the Biocontrol Bacterium *Pseudomonas fluorescens* WCS365 on the Composition of Organic Acids and Sugars in Tomato Root Exudate. *Mol. Plant Microbe Interact.* **2006,** *19*, 1121–1126.

Kamilova, F.; Lamers, G.; Lugtenberg, B. Biocontrol Strain *Pseudomonas fluorescens* WCS365 Inhibits Germination of *Fusarium oxysporum* Spores in Tomato Root Exudate as well as Subsequent Formation of New Spores. *Environ. Microbiol.* **2008,** *10*, 2455–2461.

Karlidag, H.; Esitken, A.; Turan, M.; Sahin, F. Effects of Root Inoculation of Plant Growth Promoting Rhizobacteria (PGPR) on Yield, Growth and Nutrient Element Contents of Leaves of Apple. *Sci. Hortic.-Amsterdam* **2007,** *114*, 16–20.

Karnwal, A. Production of Indole Acetic Acid by Fluorescent Pseudomonas in the Presence of l-Tryptophan and Rice Root Exudates. *J. Plant Pathol.* **2009,** *91*, 61–63.

Kaushik, R.; Saxena, A. K.; Tilak, K. V. B. R. Selection of Tn5:lacZ Mutants Isogenic to Wild Type *Azospirillum brasilense* Strains Capable of Growing at Sub-Optimal Temperature. *World J. Microbiol. Biotechnol.* **2000,** *16*, 567–570.

Kempster, V. N.; Scott, E. S.; Davies, K. A. Evidence for Systemic, Cross-Resistance in White Clover (*Trifolium repens*) and Annual Medic (*Medicago truncatula* var truncatula) Induced by Biological and Chemical Agents. *Biocontrol Sci. Technol.* **2002,** *12*(5), 615–623.

Khalid, A.; Arshad, M.; Zahir, Z. A. Factor Affecting Auxin Biosynthesis by Wheat and Rice Rhizobacteria. *Pak. J. Soil Sci.* **2001,** *21*, 11–18.

Khalid, A.; Arshad, M.; Zahir, Z. A. Growth and Yield Response of Wheat to Inoculation with Auxin Producing Plant Growth Promoting Rhizobacteria. *Pak. J. Bot.* **2003,** *35*, 483–498.

Khalid, A.; Arshad, M.; Zahir, Z. A. Screening Plant Growth-Promoting Rhizobacteria for Improving Growth and Yield of Wheat. *J. Appl. Microbiol.* **2004,** *96*, 473–480.

Khan, A. L.; Waqas, M.; Kang, S. M. Bacterial Endophyte *Sphingomonas* sp. LK11 Produces Gibberellins and IAA and Promotes Tomato Plant Growth. *J. Microbiol.* **2014,** *52*, 689–695.

Kloepper, J. W. Plant Growth Promoting Rhizobacteria as Biological Control Agents. *Soil Microbial Ecology—Applications in Agricultural and Environmental Management*; Metting, F. B. Ed.; Dekker: New York, 1993; pp 255–274.

Kloepper, J. W.; Gutierrez-Estrada, A.; Mclnroy, J. A. Photoperiod Regulates Elicitation of Growth Promotion but not Induced Resistance by Plant Growth-Promoting Rhizobacteria. *Can. J. Microbiol.* **2007,** *53*, 159–167.

Kloepper, J. W.; Leong, J.; Teintze, M.; Schroth, M. N. *Pseudomonas* Siderophores: A Mechanism Explaining Disease-Suppressive Soils. *Curr. Microbiol.* **1980,** *4*, 317–320.

Kloepper, J. W.; Schroth M. N. In *Plant Growth Promoting Rhizobacteria on Radishes*, Proceedings of the 4th International Conference on Plant Pathogenic Bacteria, Angers, 1978; pp 879–882.

Kloepper, J. W.; Tuzun, S.; Liu, L.; Wei, G. In *Plant Growth-Promoting Rhizobacteria as Inducers of Systemic Disease Resistance*; Lumsden, R. D., Waughn, J. L. Eds. Pest Management: Biologically Based Technologies; American Chemical Society Books: Washington, DC, 1993; pp 156–165.

Kohler, J.; Caravaca, F.; Roldan, A. An AM Fungus and a PGPR Intensify the Adverse Effects of Salinity on the Stability of Rhizosphere Soil Aggregates of *Lactuca sativa*. *Soil Biol. Biochem.* **2010**, *42*, 429–434.

Kumar, B.; Trivedi, P.; Pandey, A. *Pseudomonas corrugata*: A Suitable Bacterial Inoculant for Maize Grown Under Rainfed Conditions of Himalayan Region. *Soil Biol. Biochem.* **2007**, *39*, 3093–3100.

Kumar, D. P.; Anupama, P. D.; Singh, R. K.; Thenmozhi, R.; Nagasathya, A.; Thajuddin, N.; Paneerselvam, A. Evaluation of Extracellular Lytic Enzymes from Indigenous *Bacillus* Isolates. *J. Microbiol. Biotech. Res.* **2012**, *2*, 129–137.

Kumar, H.; Bajpai, V. K.; Dubey, R. C. Wilt Disease Management and Enhancement of Growth and Yield of *Cajanus cajan* (L) var. Manak by Bacterial Combinations Amended with Chemical Fertilizer. *Crop Protect.* **2010**, *29*, 591–598.

Kurek, E.; Jaroszuk-Scisel, J. Rye (*Secale cereale*) Growth Promotion by *Pseudomonas fluorescens* Strains and Their Interactions with *Fusarium culmorum* Under Various Soil Conditions. *Biol. Control* **2003**, *26*, 48–56.

Lavakush, Y. J.; Verma, J. P.; Jaiswal, D. K.; Kumar, A. Evaluation of PGPR and Different Concentration of Phosphorous Level on Plant Growth, Yield and Nutrient Content of Rice (Oryza sativa). *Ecol. Eng.* **2014**, *62*, 123–128.

Lavelle, H.; Spain, A. V. *Soil Ecology*. Kluwer Academic Publishers: Dordrecht, The Netherlands.

Lawongsa, P.; Boonkerd, N.; Wongkaew, S.; O'Gara, F.; Teaumroong, N. Molecular and Phenotypic Characterization of Potential Plant Growth-Promoting *Pseudomonas* from Rice and Maize Rhizospheres. *World J. Microbiol. Biotechnol.* **2008**, *24*, 1877–1884.

Leeman, M.; den Ouden, F. M.; van Pelt, J. A.; Dirkx, F. P. M.; Steijl, H.; Bakker, P. H. A. M.; Schippers, B. Iron Availability Affects Induction of Systemic Resistance to *Fusarium* Wilt of Radish by *Pseudomonas fluorescens*. *Phytopathology* **1996**, *86*, 149–155.

Liu, R.; Dai, M.; Wu, X.; Li, M.; Liu, X. Suppression of the Root-Knot Nematode [*Meloidogyne incognita* (Kofoid & White) Chitwood] on Tomato by Dual Inoculation with Arbuscular Mycorrhizal Fungi and Plant Growth-Promoting Rhizobacteria. *Mycorrhiza* **2012**, *22*, 289–296.

Loper, J. E.; Buyer, J. S. Siderophores in Microbial Interactions on Plant Surfaces. *Mol. Plant Microbe. Interact.* **1991**, *4*, 5–13.

Ma, W.; Guinel, F. C.; Glick, B. R. *Rhizobium leguminosarum* Bovver Viciae 1-Aminocyclopropane-1-Carboxylate Deaminase Promotes Nodulation of Pea Plants. *Appl. Environ. Microbiol.* **2003**, *69*, 4396–4402.

Ma, Y.; Rajkumar, M.; Luo, Y.; Freitas, H. Inoculation of Endophytic Bacteria on Host and Non-Host Plants—Effects on Plant Growth and Ni Uptake. *J. Hazard. Mater.* **2011a**, *195*, 230–237.

Ma, Y.; Rajkumar, M.; Vicente, J. A.; Freitas, H. Inoculation of Ni-Resistant Plant Growth Promoting Bacterium *Psychrobacter* sp. Strain SRS8 for the Improvement of Nickel Phytoextraction by Energy Crops. *Int. J. Phytoremediation* **2011b**, *13*, 126–139.

Madhaiyan, M.; Poonguzhali, S.; Ryu, J.; Sa, T. Regulation of Ethylene Levels in Canola (*Brassica campestris*) by 1-Aminocyclopropane-1-Carboxylate Deaminase-Containing *Methylobacterium fujisawaense*. *Planta* **2006**, *224*, 268–278.

Madhaiyan, M.; Poonguzhali, S.; Kang, B-G.; Lee, Y-J.; Chung, J-B.; Sa, T-M. Effect of Co-Inoculation of Methylotrophic *Methylobacterium* Oryzae with *Azospirillum brasilense* and *Burkholderia pyrrocinia* on the Growth and Nutrient Uptake of Tomato, Red Pepper and Rice. *Plant Soil* **2010**, *328*, 71–82.

Madhaiyan, M.; Poonguzhali, S.; Sa, T. Metal Tolerating Methylotrophic Bacteria Reduces Nickel and Cadmium Toxicity and Promotes Plant Growth of Tomato (*Lycopersicon esculentum* L.). *Chemosphere* **2007**, *69*, 220–228.

Mantelin, S.; Touraine, B. Plant Growth-Promoting Bacteria and Nitrate Availability: Impacts on Root Development and Nitrate Uptake. *J. Exp. Bot.* **2004**, *55*, 27–34.

Marulanda, A.; Azcón, R.; Chaumont, F.; Ruiz-Lozano, J. M.; Aroca, R. Regulation of Plasma Membrane Aquaporins by Inoculation with a *Bacillus megaterium* Strain in Maize (*Zea mays* L.) Plants Under Unstressed and Salt-Stressed Conditions. *Planta* **2010**, *232*, 533–543.

Mayak, S.; Tirosh, T.; Glick, B. R. Plant Growth-Promoting Bacteria that Confer Resistance in Tomato Plants to Salt Stress. *Plant Physiol. Biochem.* **2004a**, *42*, 565–572.

Mayak, S.; Tirosh, T.; Glick, B. R. Plant Growth-Promoting Bacteria that Confer Resistance to Water Stress in Tomato and Pepper. *Plant Sci.* **2004b**, *166*, 525–530.

Mayak, S.; Tirosh, T.; Glick, B. R. Effect of Wild-Type and Mutant Plant Growth-Promoting Rhizobacteria on the Rooting of Mung Been Cuttings. *J. Plant Growth Regul.* **1999**, *18*, 49–53.

McNeil, M.; Darvill, J.; Darvill, A.; Albersheim, P.; van Veen, R.; Hooykaas, P.; Schilperoort, R.; Dell, A. The Discernible Structural Features of the Acidic Polysaccharides Secreted by Different *Rhizobium* Species are the Same. *Carbohydr. Res.* **1986**, *146*, 307–326.

Medina, A.; Probanza, A.; Gutierrez-Mañero, F. J.; Azcon, R. Interactions of Arbuscular-Mycorrhizal Fungi and Bacillus Strains and Their Effects on Plant Growth, Microbial Rhizosphere Activity (Thymidine and Leucine Incorporation) and Fungal Biomass (Ergosterol and Chitin). *Appl. Soil Ecol.* **2003**, *22*, 15–28.

Mehnaz, S.; Lazarovits, G. Inoculation Effects of *Pseudomonas putida*, *Gluconacetobacter azotocaptans*, and *Azospirillum lipoferum* on Corn Plant Growth Under Greenhouse Conditions. *Microb. Ecol.* **2006**, *51*(3), 326–335.

Mehnaz, S.; Mirza, M. S.; Haurat, J.; Bally, R.; Normand, P.; Bano, A.; Malik, K. A. Isolation and 16S rRNA Sequence Analysis of the Beneficial Bacteria from the Rhizosphere of rice. *Can. J. Microbiol.*. **2001**, *472*, 110–117.

Mena-Violante, H.; Olalde-Portugal V. Alteration of Tomato Fruit Quality by Root Inoculation with Plant Growth-Promoting Rhizobacteria (PGPR): *Bacillus subtilis* BEB-13bs. *Sci. Hortic.-Amsterdam* **2007**, *113*, 103–106.

Meyer, J. R.; Linderman, R. G. Response of Subterranean Clover to Dual Inoculation with Vesicular–Arbuscular Mycorrhizal Fungi and a Plant Growth-Promoting Bacterium, *Pseudomonas putida*. *Soil Biol. Biochem.* **1986**, *18*, 185–190.

Meziane, H.; Van der Sluis, I.; Van Loon, L. C.; Hofte, M.; Bakker, P. A. H. M. Determinants of *Pseudomonas putida* WCS358 Involved in Inducing Systemic Resistance in Plants. *Mol. Plant Pathol.* **2005**, *6*, 177–185.

Mhadhbi, H.; Jebara, M.; Limam, F.; Aouani, M. E. Rhizobial Strain Involvement in Plant Growth, Nodule Protein Composition and Antioxidant Enzyme Activities of Chickpea–Rhizobia Symbioses: Modulation by Salt Stress. *Plant Physiol. Biochem.* **2004**, *42*, 717–722.

Mhadhbi, H.; Jebara, M.; Zitoun, A.; Limam, F.; Aouani, M. E. Symbiotic Effectiveness and Response to Mannitol-Mediated Osmotic Stress of Various Chickpea–Rhizobia Associations. *World J. Microbiol. Biotechnol.* **2008**, *24*, 1027–1035.

Minorsky, P. V. On the Inside. *Plant Physiol.* **2008**, *146*, 323–324.

Mirza, M. S.; Mehnaz, S.; Normand, P. et al. Molecular Characterization and PCR Detection of a Nitrogen Fixing Pseudomonas Strain Promoting Rice Growth. *Biol. Fertil. Soils* **2006**, *43*, 163–170.

Mishra, M.; Kumar, U.; Mishra, P. K.; Prakash, V. Efficiency of Plant Growth Promoting Rhizobacteria for the Enhancement of *Cicer arietinum* L. Growth and Germination Under Salinity. *Adv. Biol. Res.* **2010**, *4*(2), 92–96.

Muñoz-Rojas, J.; Caballero-Mellado, J. Population Dynamics of *Gluconacetobacter diazotrophicus* in Sugarcane Cultivars and Its Effect on Plant Growth. *Microb. Ecol.* **2003**, *46*, 454–464.

Muratova, A. Yu, Turkovskaya, O. V.; Antonyuk, L. P.; Makarov, O. E.; Pozdnyakova, L. I.; Ignatov, V. V. Oil-Oxidizing Potential of Associative Rhizobacteria of the genus Azospirillum. *Microbiology* **2005**, *74*, 210–215.

Murphy, J. F.; Zehnder, G. W.; Schuster, D. J.; Sikora, E. J.; Polston, J. E.; Kloepper, J. W. Plant Growth-Promoting Rhizobacterial Mediated Protection in Tomato Against Tomato Mottle Virus. *Plant Dis.* **2000**, *84*, 779–784.

Nadeem, S. M.; Zahir, Z. A.; Naveed, M.; Arshad, M. Preliminary Investigations on Inducing Salt Tolerance in Maize Through Inoculation with Rhizobacteria Containing ACC Deaminase Activity. *Can. J. Microbiol.* **2007**, *53*, 1141–9.

Nadeem, S. M.; Zahir, Z. A.; Naveed, M.; Asghar, H. N.; Arshad, M. Rhizobacteria Capable of Producing ACC-Deaminase May Mitigate The Salt Stress in Wheat. *Soil Sci. Soc. Am. J.* **2010**, *74*, 533–542.

Naveed, M.; Khalid, M.; Jones, D. L.; Ahmad, R.; Zahir, Z. A. Relative Efficacy of *Pseudomonas* spp.; Containing ACC-Deaminase for Improving Growth and Yield of Maize (*Zea mays* L.) in the Presence of Organic Fertilizer. *Pak J. Bot.* **2008**, *40*, 1243–1251.

Nehra, V.; Saharan, B. S.; Choudhary, M. Evaluation of *Brevibacillus brevis* as a Potential Plant Growth Promoting Rhizobacteria for Cotton (*Gossypium hirsutum*) Crop. *SpringerPlus*, **2016**, *5*(948), 1–10. DOI 10.1186/s40064-016-2584-8.

Noel, T. C.; Sheng, C.; Yost, C. K.; Pharis, R. P.; Hynes, M. F. *Rhizobium leguminosarum* as a Plant Growth-Promoting Rhizobacterium: Direct Growth Promotion of Canola and Lettuce. *Can. J. Microbiol.* **1996**, *42*, 279–283.

Okubara, P. A.; Bonsall, R. F. Accumulation of Pseudomonas-Derived 2, 4-Diacetylphloroglucinol on Wheat Seedling Roots is Influenced by Host Cultivar. *Biol. Control* **2008**, *46*, 322–331.

O'Neil, M. A.; Darvill, A. G.; Albersheim, P. The Degree of Esterification and Points of Substitution by *O*-Acetyl and *O*-(3-Hydroxybutanoyl) Groups in the Acidic Extracellular Polysaccharides Secreted by *Rhizobium leguminosarum* Biovars Viciae, Trifolii, and Phaseoli Are Not Related to Host Range. *J. Biol. Chem.* **1991**, *266*, 9549–9555.

Ordookhani, K.; Khavazi, K.; Moezzi, A.; Rejali, F. Influence of PGPR and AMF on Antioxidant Activity, Lycopene and Potassium Contents in Tomato. *Afr. J. Agric. Res.* **2010**, *5*, 1108–1116.

Orhan, E.; Esitken, A.; Ercisli, S.; Turan, M.; Sahin, F. Effects of Plant Growth Promoting Rhizobacteria (PGPR) on Yield, Growth and Nutrient Contents in Organically Growing Raspberry. *Sci. Hortic.-Amsterdam* **2006**, *111*, 38–43.

Park, K. S.; Kloepper, J. W. Activation of PR-1a Promoter by Rhizobacteria that Induce Systemic Resistance in Tobacco Against *Pseudomonas syringae* pv. Tabaci. *Biol. Control* **2000**, *18*, 2–9.

Parmar, N.; Dadarwal, K. R. Stimulation of Nitrogen Fixation and Induction of Flavonoid-Like Compounds by Rhizobacteria. *J. Appl. Microbiol.* **1999**, *86*, 36–44.

Pedraza, R. O.; Bellone, C. H.; de Bellone, S.; Sorte, P. M. B.; Teixeira, K. R. D. Azospirillum Inoculation and Nitrogen Fertilization Effect on Grain Yield and on the Diversity of Endophytic Bacteria in the Phyllosphere of Rice Rainfed Crop. *Eur. J. Soil Biol.* **2009**, *45*, 36–43.

Phi, Q.-T.; Yu-Mi, P.; Keyung-Jo, S.; Choong-Min, R.; Seung-Hwan, P.; Jong-Guk, K.; Sa-Youl, G. Assessment of Root-Associated *Paenibacillus polymyxa* Groups on Growth Promotion and Induced Systemic Resistance in Pepper. *J. Microbiol. Biotechnol.* **2010**, *20*, 1605–1613.

Pirlak, M.; Kose, M. Effects of Plant Growth Promoting Rhizobacteria on Yield and Some Fruit Properties of Strawberry. *J. Plant Nutr.* **2009**, *32*, 1173–1184.

Porcel, R.; Zamarreño, Á. M.; García-Mina, J. M.; Aroca, R. Involvement of Plant Endogenous ABA in *Bacillus megaterium* PGPR Activity in Tomato Plants. *BMC Plant Biol.* **2014**, *14*, 36.

Prasad, R.; Kumar, M.; Varma, A. (2015) *Role of PGPR in Soil Fertility and Plant Health in Plant-Growth-Promoting Rhizobacteria (PGPR) and Medicinal Plants Soil Biology*; Egamberdieva, D., Shrivastava, S., Varma, M.; Springer: Cham, Heidelberg, New York, Dordrecht, London, 2015; Vol. 42, pp 247–262.

Prescott, H.; Klein. Microbial Diversity in Terrestrial Environments. In *Textbook of Microbiology*; 7th Ed.; McGraw-Hill Pub. House: New York, USA, 2007.

Principe, A.; Alvarez, F.; Castro, M. G.; Zachi, L.; Fischer, S. E.; Mori, G. B.; Jofre E. Biocontrol and PGPR Features in Native Strains Isolated from Saline Soils of Argentina. *Curr. Microbiol.* **2007**, *55*, 314–322.

Raaijmakers, J. M.; Vlami, M.; de Souza, J. T. Antibiotic Production by Bacterial Biocontrol Agents. *Antonie van Leeuwenhoek* **2002**, *81*, 537–547.

Radzki, W.; Gutierrez Manero, F. J.; Algar, E. Bacterial Siderophores Efficiently Provide Iron to Iron-Starved Tomato Plants in Hydroponics Culture. *Antonie Van Leeuwenhoek* **2013**, *104*, 321–330.

Rajkumar, M.; Lee, K. J.; Lee, W. H.; Banu, J. R. Growth of *Brassica juncea* Under Chromium Stress: Influence of Siderophores and Indole-3-Acetic Acid Producing Rhizosphere Bacteria. *J. Environ. Biol.* **2005**, *26*, 693–699.

Rajkumar, M.; Ma, Y.; Freitas, H. Characterization of Metal Resistant Plant-Growth Promoting *Bacillus weihenstephanensis* Isolated from Serpentine Soil in Portugal. *J. Basic Microbiol.* **2008**, *48*, 500–508.

Rajkumar, M.; Nagendran, R.; Kui, J. L.; Wang, H. L.; Sung, Z. K. Influence of Plant Growth Promoting Bacteria and Cr (VI) on the Growth of Indian Mustard. *Chemosphere* **2006**, *62*, 741–748.

Recep, K.; Fikrettin, S.; Erkol, D.; Cafer, E. Biological Control of the Potato Dry Rot Caused by *Fusarium* Species Using PGPR Strains. *Biol. Control* **2009**, *50*, 194–198.

Reinhold-Hurek, B.; Hurek, T.; Gillis, M.; Hoste, B.; Vancanneyt, M.; Kersters, K.; Ley, J. D. *Azoarcus* gen. nov.; Nitrogen-Fixing Proteobacteria Associated with Roots of Kallar Grass (*Leptochloa fusca* (L.) Kunth), and Description of Two Species, *Azoarcus indigens* sp. nov. and *Azoarcus communis* sp. nov. *Int. J. Syst. Bacteriol.* **1993**, *43*, 574–584.

Rekha, P. D.; Lai, W.; Arun, A. B.; Young, C. Effect of Free and Encapsulated *Pseudomonas putida* CC-R2-4 and *Bacillus subtilis* CC-pg104 on Plant Growth Under Gnotobiotic Conditions. *Biores. Technol.* **2007**, *98*, 447–451.

Remans, R.; Beebe, S.; Blair, M.; Manrique, G.; Tovar, E.; Rao, I.; Croonenborghs, A.; Torres-Gutierrez, R.; El-Howeity, M.; Michiels, J.; Vanderleyden, J. Physiological and Genetic Analysis of Root Responsiveness to Auxin-Producing Plant Growth-Promoting Bacteria in Common Bean (*Phaseolus vulgaris* L.). *Plant Soil* **2008**, *302*, 149–161.

Rezzonoco, F.; Binder, C.; Defago, G.; Moenne-Loccoz, Y. The Type III Secretion System of Biocontrol *Pseudomonas fluorescens* KD Targets the Phytopathogenic Chromista *Pythium ultimum* and Promotes Cucumber Protection. *Mol. Plant Microbe Interact.* **2005**, *9*, 991–1001.

Richardson, A. E.; Barea, J. M.; McNeill, A. M.; Prigent-Combaret, C. Acquisition of Phosphorus and Nitrogen in the Rhizosphere and Plant Growth Promotion by Microorganisms. *Plant Soil* **2009**, *321*, 305–339.

Riggs, P. J.; Chelius, M. K.; Iniguez, A. L.; Kaeppler, S. M.; Triplett, E. W. Enhanced Maize Productivity by Inoculation with Diazotrophic Bacteria. *Aust. J. Plant Physiol.* **2001**, *28*, 829–836.

Robertsen, B. K.; Aman, P.; Darvill, A. G.; McNeil, M.; Albersheim, P. Host–Symbiont Interactions. V. The Structure of Acidic Extracellular Polysaccharides Secreted by *Rhizobium leguminosarum* and *Rhizobium trifolii. Plant Physiol.* **1981**, *67*, 389–400.

Rodrigues, C.; Laranjo, M.; Oliveira, S. Effect of Heat and pH Stress in the Growth of Chickpea Mesorhizobia. *Curr. Microbiol.* **2006**, *53*, 1–7.

Rodrigues, E. P.; Rodrigues, C. S.; de Oliveira, A. L. M.; Baldani, V. L.; Teixeira da Silva, J. A. *Azospirillum amazonense* Inoculation: Effects on Growth, Yield and N_2 Fixation of Rice (*Oryza sativa* L.). *Plant Soil* **2008**, *302*, 249–261.

Roesti, D.; Guar, R.; Johri, B. N.; Imfeld, G.; Sharma, S.; Kawaljeet, K.; Aragno, M. Plant Growth Stage, Fertilizer Management and Bioinoculation of Arbuscular Mycorrhizal Fungi and Plant Growth Promoting Rhizobacteria Affect the Rhizobacterial Community Structure in Rain-Fed Wheat Field. *Soil Biol. Biochem.* **2006**, *38*, 1111–1120.

Romdhane, S. B.; Trabelsi, M.; Aouani, M. E.; de Lajudie, P.; Mhamdi, R. The Diversity of Rhizobia Nodulating Chickpea (*Cicer arietinum*) Under Water Deficiency as a Source of More Efficient Inoculants. *Soil Biol. Biochem.* **2009**, *41*, 2568–2572.

Ruchi; Kapoor, R.; Kumar, A.; Kumar, A.; Patil, S.; Thapa, S.; Kaur, M. Evaluation of Plant Growth Promoting Attributes and Lytic Enzyme Production by Fluorescent *Pseudomonas* Diversity Associated with Apple and Pear. *Int. J. Scientific Res. Pub.* **2012**, *2*, 1–8.

Ryan, P. R.; Delhaize, E.; Jones, D. L. Function and Mechanism of Organic Anion Exudation from Plant Roots. *Annu. Rev. Plant Physiol. Plant Mol. Biol.* **2001**, *52*, 527–560.

Ryu, C. M.; Farag, M. A.; Hu, C. H.; Reddy, M. S.; Kloepper, J. W.; Pare, P. W. Bacterial Volatiles Induce Systemic Resistance in *Arabidopsis*. *Plant Physiol.* **2004**, *134*, 1017–1026.

Ryu, C. M.; Farag, M. A.; Hu, C. H.; Reddy, M. S.; Pare, P. W.; Kloepper, J. W. Bacterial Volatiles Promote Growth in *Arabidopsis*. *Proc. Natl. Acad. Sci. U.S.A.* **2003**, *100*, 4927–4932.

Ryu, C. M.; Kim, J.; Choi, O.; Kim, S. H.; Park, C. S. Improvement of Biological Control Capacity of *Paenibacillus polymyxa* E681 by Seed Pelleting on Sesame. *Biol. Control* **2006**, *39*, 282–289.

Sachdev, D. P.; Chaudhari, H. G.; Kasure, V. M.; Dahavale, D. D.; Chopade, B. A. Isolation and Characterization of Indole Acetic Acid (IAA) Producing *Klebsiella pneumoniae* Strains from Rhizosphere of Wheat (*Triticum aestivum*) and Their Effect on Plant Growth. *Indian J. Exp. Biol.* **2009**, *47*, 993–1000.

Safronova, V. I.; Stepanok, V. V.; Engqvist, G. L.; Alekseyev, Y. V.; Belimov, A. A. Root-Associated Bacteria Containing 1-Aminocyclopropane-1-Carboxylate Deaminase Improve Growth and Nutrient Uptake by Pea Genotypes Cultivated in Cadmium Supplemented Soil. *Biol. Fertil. Soils* **2006**, *42*, 267–272.

Sahoo, R. K.; Ansari, M. W.; Pradhan, M. Phenotypic and Molecular Characterization of Native *Azospirillum* Strains from Rice Fields to Improve Crop Productivity. *Protoplasma* **2014**, *251*, 943–953.

Sailaja, P. R.; Podile, A. R.; Reddanna, P. Biocontrol Strain of *Bacillus subtilis* AF1 Rapidly Induces Lipoxygenase in Groundnut (*Arachis hypogea*) Compared to Crown Rot Pathogen *Aspergillus niger*. *Eur. J. Plant Pathol.* **1997**, *104*, 125–132.

Saleh, S. S.; Glick, B. R. Involvement of gacS and rpoS in Enhancement of the Plant Growth-Promoting Capabilities of *Enterobacter cloacae* CAL2 and *Pseudomonas putida* UW4. *Can. J. Microbiol.* **2001**, *47*, 698–705.

Sangeeth, K. P.; Bhai, R. S.; Srinivasan, V. *Paenibacillus glucanolyticus*, A Promising Potassium Solubilizing Bacterium Isolated from Black Pepper (*Piper nigrum* L.) Rhizosphere. *J. Spices Aromat. Crops* **2012**, *21*, 118–124.

Saravanakumar, D.; Samiyappan, R. ACC Deaminase from *Pseudomonas fluorescens* Mediated Saline Resistance in Groundnut (*Arachis hypogea*) Plants. *J. Appl. Microbiol.* **2007**, *102*, 1283–1292.

Sardesai, N.; Babu, C. R. Cold Stress Induced High Molecular Weight Membrane Polypeptides are Responsible for Cold Tolerance in *Rhizobium* DDSS69. *Microbiol. Res.* **2001**, *156*, 279–284.

Sarma, R. K.; Gogoi, A.; Dehury, B.; Debnath, R.; Bora, T. C.; Saikia, R. Community Profiling of Culturable Fluorescent Pseudomonads in the Rhizosphere of Green Gram (*Vigna radiata* L.). *PLoS ONE* **2014**, *9*(10), e108378. DOI 10.1371/journal.pone.0108378.

Sekar, S.; Kandavel, D. Interaction of Plant Growth Promoting Rhizobacteria (PGPR) and Endophytes with Medicinal Plants—New Avenues for Phytochemicals. *J. Phytol.* **2010,** *2*, 91–100.

Sgroy, V.; Cassan, F.; Masciarelli, O. et al. Isolation and Characterization of Endophytic Plant Growth-Promoting (PGPB) or Stress Homeostasis-Regulating (PSHB) Bacteria Associated to the Halophyte *Prosopis strombulifera. Appl. Microbiol. Biotechnol.* **2009,** *85*, 371–381.

Shaharoona, B.; Arshad, M.; Zahir, Z. A. Effect of Plant Growth Promoting Rhizobacteria Containing ACC-Deaminase on Maize (*Zea mays* L.) Growth Under Axenic Conditions and on Nodulation in Mung Bean (*Vigna radiata* L.). *Lett. Appl. Microbiol.* **2006,** *42*, 155–159.

Shaharoona, B.; Naveed, M.; Arshad, M. Fertilizer-Dependent Efficiency of Pseudomonas for Improving Growth, Yield, and Nutrient Use Efficiency of Wheat (*Triticum aestivum* L.). *Appl. Microbiol. Biotechnol.* **2008,** *79*, 147–155.

Sharaf-Eldin, M.; Elkholy, S.; Fernandez, J. A. et al. *Bacillus subtilis* FZB24 Affects Flower Quantity and Quality of Saffron (*Crocus sativus*). *Planta Med.* **2008,** *74*, 1316–1320.

Sharma, S. K.; Johri, B. N.; Ramesh, A.; Joshi, O. P.; Prasad, S. V. S. Selection of Plant Growth-Promoting *Pseudomonas* spp. that Enhanced Productivity of Soybean-Wheat Cropping System in Central India. *J. Microbiol. Biotechnol.* **2011,** *21*, 1127–1142.

Sheng, X. F.; Jiang, C. Y.; He, L. Y. Characterization of Plant Growth-Promoting *Bacillus edaphicus* NBT and Its Effect on Lead Uptake by INDIAN mustard in A Lead-Amended Soil. *Can. J. Microbiol.* **2008,** *54*, 417–422.

Sheng, X. F.; Xia, J. J. Improvement of Rape (*Brassica napus*) Plant Growth and Cadmium Uptake by Cadmium-Resistant Bacteria. *Chemosphere* **2006,** *64*, 1036–1042.

Silo-Suh, L. A.; Lethbridge, B. J.; Raffel, S. J. Biological Activities of Two Fungi Static Antibiotics Produced by *Bacillus cereus* UW85. *Appl. Environ. Microbiol.* **1994,** *60*, 2023–2030.

Sivakumar, G.; Sharma, R. C. Induced Biochemical Changes due to Seed Bacterization by *Pseudomonas fluorescens* in Maize Plants. *Indian Phytopath.* **2003,** *56*, 134–137.

Sokolova, M. G.; Akimova, G. P.; Vaishlia, O. B. Effect of Phytohormones Synthesized By Rhizosphere Bacteria on Plants. *Prikl. Biokhim. Mikrobiol.* **2011,** *47*, 302–307.

Someya, N.; Tsuchiya, K.; Yoshida, T.; Noguchi, M. T.; Akutsu, K.; Sawada H. Co-Inoculation of an Antibiotic-Producing Bacterium and a Lytic Enzyme-Producing Bacterium for the Biocontrol of Tomato Wilt Caused by *Fusarium oxysporum* f. sp. *lycopersici. Biocontrol Sci. Technol.* **2007,** *12*, 1–6.

Soussi, M.; Santamarıa, M.; Ocana, A.; Lluch, C. Effects of Salinity on Protein and Lipo-polysaccharide Pattern in a Salt-Tolerant Strain of *Mesorhizobium ciceri. J. Appl. Microbiol.* **2001,** *90*, 476–481

Stout, M. J.; Zehnder, G. W.; Baur, M. E. Potential for the Use of Elicitors of Plant Defence in Arthropode Management Programs. *Arch. Insect. Biochem. Physiol.* **2002,** *51*, 222–235.

Sturz, A. V.; Nowak, J. Endophytic Communities of Rhizobacteria and the Strategies Required to Create Yield Enhancing Associations with Crops. *Appl. Soil Ecol.* **2000,** *15*, 183–190.

Suarez, R.; Wong, A.; Ramırez, M.; Barraza, A.; Orozco, M. C.; Cevallos, M. A.; Lara, M.; Hernandez, G.; Iturriaga, G. Improvement of Drought Tolerance and Grain Yield in

Common Bean by Overexpressing Trehalose-6-Phosphate Synthase in Rhizobia. *Mol. Plant Microbe Interact.* **2008**, *21*, 958–966.

Suslow, T. V.; Schroth, M. N. Rhizobacteria of Sugar Beets: Effects of Seed Application and Root Colonization on Yield. *Phytopathology* **1982**, *72*, 199–206.

Tan, K. Z.; Radziah, O.; Halimi, M. S.; Khairuddin, A. R.; Shamsuddin, Z. H. Assessment of Plant Growth-Promoting Rhizobacteria (PGPR) and Rhizobia as Multi-Strain Biofertilizer on Growth and N₂ Fixation of Rice Plant. *Aust. J. Crop Sci.* **2015**, *9*(12), 1257–1264.

Tan, K. Z.; Radziah, O.; Halimi, M. S.; Khairuddin, A. R.; Habib, S. H.; Shamsuddin, Z. H. Isolation and Characterization of Rhizobia and Plant Growth-Promoting Rhizobacteria (PGPR) and Their Effects on Growth of Rice Seedlings. *Am. J. Agric. Biol. Sci.* **2014**, *9*(3), 342–360.

Tank, N.; Saraf, M. Salinity-Resistant Plant Growth Promoting Rhizobacteria Ameliorates Sodium Chloride Stress on Tomato Plants. *J. Plant Interact.* **2010**, *5*, 51–58.

Tank, N.; Saraf, M. Enhancement of Plant Growth and Decontamination of Nickel-Spiked Soil Using PGPR. *J. Basic Microbiol.* **2009**, *49*, 195–204.

Tavasolee, A.; Aliasgharzad, N.; SalehiJouzani, G.; Mardi, M.; Asgharzadeh, A. Interactive Effects of Arbuscular Mycorrhizal Fungi and Rhizobial Strains on Chickpea Growth and Nutrient Content in Plant. *Afr. J. Biotechnol.* **2011**, *10*, 7585–7591.

Tejera, N.; Lluch, C.; Martínez-Toledo, M. V. Isolation and Characterization of *Azotobacter* and *Azospirillum* Strains from the Sugarcane Rhizosphere. *Plant Soil* **2005**, *270*, 223–232.

Thakuria, D.; Taleekdar, N. C.; Goswami, C.; Hazarika, S.; Boro, R. C.; Khan, M. R. Characterization and Screening of Bacteria from Rhizosphere of Rice Grown in Acidic Soils of Assam. *Curr. Sci.* **2004**, *86*(7), 978–985.

Tilak, K. V.; Ranganayaki, N. Synergistic Effects of Plant-Growth Promoting Rhizobacteria and Rhizobium on Nodulation and Nitrogen Fixation by Pigeon Pea (*Cajanus cajan*). *Eur. J. Soil Sci.* **2006**, *57*, 67–71.

Timmusk, S.; Nicander, B.; Granhall, U.; Tillberg, E. Cytokinin Production by *Paenibacillus polymyxa*. *Soil Biol. Biochem.* **1999**, *31*, 1847–1852.

Tisdale, S. L.; Nelson, W. L. (1975). *Soil Fertility and Fertilizers*, 3rd edn. Macmillan, New York, p 694.

Tripathi, M.; Munot, H. P.; Shouch, Y.; Meyer, J. M.; Goel, R. Isolation and Functional Characterization of Siderophore-Producing Lead- and Cadmium-Resistant *Pseudomonas putida* KNP9. *Curr. Microbiol.* **2005**, *5*, 233–237.

Trivedi, P.; Pandey, A. Recovery of Plant Growth-Promoting Rhizobacteria from Sodium Alginate Beads After 3 Years Following Storage at 4°C. *J. Ind. Microbiol. Biotechnol.* **2008**, *35*, 205–209.

Uma, C.; Sivagurunathan, P.; Sangeetha, D. Performance of Bradyrhizobial Isolates Under Drought Conditions. *Int. J. Curr. Microbiol. Appl. Sci.* **2013**, *2*, 228–232.

Upadhyay, S. K.; Singh, J. S.; Singh, D. P. Exopolysaccharide-Producing Plant Growth-Promoting Rhizobacteria Under Salinity Condition. *Pedosphere* **2011**, *21*, 214–222.

Valverde, A.; Burgos, A.; Fiscella, T.; Rivas, R.; Velazquez, E.; Rodriguez-Barrueco, C.; Cervantes, E.; Chamber, M.; Igual, J. M. Differential Effects of Coinoculations with *Pseudomonas jessenii* PS06 (a Phosphate-Solubilizing Bacterium) and *Mesorhizobium ciceri* C-2/2 Strains on the Growth and Seed Yield of Chickpea Under Greenhouse and Field Conditions. *Plant Soil* **2006**, *287*, 43–50.

Van, V. T.; Berge, O.; Ke, S. N.; Balandreau, J.; Heulin, T. Repeated Beneficial Effects of Rice Inoculation with a Strain of Burkholderia Vietnamiensis on Early and Late Yield Components in Low Fertility Sulphate Acid Soils of Vietnam. *Plant Soil* **2000**, *218*, 273–284.

Velazhahan, R.; Samiyappan, R.; Vidhyasekaran, P. Relationship Between Antagonistic Activities of *Pseudomonas fluorescens* Isolates Against *Rhizoctonia solani* and Their Production of Lytic Enzymes. *Z. Pflanz. Pflanzen.* **1999**, *106*, 244–250.

Vessey, J. K. Plant Growth Promoting Rhizobacteria as Biofertilizers. *Plant Soil* **2003**, *255*, 571–586.

Vidal, C.; Chantreuil, C.; Berge, O.; Maure, L.; Escarree, J.; Bena, G.; Brunel, B.; Cleyet-Marel, J. C. *Mesorhizobium metallidurans* sp. nov.; A Metal Resistant Symbiont of *Anthyllis vulneraria* Growing on Metallicolous Soil in Languedoc France. *Int. J. Syst. Evol. Microbiol.* **2009**, *59*, 850–855.

Vivas, A.; Azcon, R.; Biro, B.; Barea, J. M.; Ruiz-Lozano, J. M. Influence of Bacterial Strains Isolated from Lead-Polluted Soil and Their Interactions with Arbuscular Mycorrhizae on the Growth of *Trifolium pratense* L. Under Lead Toxicity. *Can. J. Microbiol.* **2003**, *49*, 577–588.

Vivas, A.; Biro, B.; Ruiz-Lozano, J. M.; Barea, J. M.; Azcon, R. Two Bacterial Strains Isolated from a Zn-Polluted Soil Enhance Plant Growth and Mycorrhizal Efficiency Under Zn Toxicity. *Chemosphere* **2006**, *52*, 1523–1533.

Vivekananthan, R.; Ravi, M.; Ramanathan, A.; Samiyappan, R. Lytic Enzymes Induced by *Pseudomonas fluorescens* and Other Biocontrol Organisms Mediate Defence Against the Anthracnose Pathogen in Mango. *World J. Microbiol. Biotechnol.* **2004**, *20*, 235–244.

Wang, E. T.; Tan, Z. Y.; Guo, X. W.; Rodriguez-Duran, R.; Boll, G.; Martinez-Romero, E. Diverse Endophytic Bacteria Isolated from a Leguminous Tree *Conzattia* Multiflora Grown in Mexico. *Arch. Microbiol.* **2006**, *186*, 251–259.

Wani, P. A.; Khan, M. S. Bioremediation of Lead by a Plant Growth Promoting *Rhizobium* Species RL9. *Bacteriol. J.* **2012**, *2*, 66–78

Wani, P. A.; Khan, M. S. Nickel Detoxification and Plant Growth Promotion by Multi Metal Resistant Plant Growth Promoting *Rhizobium* Species RL9. *Bull. Environ. Contam. Toxicol.* **2013**, *91*, 117–124.

Wani, P. A.; Khan, M. S.; Zaidi, A. Chromium-Reducing and Plant Growth Promoting Mesorhizobium Improves Chickpea Growth in Chromium-Amended Soil. *Biotechnol. Lett.* **2008**, *30*, 159–163

Wani, P. A.; Khan, M. S. *Bacillus* Species Enhance Growth Parameters of Chickpea (*Cicer arietinum* L.) in Chromium Stressed Soils. *Food Chem. Toxicol.* **2010**, *48*, 3262–3267.

Wani, P. A.; Khan, M. S.; Zaidi, A. Effect of Metal Tolerant Plant Growth Promoting *Bradyrhizobium* sp. (vigna) on Growth, Symbiosis, Seed Yield and Metal Uptake by Greengram Plants. *Chemosphere* **2007a**, 70, 36–45.

Wani, P. A.; Khan, M. S.; Zaidi, A. Co Inoculation of Nitrogen Fixing and Phosphate Solubilizing Bacteria to Promote Growth, Yield and Nutrient Uptake in Chickpea. *Acta Agron. Hung.* **2007b**, 55, 315–323.

Wani, P. A.; Khan, M. S.; Zaidi, A. Synergistic Effect of the Inoculation with Nitrogen-Fixing and Phosphate-Solubilizing Rhizobacteria on Performance of Field-Grown Chickpea. *J. Plant Nutr. Soil Sci.* **2007c**, *170*, 283–287.

Wei, L.; Kloepper, J. W.; Tuzun, S. Induced Systemic Resistance to Cucumber Diseases and Increased Plant Growth by Plant Growth Promoting Rhizobacteria Under Field Conditions. *Phytopathology* **1996**, *86*, 221–224.

Weller, D. M.; Raaijmakers, J. M.; Gardener, B. B.; Thomashow, L. S. Microbial Populations Responsible for Specific Soil Suppressiveness to Plant Pathogens. *Annu. Rev. Phytopathol.* **2002**, *40*, 309–348.

Whipps, J. M. Microbial Interactions and Biocontrol in the Rhizosphere. *J. Exp. Biol.* **2001**, *52*, 487–511.

Wu, C. H.; Wood, T. K.; Mulchandani, A.; Chen, W. Engineering Plant–Microbe Symbiosis for Rhizoremediation of Heavy Metals. *Appl. Environ. Microbiol.* **2006**, *72*, 1129–1134.

Yanni, Y.; Rizk, R.; Abd-El Fattah, F. The Beneficial Plant Growth-Promoting Association of *Rhizobium leguminosarum* bv. Trifolii with Rice Roots. *Aust. J. Plant Physiol.* **2001**, *28*, 845–870.

Yao, A. V.; Bochow, H.; Karimov, S.; Boturov, U.; Sanginboy, S.; Sharipov, A. K. Effect of FZB 24® *Bacillus subtilis* as a Biofertilizer on Cotton Yields in Field Tests. *Arch. Phytopathol. Plant Protect.* **2006**, *39*, 323–328.

Yao, L.; Wu, Z.; Zheng, Y.; Kaleem, I.; Li, C. Growth Promotion and Protection Against Salt Stress by *Pseudomonas putida* Rs-198 on Cotton. *Eur. J. Soil Biol.* **2010**, *46*(1), 49–54.

Yao, T.; Yasmin, S.; Hafeez, F. Y. Potential Role of Rhizobacteria Isolated from Northwestern China for Enhancing Wheat and Oat Yield. *J. Agric. Sci.* **2008**, *146*, 49–56.

Younis, M. Responses of *Lablab purpureus–Rhizobium* Symbiosis to Heavy Metals in Pot and Field Experiments. *World J. Agric. Sci.* **2007**, *3*, 111–122.

Yue, H.; Mo, W.; Li, C.; Zheng, Y.; Li, H. The Salt Stress Relief and Growth Promotion Effect of RS-5 on cotton. *Plant Soil* **2007**, *297*, 139–145.

Zahir, Z. A.; Abbas, S. A.; Khalid, M.; Arshad, M. Substrate Dependent Microbially Derived Plant Hormones for Improving Growth of Maize Seedlings. *Pak. J. Biol. Sci.* **2000**, *3*, 289–291.

Zahir, Z. A.; Ghani, U.; Naveed, M.; Nadeem, S. M.; Asghar, H. N. Comparative Effectiveness of *Pseudomonas* and *Serratia* sp. Containing ACC-Deaminase for Improving Growth and Yield of Wheat (*Triticum aestivum* L.) Under Salt-Stressed Conditions. *Arch. Microbiol.* **2009**, *191*, 415–424.

Zahir, Z. A.; Asghar, H. N.; Akhtar, M. J.; Arshad, M. Precursor (L-Tryptophan)-Inoculum (*Azotobacter*) Interaction for Improving Yields and Nitrogen Uptake of Maize. *J. Plant Nutr.* **2005**, *28*, 805–817.

Zahir, Z. A.; Shah, M. K.; Naveed, M.; Akhter, M. J. Substrate Dependent Auxin Production by *Rhizobium phaseoli* Improves the Growth and Yield of *Vigna radiata* L. Under Salt Stress Conditions. *J. Microbiol. Biotechnol.* **2010**, *20*, 1288–1294.

Zaidi, S.; Usmani, S.; Singh, B. R.; Musarrat, J. Significance of *Bacillus subtilis* Strain SJ 101 as a Bioinoculant for Concurrent Plant Growth Promotion and Nickel Accumulation in *Brassica juncea*. *Chemosphere* **2006**, *64*, 991–997.

Zakria, M.; Njoloma, J.; Saeki, Y.; Akao, S. Colonization and Nitrogen-Fixing Ability of *Herbaspirillum* sp Strain B501 gfp1 and Assessment of Its Growth-Promoting Ability in Cultivated Rice. *Microbes Environ.* **2007,** *22,* 197–206.

Zehnder, G. W.; Murphy, J. F.; Sikora, E. J.; Kloepper, J. W. Application of Rhizobacteria for Induced Resistance. *Eur. J. Plant Pathol.* **2001,** *107,* 39–50.

Zhang, H.; Sekiguchi, Y.; Hanada, S.; Hugenholtz, P.; Kim, H.; Kamagata, Y.; Nakamura, K. *Gemmatimonas aurantiaca* gen. nov.; sp. nov.; A Gram-Negative, Aerobic, Polyphosphate Accumulating Microorganism, the First Cultured Representative of the New Bacterial Phylum Gemmatimonadetes phyl. nov. *Int. J. Syst. Evol. Microbiol.* **2003,** *53,* 1155–1163.

CHAPTER 2

BENEFICIAL USES AND APPLICATIONS OF PLANT GROWTH-PROMOTING RHIZOBACTERIA IN SUSTAINABLE AGRICULTURE

PRADEEP KUMAR[1], MADHU KAMLE[1*],
PAWAN KUMAR MAURYA[2], and RAVI KANT SINGH[4]

[1]Department of Forestry, North Eastern Regional Institute of Science and Technology, Nirjuli-791109, Arunachal Pradesh, India

[2]Department of Biochemistry, Central University of Haryana, Jant-Pali, Mahendergarh (Haryana)-123031, India

[3]Amity Institute of Biotechnology, Amity University Chhattisgarh, Raipur, C.G.-493225, India

*Corresponding author. E-mail: madhu.kamle18@gmail.com

2.1 INTRODUCTION

The plant growth–promoting rhizobacteria (PGPR) is characterized by the following inherent distinctiveness: (1) they must be proficient to colonize the root surface; (2) they must survive, multiply, and compete with other microbiota, at least for the time needed to express their plant growth promotion/protection activities; and (3) they must promote plant growth (Kloepper, 1994). About 2–5% of the rhizobacteria, when reintroduced by plant inoculation in a soil-containing competitive microflora, exert a beneficial effect on plant growth and called PGPR (Kloepper and Schroth, 1978). In accordance with Vessey (2003), soil bacterial species expanding in plant rhizosphere grow around plant tissues, stimulate plant growth by a plethora of mechanisms, and are collectively known as PGPR. PGPRs based on their functional activities are classified by Somers et al. (2004) as (1) biofertilizers (increasing the availability of nutrients to plant), (2) phytostimulators (plant growth promotion, generally through phytohormone), (3) rhizoremediators (degrading organic pollutants), and (4) biopesticides (controlling diseases, mainly by the production of anti-biotics and antifungal metabolites) (Antoun and Prevost, 2005). Gener-ally, a single PGPR will often reveal multiple modes of action including biological control (Kloepper, 2003; Vessey, 2003). Furthermore, Gray and Smith (2005) have recently shown that the PGPR associations range in the degree of bacterial proximity to the root and intimacy of association. In general, these can be separated into extracellular (ePGPR), existing in the rhizosphere, on the rhizoplane, or in the spaces between cells of the root cortex, and intracellular (iPGPR), which exist inside root cells, generally in specialized nodular structures (Figueiredo et al., 2011). Some examples of ePGPR are like *Agrobacterium, Arthrobacter, Azotobacter, Azospi-rillum, Bacillus, Burkholderia, Caulobacter, Chromobacterium, Erwinia, Flavobacterium, Micrococcous, Pseudomonas, Serratia*, etc. (Bhattacha-ryya and Jha, 2012). Similarly, the iPGPR includes *Allorhizobium, Azorhi-zobium, Bradyrhizobium, Mesorhizobium,* and *Rhizobium* of the family Rhizobiaceae. Most of the rhizobacteria belonging to this group are Gram-negative rods with a lower proportion being Gram-positive rods, cocci, or pleomorphic (Bhattacharyya and Jha, 2012). Moreover, numerous actino-mycetes are also one of the major components of rhizosphere microbial communities displaying marvelous plant growth.

2.2 SUSTAINABLE AGRICULTURE AND PLANT GROWTH–PROMOTING RHIZOBACTERIA

Sustainable agriculture is a comprehensive concept which embraces advances in agricultural management practices and technology with the ever-increasing food and feed demands for increased population. Conventional agricultural fails to meet the current agriculture produc-tion demands. Efficient and potential soil microbiota with implication of PGPR is an alternative for sustainable agricultural practices (Singh et al., 2011). If used effectively, microbial communities can significantly benefit the agriculture produced with increased yield. PGPRs are excel-lent model systems that provide novel genetic constituents and bioactive compounds with multifaceted use in environment-friendly agriculture.

Current and future progress in consideration of PGPR diversity, colonization ability, mechanisms of interactions (direct or indirect), formulation, and application could facilitate their development as the reliable biofertilizers and biocontrol agents (BCAs) in management of sustainable agriculture with innate genetic potential. In the expansive sense, PGPR include the N_2-fixing rhizobacteria that colonize the rhizosphere, providing N to plants in addition to the well characterized legume rhizobia symbioses (Singh et al., 2011).

A narrow zone of soil directly surrounding the root system is referred to as rhizosphere and commonly known as hot spot of microbial activities (Walker et al., 2003). However, the term rhizobacteria includes a group of rhizosphere bacteria competent in colonizing the root environments (Kloepper et al., 1991). Soil enriched with microscopic life forms including bacteria, fungi, actinomycetes, protozoa, and algae; among these, 95% are bacteria. Number of culturable bacterial cells in soil is generally about 1% of the total number present in soil. The number and the type of bacteria that are found in different soils are influenced by the soil condition including temperature, moisture, and the presence of salt and other chemicals as well as by the number and type of plants found in the soil. In general, bacteria are not evenly distributed in the soil. Bacterial concentration found near the root rhizosphere is much higher than the rest of the soil. This is because of the presence of nutrients including sugars, amino acids, organic acids, and other small molecules from plant root exudates that may account for up to a third of the carbon fixed by a plant (Glick, 2012). Plants roots synthesize, accumulate, and secrete diverse compounds and these compounds act as chemoattractant for a vast number of heterogeneous, diverse, and actively metabolizing soil microbial communities (Ahemad and Kibret, 2014) and supply biological fixed nitrogen. PGPRs affect plant growth through indirect mechanisms, such as suppression of bacterial, fungal, and nematode pathogens (biocontrol) by production of various metabolites, induce systemic resistance, and compete with the pathogen for nutrients and colonization of space. Number of plant growth–promoting bacteria (PGPB) that occupy the endo-rhizosphere that reside in several plants like tomato, brinjal, chili, and capsicum roots has been isolated and their growth-promoting effects proved (Sukhada, 1999). These bacteria present particularly near roots affect the plants growth in one of the three ways by interacting between soil bacteria and plants (Lynch, 1990), they are either beneficial, harmful, or sometimes neutral. While that these PGPR bacteria which may change its mode of interaction with changing conditions. They facilitate plant growth by providing either fixed nitrogen or phosphorus; compounds are often present in only limited amounts in many soils and are unlikely to provide any benefit to plants when significant amount of chemical fertilizer is added to the soil. Additionally, bacteria affect different plants incompatibly, as in the case of IAA (indole acetic acid) overproducing mutant *Pseudomonas fluorescens* BSP53a stimulated root development in blackcurrant cuttings while inhibiting the development of roots in cherry cuttings (Dubeikovsky et al., 1993). This indicates that blackcurrant cuttings contained a suboptimal level of IAA enhanced by the presence of mutant *P. fluorescens* BSP53a. Apart of these forewarnings, it is usually a direct matter to decide whether a PGPR either promotes or inhibits plant growth including free living, that form specific symbiotic relationship with plants (e.g., *Rhizobia* spp. and *Frankia* spp.), endophytic bacteria that are colonized in some or a portion of a plants interior tissues, and cyanobacteria (BGA = blue green algae). Yet the differences between these bacteria are that they all utilize the same mechanisms. PGPB may promote plant growth directly usually by either facilitating resource acquisition or modulating plant hormone levels, or indirectly by decreasing the inhibitory effects of various pathogenic agents on plant growth and development by acting as biocontrol bacteria (Glick, 1995).

2.3   MECHANISM FOR PLANT GROWTH PROMOTION

Advantage of PGPRs is increasing rapidly in order to utilize them as commercial eco-friendly biofertilizers. PGPRs change the complete microbial community structures present in the rhizosphere and thus efficiently enhanced the plant growth (Kloepper and Schroth, 1981). Other promising mechanisms of PGPRs for plant growth promotion include associative nitrogen fixation, suppression of disease causing plant pathogens, competition with pathogenic microorganisms, production of plant growth regulators (PGRs), lowering ethylene level in root cells, solubilization of nutrients, promotion of mycorrhizal functioning, decreasing pollutant toxicity, etc. (Glick et al., 1999). PGPR-mediated induced systemic resistance (ISR) is an important mechanism of biological disease control. They can hold soil aggregates, creating channels through which roots grow, soil fauna move, and water percolates. The extent of plant growth promotion envisages vast number of traits, regulated in complex ways. The gene products of both the *nif* and the *fix* genes involved in nitrogen fixation are well document in many PGPR. Soil microorganisms mainly fungi possessing phytase (myo-inositol hexakisphosphate phosphohydrolase) enzyme that catalyzes the hydrolysis of phytic acid (myo-inositol hexakisphosphate) an indigestible, organic form of phosphorus that is found in grains and oil seeds and releases a usable form of inorganic phosphorus whose activity contribute to plant phosphorus nutrition. Another beneficial effect due to bacterial phytase activity in the rhizosphere is elimination of chelate-forming phytate, which binds nutritionally important minerals. Siderophores help by chelating iron from the rhizosphere to make it bioavailable. Some PGPR release a blend of volatile components that promote growth.

Generally, PGPR promote plant growth and development either directly and indirectly mechanisms depicted in Figure 2.1 (Ortiz-Castro et al., 2009; Glick, 2012). The direct mode includes biological nitrogen fixation, phytohormone production (auxins, cytokinins, and gibberellins), minerals (like phosphorus and iron) solubilization, production of siderophores and enzymes, induction of systemic resistance, whereas the indirect modes includes biocontrol, antibiotics production, chelation's of available Fe in rhizosphere, synthesis of extracellular enzymes to hydrolyze the fungal cell wall and competition for niches within rhizosphere (Zahir et al., 2004; van Loon, 2007). *P. fluorescens* and *Bacillus subtilis* PGPR strains recorded as the most promising candidates of indirect stimulation (Damayanti et al., 2007). Besides, nitrogen transformation, increasing bioavailability of phosphate, iron acquisition, exhibition of specific enzymatic activity, and plant protection from harmful pathogens with the production of antibiotics can also successfully improve the quality of crops in agriculture (Spaepen et al., 2007).

2.3.1   DIRECT MECHANISM

2.3.1.1   NITROGEN FIXATION

Nitrogen is the most essential element for all the form of life and it is a vital nutrient for the plant growth and productivity. Though there is 78% of nitrogen present in atmosphere, it remain unavailable to the growing plant. Atmospheric N_2 is converted into plant utilizable form ammonia by

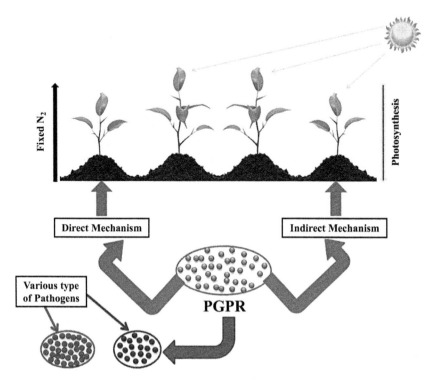

FIGURE 2.1   (See color insert.) Direct and indirect mechanisms of plant growth–promoting rhizobacteria (PGPR) on plant. (*Source*: Picture adapted from García-Fraile et al., 2015)

biological nitrogen fixation microorganisms by using a complex enzyme system known as nitrogenase (Kim and Rees, 1994). Nitrogen-fixing microorganisms fix biological nitrogen at mild temperature, which are widely distributed in nature (Raymond et al., 2004). These nitrogen fixations by PGPR have ability to fix atmospheric nitrogen and provide it to the plants by two mechanisms, they are (1) symbiotic and (2) nonsymbiotic. Symbiotic nitrogen fixation is mutualistic relationship between microbe and plants. Microbe forms a nodule in plant roots where nitrogen fixation occurs. *Rhizobia* are a vast group of rhizobacteria that have the ability to lay symbiotic interactions by the colonization and formation of root nodules with leguminous plants, where nitrogen is fixed to ammonia and made available for the plant (Ahemad and Kibret, 2014). Most widely present PGPRs as symbionts are *Rhizobium, Bradyrhizobium, Sinorhizo-bium*, and *Mesorhizobium* with leguminous plants. *Frankia* (Gram-positive) member of actinomycetes family is associated with broad-spectrum plants particularly nonleguminous plants in symbiotic relationship. Nonsymbiotic nitrogen fixation is carried out by free-living diazotrophs and this can stimulate nonleguminous plants growth such as radish and rice. These nitrogen fixing rhizospheric bacteria belongs to genera including *Azoarcus, Azobacter, Acetobacter, Azospirillum, Burkhold-eria, Diazotrophicus, Enterobactors, Gluconacetobacter, Pseudomonas*, and cyanobacteria, namely, Anabaena and Nostoc (Vessey, 2003; Bhattacharyya and Jha, 2012). Nitrogen fixation (*nif*) genes are found in both symbiotic and free living systems. Nitrogenase (*nif*) genes include structural genes, involved in activation of the Fe protein, iron molybdenum cofactor biosynthesis, electron donation, and regulatory genes required for the synthesis and function of the enzyme. These PGPRs with biological nitrogen fixation provide an integrated approach for disease management, growth promotion activity and maintain the nitrogen level in agricultural soil.

2.3.1.2 PHOSPHATE SOLUBILIZING BACTERIA

Phosphorus (P) is the second most important key element in plant growth and nutrition. It plays an important role in nearly all major metabolic processes of plants including photosynthesis, energy transfer, signal transduction, macromolecular biosynthesis, and respirations (Khan et al., 2010). Phosphorus (P) minerals are abundantly available in soils as organic as well as inorganic forms (Khan et al., 2009). The movement of P takes place in soil, as shown in Figure 2.2. Plants are unable to utilize phosphate because 95–99% phosphate present in insoluble, immobilized and precipitated form (Pandey and Maheshwari, 2007). Plants utilize phosphate only in two soluble forms either monobasic (H_2PO_4) or dibasic (HPO_4^{2-}) ions (Bhattacharyya and Jha, 2012). PGPR and phosphate solubilizing bacteria (PSB) are presented in the soil employs different strategies and mechanisms to make use of unavailable forms of phosphorus and in turn, also help in making phosphorus available for plants to absorb (Sharma et al., 2013). The key mechanisms employed for solubilizing phosphate by PGPR-PSB are:

1. Release of complex or mineral dissolving compounds, for example, organic acid anions, protons, hydroxyl ions, CO_2
2. Liberation of extracellular enzymes (biochemical phosphate mineralization)
3. Release of phosphate during substrate degradation (biological phosphate mineralization).

These PSB includes the following genera: *Arthrobacter, Bacillus, Beijerinckia, Burkholderia, Enterobacter, Erwinia, Flavobacterium, Microbacterium, Pseudomonas, Rhizobium, Rhodococcus, Serratia*, etc. All these PSB have great attention as soil inoculums to improve plant growth and yield (Bhattacharyya and Jha, 2012) alone or in combination with other rhizosphere microbes (Zaidi et al., 2009).

2.3.1.3 SIDEROPHORES PRODUCTION

Iron is one of the fourth most abundant bulk minerals present on the earth, in aerobic soils yet it is unavailable in the soil for plants. Iron is not readily

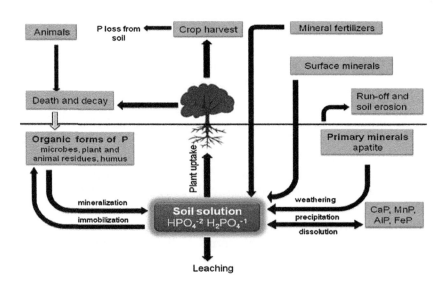

FIGURE 2.2 (See color insert.) Movement of phosphorus in soil. (Reprinted from Ahemad and Kibret, 2014. © 2004 with permission from Elsevier.)

assimilated by either bacteria or plants because ferric ion (Fe^{3+}) is the predominant form in nature, only sparingly soluble so that the amount of iron available for assimilation by living organisms is extremely low (Ma, 2005). In aerobic soils, both microorganisms and plants require high level of iron, and obtaining sufficient iron is even more problematic in the rhizosphere where plants, bacteria, and fungi compete for iron (Loper and Buyer, 1991; Guerinot and Yang, 1994). Microorganisms evolved specialized mechanisms for the assimilation of these iron including the production of low-molecular-weight iron-chelating compound known as siderophores, which transport this element in to their cells (Schwyn and Neilands, 1987; Arora et al., 2013). These siderophores divided into three families depend on the characteristics functional group hydroxa-mates, catecholates, and carboxylates. Currently, more than 500 different types of siderophores are known; among which, 270 have been structur-ally characterized (Cornelis, 2010). Siderophores have been associated with both direct and indirect enhancement of plant growth by PGPR. Direct benefits of bacterial siderophores on plant growth demonstrate by radiolabeled ferric siderophores as a sole source of iron. This indicated that plants are able to take up the labeled iron by large number of PGPR including *Aeromonas* spp., *Azadirachta* spp., *Azotobacter* spp., *Bacillus* spp., *Burkholderia* spp., *Pseudomonas* spp., *Rhizobium* spp., *Serratia* spp., and *Streptomyces* spp. (Sujatha and Ammani, 2013) and enhanced chlorophyll level compared to un inoculated plants (Sharma et al., 2003). Siderophore-producing PGPRs recognized efficient BCA against patho-gens. These siderophores are specifically ferric ion chelator but can also bind with other metals. Thus, heavy metal pollution is largely influenced by siderophores. Siderophores as BCAs are gaining commercial benefits as safe and self-replicating circumvents repeated application but target organisms do not develop resistance (Sayyed et al., 2004).

2.3.1.4 PHYTOHORMONES PRODUCTION

2.3.1.4.1 Indole Acetic Acid

PGPRs synthesize and export phytohormone, also called PGRs, that are synthesized in defined organs of plant and can be translocated to other sites, where these trigger specific biochemical, physiological, and morphological role in plant growth and development (Hayat et al., 2011). PGRs are organic in nature, promote physiological processes of plants even at low concentrations, and take part in the development of tissues where they are produced. Among the PGPRs, gibberellins, cytokinins, auxins, abscisic acid, and ethylene well documented. Up to 80% of the rhizobacteria can synthesize IAA colonized the seed or root surfaces is proposed to act in conjunction with endogenous IAA in plant to stimulate cell proliferation and enhance the host's uptake of minerals and nutrients from the soil (Vessey, 2003). IAA affects plant cell division, extension, and differentiation; stimulates seed and tuber germination; increases the rate of xylem and root development; controls processes of vegetative growth; initiates lateral and adventitious root formation; mediates responses to light, gravity, and florescence; affects photosynthesis, pigment formation, biosynthesis of various metabolites, and resistance to stressful conditions (Spaepen and Vanderleyden, 2011). Tryptophan (amino acid) commonly found in root exudates, identified as main precursor molecule for biosynthesis of IAA in bacteria (Shilev, 2013). The biosynthesis of IAA by PGPR involves formation via indole-3-pyruvic acid and indole-3-acetic aldehyde, which is the most common mechanism in bacteria like *Pseudomonas*, *Rhizobium*, *Bradyrhizobium*, *Agrobacterium*, *Enterobacter*, and *Klebsiella* (Kang et al., 2010). Root-growth promotion by the free-living PGPRs, for example, *Alkaligenes faecalis, Enterobacter cloacae, Acetobacter dizotrophicous*, species of *Azospirillum, Pseudomonas*, and *Xanthomonas* have been related to low level of IAA secretion. However, phytohormones production from microbial sources are more effective due to the reason that the threshold between inhibitory and stimulatory levels of chemically produced hormones is low, while microbial hormones are more effective by virtue of their continuous slow release.

2.3.1.4.2 Cytokinins, Gibberellin, and Ethylene

Cytokinins and gibberellins are very important phytohormones present in low amount in biological samples that enhance cell division, root developments, root formations, involve in photosynthesis and chlorophyll differentiations (Frankenberger and Arshad 1995; Vessey, 2003). They are also known to induce opening of stomata, suppression of auxin-induce apical dominance and inhibit senescence of plant organs, especially in leaves

(Crozier et al., 2001). More than 30 growth-promoting compounds of cyto-kinin group are reported produced by plant associated PGPR. *Azotobacter* spp., *Rhizobium* spp., *Pantoea agglomerans*, *Rhodospirillum rubrum*, *P. fluorescens*, *Bacillus subtilis*, and *Paenibacillus polymyxa* produce either cytokinins, gibberellins, or both for plant growth promotion (Kang et al., 2010). Some strains of phytopathogens can also synthesize cytokinins. However, PGPR produce lower cytokinin levels as compared to phyto-pathogens so that the effect of PGPR on plant growth is stimulatory, while on pathogens is inhibitory.

Ethylene is another a key phytohormone that plays large number of biological activities including promoting root initiation, inhibiting root elongation, fruit ripening, lower wilting, seed germination, leaf abscis-sion, activating the synthesis of other plant hormones (Glick et al., 2007). The high concentration of ethylene induces defoliation and other cellular processes that may lead to reduce crop performance (Bhattacharyya and Jha, 2012). The enzyme, 1-aminocyclopropane-1-carboxylic acid oxidase (ACCO) is a prerequisite for ethylene production. Iqbal et al. (2012) reported to improved nodule number, nodule dry weight, fresh biomass, grain yield, straw yield, and nitrogen content in grains of lentil results in lowering of the ethylene production via inoculation with plant growth–promoting strains of *Pseudomonas* spp. containing ACC deaminase along with *R. leguminosarum*. Currently, many bacterial strains exhibiting ACC deaminase activity recognized in a wide range of genera.

2.3.2 INDIRECT MECHANISM

The biocontrol bacteria have ability to indirectly promote the growth has been the source of considerable interest both in terms of (1) development of underlying mechanisms used by biocontrol bacteria and (2) utilization of these bacteria commercially instead of chemical pesticides. These two objectives are complementary to each other by effective implementation of the bacterial strains as BCAs (Glick, 2012). These PGPRs are a great source toward promising, sustainable and environment-friendly agriculture and maintains fertility of soil and plant growth indirectly. This approach leads to inspire a wide range of exploitation of PGPR led to reducing the need for agrochemicals (fertilizers and pesticides) for improve soil fertility by a variety of mechanisms that via production of antibiotics, siderophores, hydrogen cyanide (HCN), hydrolytic enzymes, etc. (Lugtenberg and Kamilova, 2009; Tariq et al., 2014). The production of antibiotics is considered to be one of the most powerful and studied biocontrol mechanisms of PGPR against phytopathogens that have become increasingly better understood over the past two decades. The synthesis of a range of different type of antibiotics is the PGPB traits that are most often associated with the ability of the bacterium to prevent proliferation of plant pathogen (Compant et al., 2005; Mazurier et al., 2009).

2.3.2.1   ANTIBIOSIS AND LYTIC ENZYMES

The production of antibiotics considered the most powerful and studied biocontrol mechanisms of PGPR against phytopathogens and become increasingly better understood over the past two decades (Shilev, 2013). Many antibiotics identified, compounds such as amphisin, 2,4-diacetylphloroglucinol (DAPG), oomycin A, phenazine, pyoluteorin, pyrrolnitrin, tensin, tropolone, and cyclic lipopeptides produced by *Pseudomonads* (Loper and Gross, 2007) and oligomycin A, kanosamine, zwittermicin A, and xanthobaccin produced by *Bacillus* spp., *Streptomyces* spp., and *Stenotrophomonas* spp. to prevent the proliferation of fungal plant pathogens (Compant et al., 2005). Many of these antibiotics together with their specificity and mode of action studied in detail, and some of the biocontrol strains are commercially available. One problem with depending too much on antibiotic-producing bacteria as BCAs is that with the increased use of these strains, some phytopathogens may develop resistance to specific antibiotics. To overcome this problem, few researchers suggest utilization of biocontrol strains that synthesize HCN as well as one or more antibiotics (Glick, 2012). In soils, antibiotic DAPG is produced by *Pseudomonas* spp. which was reported for biocontrol of fungal disease in wheat caused by the *Gaeumanomyces graminis* var. *tritici* (de Souza et al., 2003). Bacterization of wheat seeds with *P. fluorescens* strains producing the antibiotic phenazine-1-carboxylic acid (PCA) results in significant suppression of 60% in field trials (Weller, 2007). *Bacillus amyloliquefaciens* is known for lipopeptide and polyketide production for biocontrol activity and PGPR against soilborne pathogens (Ongena and Jacques, 2008). Apart from antibiotic production, some rhizobacteria are also capable of producing volatile compound known as HCN for biocontrol of black root rot of tobacco caused by *Thielaviopsis basicola* (Sacherer et al., 1994). Production of DAPG and HCN by *Pseudomonas* contribute biocontrol of bacterial canker of tomato (Lanteigne et al. 2012). This approach is effective because while HCN may not have sufficient biocontrol activity alone, it appears to act synergistically with bacterial encoded antibiotics. Some biocontrol bacteria produce enzymes including chitinases, cellulases, β-1,3 glucanases, proteases, and lipases that can lyse a portion of the cell walls of many pathogenic fungi (Hayat et al., 2010; Joshi et al., 2012). These enzymes have biocontrol activity against a range of pathogenic fungi including *Botrytis cinerea*, *Sclerotium rolfsii*, *Fusarium oxysporum*, *Phytophthora* spp., *Rhizoctonia solani*, and *Pythium ultimum* (Singh et al, 1999; Kim et al., 2008; Nadeem et al., 2013; Upadyay et al., 2012). Many researchers claimed the effectiveness of PGPR as efficient BCAs including *P. fluorescens* CHA0 suppresses black root rot of tobacco caused by the fungus *T. basicola* (Voisard et al., 1989), *Pseudomonas putida* against *Macrophomina phaseolina* in chickpea and *Azotobacter chroococcum* against *F. oxysporum* in *Sesamum indicum*, respectively, in field condition (Maheshwari et al., 2012).

2.3.2.2   SIDEROPHORE

Some bacterial strains that do not employ any other means of biocontrol can act as BCAs by producing siderophores. Siderophores from PGPB can prevent some phytopathogens from acquiring a sufficient amount of iron thereby limiting their ability to proliferate (Dowling et al., 1996; Kloepper et al., 1980). This mechanism is effective because biocontrol PGPB produce siderophores that have a much greater affinity for iron than do fungal pathogens (Schippers et al., 1987). Conversely, fungal pathogens are unable to proliferate in the rhizosphere of the roots of the host plant because of lack of iron (O'Sullivan and O'Gara, 1992). In this model, the biocontrol PGPB effectively out-compete fungal pathogens for available iron. Siderophore-producing PGPR can prevent the proliferation of pathogenic microorganisms by sequestering Fe^{3+} in the area around the root (Mehnaz, 2013). These siderophores bind with ferric ion and make siderophore–ferric complex, which subsequently binds with iron limitation dependent receptors at the bacterial cell surface. The ferric ion is subsequently released and active in the cytoplasm as ferrous ion. Many plants can use various bacterial siderophores as iron sources, although the total concentrations are probably too low to contribute substantially to plant iron uptake. Various studies showed isolation of siderophore producing bacteria belonging to the *Bradyrhizobium, Pseudomonas, Rhizobium, Serratia,* and *Streptomyces* genera from the rhizosphere (Kuffner et al., 2008).

2.3.2.3   INDUCE SYSTEMIC RESISTANCE (ISR)

ISR defined as a physiological state of enhanced defensive capacity elicited in response to specific environmental stimuli and consequently the plant's innate defense system effective against subsequent biotic challenges (Avis et al., 2008). This is similar to systemic acquired resistance (SAR) that occurs when plants activate their defense mechanisms in response to infection by a pathogenic agent (Pieterse et al., 2009). ISR does not target specific pathogens but involves jasmonate and ethylene signaling within the plant, and these hormones stimulate the host plant's defense responses against a variety of plant pathogens (Glick, 2012). ISR does not require any direct interaction between the resistance-inducing PGPR and the pathogen (Bakker et al., 2007). Fungal, bacterial, and viral diseases occur in some instances however, damage caused by insects and nematodes can be reduced after application of PGPR (Naznin et al., 2012). Many individual bacterial components induce ISR such as lipopolysaccharides (LPS), flagella, siderophores, cyclic lipopeptides, 2,4-diacetylphloroglucinol, homoserine lactones, and volatiles like, acetoin, and 2,3-butanediol (Doornbos et al., 2012).

2.4   PGPR AS BIOCONTROL AGENT

PGPR gained worldwide attention and expansion for eco-friendly smart agriculture benefits. PGPRs including pathogenic, symbiotic, and free living rhizobacterial species are reported to produce IAA and gibberllic acid in the rhizospheric soil and thereby plays a significant role in increasing the root surface area and number of root tips in many plants (Han et al. 2005; Kloepper et al., 2007). Rhizobacteria are the most widely used in association with plant rhizosphere and are present in all agroecosystems (Kloepper and Schroth, 1978). Many effective strains of bacteria considered ideal BCAs because of the rapid growth, user friendly and aggressive colonization of the rhizosphere. Use of PGPRs as BCA of soilborne fungal pathogens is an alternative or complementary strategy to mitigate physical and chemical disease management practices (Berg and Smalla, 2009). PGPR indirectly enhanced growth via suppressing the pathogen by producing chemicals that inhibit the growth and invasion of plant pathogens. PGPR reported beneficial to the plants like tomatoes and peppers growing on water-deficient soils for conferring resistance to water stress conditions (Aroca and Ruiz-Lozano, 2009) as well as disease control. For sustainable biocontrol, the mode of action of the PGPRs strains will be a determining factor in the type of disease control strategy implemented. Most biocontrol strains of PGPR produce one or several groups of antibiotics, which inhibit fungal pathogens (Haas and Defago, 2005). Antibiotics produced by these biocontrol PGPRs reduce or suppress soilborne infections of cereal crops including wheat, rice, maize, chickpea, and barley (Raaijmakers et al., 2002). Genetic analysis of many biocontrol strains of *Pseudomonas* spp. indicates that there is a positive correlation between disease suppression and antibiotic production (Vincent et al., 1991). The potential uses of antibiotic-producing PGPR as BCAs are reported in many cereals including maize, sorghum, rice, and chickpea. In maize for instance, *Fusarium verticilloides*, causing root rot and yield loss, is significantly suppressed by the application of *Bacillus amyloliquifaciens* during seed treatment (Pereira et al., 2009). A potential biocontrol strain reported, *Bacillus* spp. L324-92, with a broad-spectrum inhibitory activity against root rot caused by *R. solani*, *Pythium irregulare*, and *P. ultimum* (Kim et al., 1997). The study on auxin synthesizing rhizobacteria (Spaepen et al., 2007) as phytohormone producer demonstrates that rhizobacteria can synthesize IAA from tryptophan by different pathways, although the general mechanism of auxin synthesis concentrated on the tryptophan-independent pathways. The phytopathogenic bacteria rather use the indole acetamide pathway to synthesize IAA implicated earlier in the tumor induction in plants.

2.5 COMMERCIALIZATION OF PGPR AND MAJOR CONCERN

So far, with the limited understanding of PGPR-Plant interaction, a several number of these bacteria nevertheless accepted commercially as adjuncts to agricultural practices (Lucy et al., 2004; Banerjee et al., 2006). Commercialized PGPR strains include *Agrobacterium radiobacter, Azospirillum brasilense, Azospirillum lipoferum, Azotobacter chroococcum, Bacillus fimus, Bacillus licheniformis, Bacillus megaterium, Bacillus mucilaginous, Bacillus pumilus, Bacillus* spp., *Bacillus subtilis, Bacillus subtilis* var. *amyloliquefaciens*, etc. *Burkholderia cepacia, Delfitia acidovorans, Paenobacillus macerans, Pantoea agglomerans, Pseudomonas aureofaciens, Pseudomonas chlororaphis, P. fluorescens, Pseudomonas solanacearum, Pseudomonas* spp., *Pseudomonas syringae, Serratia entomophilia, Streptomyces griseoviridis, Streptomyces* spp., *Streptomyces lydicus*, and various species of *Rhizobia*. However, PGPR-inoculated crops represent only a small fraction of current worldwide agricultural practice (Glick, 2012). For the commercialization of PGPR bacterial strain, there are different issues that are needed to be taken into consideration, which are as follows:

1. Determination of traits most important for efficacious functioning and subsequent selection of PGPR strains with appropriate biological activities;
2. Consistency among regulatory agencies in different countries regarding which strains released in the environment, under what conditions genetically engineered strains are suitable for environmental use;
3. A better understanding of the advantages and disadvantages of using rhizosphere versus endophytic bacteria;
4. Selection of PGPR strains that function optimally under specific environmental conditions (e.g., those that work well in warm and sandy soils versus organisms better adapted to cool and wet environments);
5. Development of effective implications of PGPR to plants in various settings (e.g., in the field versus in the greenhouse);
6. A better understanding of the potential interactions between PGPR and mycorrhizae and other soil fungi.

Moreover, commercial success of PGPR strains requires economical and viable market demand, consistent and broad-spectrum action, safety and stability, longer shelf life, low capital costs, and easy availability of career materials (Gupta et al., 2015).

2.6   FUTURE PROSPECTS AND CHALLENGES

PGPR strains have diverse and worthwhile applications as bioinoculants, biofertilizers, and BCAs, with practical potential proved as a boon in environment-friendly sustainable agriculture. With the application of biotechnological approach to improve the efficacy of PGPR strains using genetic engineered crop production for sustainable agriculture. The applications of rhizosphere soil in agricultural crops with desirable bacterial populations have established considerable promises in both the laboratory and greenhouse experiment. Improved understanding of mode of action by which PGPRs promote plant growth can lead to expanded exploitation of these as potential "biofertilizers" to reduce the negative environmental effects associated with the food and fiber production. An effort of application of genetically engineered PGPRs to remediate complex contaminated soil (Denton, 2007) and thereby increasing crop productivity in agriculture is another attractive idea of research in recent decade. The rhizobacteria community can be specifically engineered to target various pollutants at cocontaminated sites to provide customized rhizoremediation system (Wu et al., 2006). Recent progress of molecular biology and biotechnology in the understanding of rhizobacteria interactions with the nodules of crop plants will encourage a suitable area of research in PGPR mechanisms relating to rhizosphere colonization. Inoculation of rhizobacteria population in genetically engineered Arabidopsis for removing the cadmium and lead proved successful. Application of PGPRs in transgenic crops proved promising for future sustainable agriculture (Ali et al., 2010) in advancing rhizoremediation technologies. Efforts are under progress on the production of transgenic plants that can increase remediation efficiency by expressing a particular PGPR protein (Zhuang et al., 2007). Genomic tinkering of naturally occurring PGPR strains with effective genes (Nakkeeran et al., 2005) could lead to accentuated expression of genomic products and thereby alleviating the attack of both pests and diseases on field crops that would further facilitate for better introduction of a single bacterium with multiple modes of action to benefit the growers. Thus, future success of industries producing microbial inoculants, especially PGPRs, will depend on innovative business management, product marketing, extension education, and extensive research. Further optimization is required for better fermentation and formulation processes of effective PGPR strains to introduce in agriculture.

2.7   SUMMARY

Rhizosphere promotes the growth and development of several classified and nonclassified microorganism; among them, several organisms stimulate the growth of plants. Many microbes colonize in the roots of monocots and dicots and enhance growth by direct and indirect mechanisms such microbes known as PGPR. PGPR assist the plant growth directly by modifying root functioning, improve resources acquisition (nitrogen, phosphorus, and essential minerals), and control plant hormones level. Indirectly support plant growth and suppression of many phytopathogens as biocontrol and influence the physiology of the entire plants. In current perspective, environment-friendly microbial input is the need of the day that comprises plant growth–promoting rhizobacteria in agriculture for restructuring the crop rhizospheres for improved and sustainable nutrient supply in the soils and increasing the health and yield of crops. The two major constraints in agriculture are enhanced crop production and crop protection against diseases. Sustainable approaches not aimed solely to capitalize on short-term production but rather those that consider long-standing production expansions, ecology of agricultural systems, and user-friendly farmer profitability. Microbial inoculants have great potential to provide holistic health and sustainable crop yields. Maintaining biodiversity of PGPR in soil could be an important component of environment-friendly sustainable agriculture strategies. PGPR proved to be an attractive alternative to avoid global dependence on lethal agricultural chemicals disrupt the agro-ecosystems. In this chapter, we address the insight knowledge and future prospects of the PGPR their mode of actions, recent developments effects on plant physiology, and morphology induced by PGPR.

ACKNOWLEDGMENTS

Authors Pradeep Kumar and Madhu Kamle would like to thank Ben-Gurion University of the Negev, Israel. We would like to especially acknowledge Professor Ariel Kushmaro (The Department of Biotechnology Engineering), Professor Arieh Zaritsky (Faculty of Natural Sciences), Professor Yaron Sitrit (French Associates Institute for Agriculture and Biotechnology of Drylands) Ben-Gurion University of the Negev,

Israel, and Professor Eitan Ben-Dov, Achava Academic College, Israel for their support and motivations.

KEYWORDS

- **biocontrol agents**
- **induced systemic resistance**
- **PGPR**
- **plant growth regulators**
- **plant rhizosphere**
- **soil microorganisms**
- **sustainable agriculture**

REFERENCES

Ahemad, M.; Kibret, M. Mechanisms and Applications of Plant Growth Promoting Rhizobacteria: Current Perspective. *J. King Saud Univ.-Sci.* **2014**, *26*, 1–20.

Ali, K.; Shamsuddin; Zulkifli, H. Phytoremediation of Heavy Metals with Several Efficiency Enhancer Methods. *Afr. J. Biotechnol.* **2010**, *9*(25), 3689–3698.

Antoun, H.; Prevost, D. Ecology of Plant Growth Promoting Rhizobacteria. In: *PGPR: Biocontrol and Biofertilization*; Siddiqui, Z. A., Ed.; Springer: Dordrecht, 2005; pp 1–38.

Aroca, R.; Ruiz-Lozano, J. M. Induction of Plant Tolerance to Semi-Arid Environments by Beneficial Soil Microorganisms—A Review. In *Climate Change, Intercropping, Pest Control and Beneficial Microorganisms, Sustainable Agriculture Reviews*; Lichtouse, E., Ed.; Springer: The Netherlands, 2009; pp 121–135.

Arora, N. K.; Tewari, S.; Singh, R. Multifaceted Plant-Associated Microbes and Their Mechanisms Diminish the Concept of Direct and Indirect PGPRs. In *Plant Microbe Symbiosis: Fundamentals and Advances*; Arora, N. K., Ed.; Springer: The Netherlands; 2013; pp 411–449.

Avis, T. J.; Gravel, V.; Antoun, H.; Tweddell, R. J. Multifaceted Beneficial Effects of Rhizosphere Microorganisms on Plant Health and Productivity. *Soil Biol. Biochem.* **2008**, *40*, 1733–1740.

Bakker, P. A. H. M.; Pieterse, C. M. J.; van Loon, L. C. Induced Systemic Resistance by Fluorescent *Pseudomonas* spp. *Phytopathology* **2007**, *97*(2), 239–243.

Banerjee, M. R.; Yesmin, L.; Vessey, J. K. Plant-growth-promoting Rhizobacteria as Biofertilizers and Biopesticides. In: Rai M. K. Ed. *Handbook of Microbial Biofertilizers*; Food Products Press: Binghamton, NY, USA, 2006; pp 137–181.

Berg, G.; Smalla, K. Plant Species and Soil Type Cooperatively Shape the Structure and Function of Microbial Communities in the Rhizosphere. *FEMS Microbiol. Ecol.* **2009**, *68*, 1–13.

Bhattacharyya, P. N.; Jha, D. K. Plant Growth-Promoting Rhizobacteria (PGPR): Emergence in Agriculture. *World J. Microbiol. Biotechnol.* **2012**, *28*, 1327–1350.

Compant, S.; Reiter, B.; Sessitsch, A.; Nowak, J.; Clément, C.; Barka, E. D. Endophytic Colonization of *Vitis vinifera* L. by Plant Growth-Promoting Bacterium *Burkholderia* sp. Strain PSJN. *Appl. Environ. Microbiol.* **2005**, *71*, 1685–1693.

Cornelis, P. Iron Uptake and Metabolism in *Pseudomonads*. *Appl. Microbiol. Biotechnol.* **2010**, *86*, 1637–1645.

Crozier, A.; Kamiya, Y.; Bishop, G.; Yokota, T. Biosynthesis of Hormones and Elicitors Molecules. In *Biochemistry and Molecular Biology of Plants*; Buchanan, B. B., Grussem, W., Jones, R. L., Eds.; American Society of Plant Biologists: Rockville, 2001; pp 850–900.

Damayanti, T. A.; Pardede, H.; Mubarik, N. R. Utilization of Root-colonizing Bacteria to Protect Hot-pepper Against Tobacco Mosaic Tobamovirus. *Hayati. J. Biosci.* **2007**, *14*, 105–109.

de Souza, J. T.; Weller, D. M.; Raaijmakers, J. M. Frequency, Diversity and Activity of 2,4-Diacetylphloroglucinol Producing Fluorescent *Pseudomonas* spp. in Dutch Take-All Decline Soils. *Phytopathology* **2003**, *93*, 54–63.

Denton, B. Advances in Phytoremediation of Heavy Metals Using Plant Growth Promoting Bacteria and Fungi. MMG 445. *Basic Biotechnol.* **2007**, *3*, 1–5.

Doornbos, R. F.; van Loon, L. C.; Peter, A. H. M.; Bakker, A. Impact of Root Exudates and Plant Defense Signaling on Bacterial Communities in the Rhizosphere. *Rev. Sustain. Dev.* **2012**, *32*, 227–243.

Dowling, D. N.; Sexton, R.; Fenton, A.; et al. Iron Regulation in Plant-Associated *Pseudomonas fuorescens* M114: Implications for Biological Control. In *Molecular Biology of Pseudomonads*. Nakazawa, T., Furukawa, K., Haas, D., and Silver, S., Eds.; American Society for Microbiology Press: Washington, DC, USA, 1996; pp 502–511.

Dubeikovsky, A. N.; Mordukhova, E. A.; Kochetkov, V. V.; Polikarpova, F. Y.; Boronin, A. M. Growth Promotion of Blackcurrant Softwood Cuttings by Recombinant Strain *Pseudomonas fuorescens* BSP53a Synthesizing an Increased Amount of Indole-3-Acetic Acid. *Soil Biol. Biochem.* **1993**, *25*(9), 1277–1281.

Figueiredo, M. V. B.; Seldin, L.; Araujo, F. F.; Mariano, R. L. R. Plant Growth Promoting Rhizobacteria: Fundamentals and Applications. In: *Plant Growth and Health Promoting Bacteria;* Maheshwari, D. K., Ed.; Springer-Verlag: Berlin, Heidelberg, 2011; pp 21–42.

Frankenberger, W. T. Jr.; Arshad, M. *Photohormones in Soil: Microbial Production and Function*; Marcel Dekker Inc.: New York, USA, 1995, p 503.

García-Fraile, P.; Menéndez, E.; Rivas, R. Role of Bacterial Biofertilizers in Agriculture and Forestry. *Bioengineering* **2015**, *2*(3), 183–205.

Glick, B. R. The Enhancement of Plant Growth by Free-Living Bacteria. *Can. J. Microbiol.* **1995**, *41*(2), 109–117.

Glick, B. R.; Patten, C. L.; Holguin, G.; Penrose, G. M. *Biochemical and Genetic Mechanisms Used by Plant Growth Promoting Bacteria*. Imperial College Press: London, 1999.

Glick, B. R. Plant Growth-Promoting Bacteria: Mechanisms and Applications. *Scientifica* **2012**, *2012*, 963401.

Glick, B. R.; Todorovic, B.; Czarny, J.; Cheng, Z.; Duan, J. Promotion of Plant Growth by Bacterial ACC Deaminase. *Crit. Rev. Plant Sci.* **2007**, *26*, 227–242.

Gray, E. J.; Smith, D. L. Intracellular and Extracellular PGPR: Commonalities and Distinctions in the Plant–Bacterium Signaling Processes. *Soil Biol. Biochem.* **2005**, *37*, 395–412.

Guerinot, M. L.; Ying, Y. Iron: Nutritious, Noxious, and Not Readily Available. *Plant Physiol.* **1994,** *104*(3), 815–820.

Haas, D.; Defago, G. Biological Control of Soil-Borne Pathogens by Fluorescent Pseudo-monads. *Nat. Rev. Microb.* **2005,** *3*(4), 307–319.

Han, J.; Sun, L.; Dong, X.; Cai, Z.; Sun, X.; Yang, H.; Wang, Y.; Song, W. Characterization of a Novel Plant Growth-Promoting Bacteria Strain *Delftia tsuruhatensis* HR4 Both as a Diazotroph and a Potential Biocontrol Agent against Various Plant Pathogens. *Syst. Appl. Microbiol.* **2005,** *28*(1), 66–76.

Hayat, R.; Safdar Ali, S.; Amara, U.; Khalid, R.; Ahmed, I. Soil Beneficial Bacteria and Their Role in Plant Growth Promotion: A Review. *Ann. Microbiol.* **2010,** *60,* 579–598.

Hayat, R.; Ali, S.; Amara, U.; Khalid, R.; Ahmed, I. Soil Beneficial Bacteria and Their Role in Plant Growth Promotion: A Review. *Ann. Microbiol.* **2011,** *60*, 579–598.

Iqbal, M. A.; Khalid, M.; Shahzad, S. M.; Ahmad, M.; Soleman, N.; Akhtar, N. Integrated Use of *Rhizobium leguminosarum*, Plant Growth Promoting Rhizobacteria and Enriched Compost for Improving Growth, Nodulation and Yield of Lentil (*Lens culinaris* Medik). *Chilean J. Agric. Res.* **2012,** *72*, 104–110.

Joshi, M.; Shrivastava, R.; Sharma, A. K.; Prakash, A. Screening of Resistant Verities and Antagonistic *Fusarium oxysporum* for Biocontrol of *Fusarium* Wilt of Chilli. *Plant Pathol. Microbiol.* **2012,** *3*, 134.

Kang, B. G.; Kim, W. T.; Yun, H. S.; Chang, S. C. Use of Plant Growth-Promoting Rhizo-bacteria to Control Stress Responses of Plant Roots. *Plant Biotechnol. Rep.* **2010,** *4,*179–183.

Khan, M. S.; Zaidi, A.; Wani, P. A.; Oves, M. Role of Plant Growth Promoting Rhizobac-teria in the Remediation of Metal Contaminated Soils. *Environ. Chem. Lett.* **2009,** *7*, 1–19.

Khan, M. S.; Zaidi, A.; Ahemad, M.; Oves, M.; Wani, P. A. Plant Growth Promotion by Phosphate Solubilizing Fungi—Current Perspective. *Arch. Agron. Soil. Sci.* **2010,** *56*, 73–98.

Kim, J.; Rees, D. C. Nitrogenase and Biological Nitrogen Fixation. *Biochemistry* **1994,** *33*, 389–397.

Kim, D. S.; Cook, R. J.; Weller, D. M. *Bacillus* sp. L324-92 for Biological Control of Three Root Diseases of Wheat Grown Reduced Tillage. *Phytopathology* **1997,** *87*, 551–558.

Kim, Y. C.; Jung, H.; Kim, K. Y.; Park, S. K. An Effective Biocontrol Bioformulation Against *Phytophthora* Blight of Pepper Using Growth Mixtures of Combined Chitino-lytic Bacteria Under Different Field Conditions. *Eur. J. Plant. Pathol.* **2008,** *102*(4), 373–382.

Kloepper, J. W.; Schroth, M. N. In *Plant Growth-Promoting Rhizobacteria on Radishes,* Proceedings of the 4th International Conference on Plant Pathogenic Bacteria. Gilbert-Clarey, Tours, 1978; pp 879–882.

Kloepper, J. W.; Gutierrez-Estrada, A.; Mclnroy, J. A. Photoperiod Regulates Elicitation of Growth Promotion but not Induced Resistance by Plant Growth-Promoting Rhizobac-teria. *Can. J. Microbiol.* **2007,** *53*(2), 159–167.

Kloepper, J. W.; Leong, J.; Teintze, M.; Schroth, M. N. Enhanced Plant Growth by Sidero-phores Produced by Plant Growth-Promoting Rhizobacteria. *Nature* **1980,** *286*(5776), 885–886.

Kloepper, J. W.; Schroth M. N. Relationship of In Vitro Antibiosis of Plant Growth Promoting Rhizobacteria to Plant Growth and the Displacement of Root Microflora. *Phytopathology* **1981,** *71*, 1020–1024.

Kloepper, J. W.; Zablotowick, R. M.; Tipping, E. M.; Lifshitz, R. Plant Growth Promotion Mediated by Bacterial Rhizosphere Colonizers. In *The Rhizosphere and Plant Growth*; Keister, D. L., Cregan, P. B., Eds.; Kluwer Academic Publishers: Dordrecht, Netherlands, 1991; pp 315–326.

Kloepper, J. W. Plant Growth-promoting Rhizobacteria (Other Systems). In: *Azospirillum/ Plant Associations*; Okon, Y., Ed.; CRC Press: Boca Raton, FL, USA, 1994; pp 111–118.

Kloepper, J. W. A Review of Mechanisms for Pplant Growth Promotion by PGPR In: *Abstracts and Short Papers*; Reddy, M. S., Anandaraj, M., Eapen, S. J., Sarma, Y. R., Kloepper, J. W., Eds.; 6th International PGPR Workshop, 5–10 October 2003, Indian Institute of Spices Research, Calicut, India, 2003, pp 81–92.

Kuffner, M.; Puschenreiter, M.; Wieshammer, G.; Gorfer, M.; Sessitsch, A. Rhizosphere Bacteria Affect Growth and Metal Uptake of Heavy Metal Accumulating Willows. *Plant Soil.* **2008,** *304*, 35–44.

Lanteigne, C.; Gadkar, V. J.; Wallon, T.; Novinscak, A.; Filion, M. Production of DAPG and HCN by *Pseudomonas* sp. LBUM300 Contributes to the Biological Control of Bacterial Canker of Tomato. *Phytopathology* **2012,** *102*, 967–973.

Loper, J. E.; Buyer, J. S. Siderophores in Microbial Interactions on Plant Surfaces. *Mol. Plant-Microbe Interact.* **1991,** *4*, 5–13.

Loper, J. E.; Gross, H. Genomic Analysis of Antifungal Metabolite Production by *Pseudomonas fluorescens* Pf-5. *Eur. J. Plant Pathol.* **2007,** *119*, 265–278.

Lucy, M.; Reed, E.; Glick. R. B. Applications of Free Living Plant Growth-promoting Rhizobacteria. *Antonie van Leeuwenhoe* **2004,** *86*(1), 1–25.

Lugtenberg, B.; Kamilova, F. Plant-Growth-Promoting Rhizobacteria. *Annu. Rev. Microbiol.* **2009,** *63*, 541–556.

Lynch, J. M., Ed. *The Rhizosphere*; Wiley-Interscience: Chichester, UK, 1990.

Ma, J. F. Plant Root Responses to Three Abundant Soil Minerals: Silicon, Aluminum and Iron. *Crit. Rev. Plant Sci.* **2005,** *24*(4), 267–281.

Maheshwari, D. K.; Dubey, R. C.; Aeron, A.; Kumar, B.; Kumar, S.; Tewari, S.; Arora, N. K. Integrated Approach for Disease Management and Growth Enhancement of *Sesamum indicum* L. Utilizing *Azotobacter chroococcum* TRA2 and Chemical Fertilizer. *World J. Microbiol. Biotechnol.* **2012,** *28*, 3015–3024.

Mazurier, S.; Corberand, T.; Lemanceau, P.; Raaijmakers, J. M. Phenazine Antibiotics Produced by Fluorescent *Pseudomonads* Contribute to Natural Soil Suppressiveness to *Fusarium* Wilt. *ISME J.* **2009,** *3*(8), 977–991.

Mehnaz, S. Secondary Metabolites of *Pseudomonas aurantiaca* and Their Role in Plant Growth Promotion. In *Plant Microbe Symbiosis: Fundamentals and Advances*; Arora, N. K., Ed.; Springer: India, 2013, pp 373–394.

Nadeem, S. M.; Naveed, M.; Zahir, Z. A.; Asghar, H. N. Plant-Microbe Interactions for Sustainable Agriculture: Fundamentals and Recent Advances In *Plant Microbe Symbiosis: Fundamentals and Advances*; Arora, N. K., Ed.; Springer: India, 2013, pp 51–103.

Nakkeeran, S., Fernando, W. G. D.; Siddiqui, Z. A. Plant Growth Promoting Rhizobacteria Formulations and Its Scope in Commercialization for the Management of Pests and Diseases. In *PGPR: Biocontrol and Biofertilization*; Springer: Dordrecht, 2005, pp 257–296.

Naz111, II. A.; Kimura, M.; Miyazawa, M.; Hyakumachi, M. Analysis of Volatile Organic Compounds Emitted by Plant Growth Promoting Fungus *Phoma* sp. GS8-3 for Growth Promotion Effects on Tobacco. *Microbe Environ.* **2012**, *28*, 42–49.

Ongena, M.; Jacques, P. *Bacillus lipopeptides*: Versatile Weapons for Plant Disease Biocontrol. *Trends Microbiol.* **2008**, *16*, 115–125.

Ortiz-Castro, R.; Contreras-Cornejo, H. A.; Macias-Rodriguez, L.; Lopez-Bucio, J. The Role of Microbial Signals in Plant Growth and Development. *Plant Signal Behav.* **2009**, *4*, 701–712.

O'Sullivan, D. J.; O'Gara, F. Traits of Fluorescent *Pseudomonas* spp. Involved in Suppression of Plant Root Pathogens. *Microbiol. Rev.* **1992**, *56*(4), 662–676.

Pandey, P.; Maheshwari, D. K. Two Species Microbial Consortium for Growth Promotion of *Cajanus Cajan*. *Curr. Sci.* **2007**, *92*, 1137–1142.

Pereira, P.; Nesci, A.; Etcheverrg, M. G. Efficacy of Bacterial Seed Treatments for the Control of *Fusarium verticilliodes* in Maize. BioControl **2009**, *54*,103–111.

Pieterse, C. M. J.; Leon-Reyes, A.; Van der Ent, S.; VanWees, S. C. Networking by Small-Molecule Hormones in Plant Immunity. *Nat. Chem. Biol.* **2009**, *5*, 308–316.

Raaijmakers, J. M.; Vlami, M.; de Souza, J. T. Antibiotic Production by Bacterial Biocontrol Agent. *Anton van Leeuwenhoek* **2002**, *81*, 537–547.

Raymond, J.; Siefert, J. L.; Staples, C. R.; Blankenship, R. E. The Natural History of Nitrogen Fixation. *Mol. Biol. Evol.* **2004**, *21*, 541–554.

Sacherer, P.; Défago, G.; Haas, D. Extracellular Protease and Phospholipase C Are Controlled by the Global Regulatory Gene *gacA* in the Biocontrol Strain *Pseudomonas fluorescens* CHA0. *FEMS Microbiol. Lett.* **1994**, *116*, 155–160.

Sayyed, R. Z.; Naphade, B. S.; Chincholkar, S. B. Ecologically Competent Rhizobacteria for Plant Growth Promotion and Disease Management. In *Recent Trends in Biotechnology*; Rai, M. K., Chikhale, N. J., Thakare, P. V., Wadegaonkar, P. A., Ramteke, A. P., Eds.; Scientific Publisher: Jodhpur (Raj.), India, 2004, pp 1–16.

Schippers, B.; Bakker, A. W.; Bakker, A. H. M. Interactions of Deleterious and Beneficial Rhizosphere Microorganisms and the Effect of Cropping Practice. *Annu. Rev. Phytopathol.* **1987**, *25*, 339–358.

Schwyn, B.; Neilands, J. B. Universal Chemical Assay for the Detection and Determination of Siderophores. *Anal. Biochem.* **1987**, *160*, 47–56.

Sharma, A.; Johri, B. N.; Sharma, A. K.; Glick, B. R. Plant Growth-Promoting Bacterium *Pseudomonas* sp. strain GRP3 Influences Iron Acquisition in Mung Bean (*Vigna radiata* L. Wilzeck). *Soil Biol. Biochem.* **2003**, *35*, 887–894.

Sharma, S. B.; Sayyed, R. Z.; Trivedi, M. H.; Gobi, T. A. Phosphate Solubilizing Microbes: Sustainable Approach for Managing Phosphorus Deficiency in Agricultural Soils. *Springerplus* **2013**, *2*, 587.

Shilev, S. Soil Rhizobacteria Regulating the Uptake of Nutrients and Undesirable Elements by Plants. In *Plant Microbe Symbiosis: Fundamentals and Advances*; Arora, N. K., Ed.; Springer: India, 2013, pp 147–150.

Singh, J. S.; Pandey, V. C.; Singh, D. P. Efficient Soil Microorganisms: A New Dimension for Sustainable Agriculture and Environmental Development. *Agric. Ecosyst. Environ.* **2011**, *140*, 339–353.

Singh, P. P.; Shin, Y. C.; Park, C. S.; Chung, Y. R. Biological Control of *Fusarium* Wilts of Cucumber by Chitinolytic Bacteria. *Phytopathology* **1999**, *89*(1), 92–99.

Somers, E.; Vanderleyden, J.; Srinivasan, M. Rhizosphere Bacterial Signalling: A Love Parade Beneath Our Feet. *Crit. Rev. Microbiol.* **2004**, *30*, 205–240.

Spaepen, S.; Vanderleyden, J. Auxin and Plant–Microbe Interactions. *Cold Spring Harb. Perspect. Biol.* **2011**, *3*, a001438.

Spaepen, S.; Vanderleyden, J.; Remans, R. Indole-3-Acetic Acid in Microbial and Microorganism–Plant Signaling. In *FEMS Microbiol Rev*; Unden, F., Ed.; Blackwell Publishing Ltd.: New York, 2007; pp 1–24.

Sujatha, N.; Ammani, K. Siderophore Production by the Isolates of Fluorescent *Pseudomonads. Int. J. Cur. Res. Rev.* **2013**, *5*, 1–7.

Sukhada. M. Biofertilizers for Horticultural Crops. *Indian Hort.* **1999**, *44*(1), 32–35.

Tariq, M.; Hameed, S.; Yasmeen, T.; Zahid, M.; Zafar, M. Molecular Characterization and Identification of Plant Growth Promoting Endophytic Bacteria Isolated from the Root Nodules of Pea (*Pisum sativum* L.). *World J. Microbiol. Biotechnol.* **2014**, *30*, 719–725.

Upadyay, S. K.; Maurya, S. K.; Singh, D. P. Salinity Tolerance in Free Living Plant Growth Promoting Rhizobacteria. *Ind. J. Sci. Res.* **2012**, *3*, 73–78.

Van Loon, L. C. Plant Responses to Plant Growth Promoting Bacteria. *Eur. J. Plant Pathol.* **2007**, *119*, 243–254.

Vessey, J. K. Plant Growth Promoting Rhizobacteria as Biofertilizers. *Plant Soil* **2003**, *255*, 571–586.

Vincent, M. N.; Harrison, L. A.; Brackin, J. M.; Kovacevich, P. A.; Mukerji, P.; Weller, D. M.; Pierson, E. A. Genetic Analysis of the Antifungal Activity of a Soilborne *Pseudomonas aureofaciens* Strain. *Appl. Environ. Microbiol.* **1991**, *57*, 2928–2934.

Voisard, C.; Keel, C.; Haas, D.; Dèfago, G. Cyanide Production by *Pseudomonas fluorescens* Helps Suppress Black Root Rot of Tobacco Under Gnotobiotic Conditions. *EMBO J.* **1989**, *8*, 351–358.

Walker, T. S.; Bais, H. P.; Grotewold, E.; Vivanco, J. M. Root Exudation and Rhizosphere Biology. *Plant Physiol.* **2003**, *132*, 44–51.

Weller, D. M. *Pseudomonas* Biocontrol Agents of Soilborne Pathogens: Looking Back Over 30 Years. *Phytopathology* **2007**, *97*, 250–256.

Wu, C. H.; Wood, T. K.; Mulchandani, A.; Chen, W. Engineering Plant-Microbe Symbiosis for Rhizoremediation of Heavy Metals. *Appl. Environ. Microbiol.* **2006**, *72*(2), 1129–1134.

Zaidi, A.; Khan, M. S.; Ahemad, M.; Oves, M. Plant Growth Promotion by Phosphate Solubilizing Bacteria. *Acta Microbiol. Immunol. Hung.* **2009**, *56*, 263–284.

Zahir, Z. A.; Muhammad, A.; Frankenberger, W. T. Jr. Plant Growth Promoting Rhizobacteria: Applications and Perspectives in Agriculture. *Adv. Agron* **2004**, *81*, 97–168.

Zhuang, X.; Chen, J.; Shim, H.; Bai, Z. New Advances in Plant Growth-Promoting Rhizobacteria for Bioremediation. *Environ. Int.* **2007**, *33*(3), 406–413.

PART 2

Microbiology for Crop Disease Management and Pathogenic Control in Sustainable Environments

CHAPTER 3

MICROBIAL CONTROL: A POTENTIAL SOLUTION FOR PLANT DISEASE MANAGEMENT IN SUSTAINABLE ENVIRONMENTS AND AGRICULTURE

DEEPAK KUMAR VERMA[1*], SHIKHA SRIVASTAVA[2], BALARAM MOHAPATRA[3*], RESHMA PRAKASH[4], VIPUL KUMAR[5*], DIGANGGANA TALUKDAR[6*], RONI YULIANTO[7], ABHAY K. PANDEY[8], ADESH KUMAR[9], ALI TAN KEE ZUAN[10], ARPANA H. JOBANPUTRA[11], HUEY-MIN HWANG[12*], MAMTA SAHU[13], and BAVITA ASTHIR[14]

[1]Agricultural and Food Engineering Department, Indian Institute of Technology, Kharagpur 721302, West Bengal, India

[2]Department of Botany, Deen Dayal Upadhyay Gorakhpur University, Gorakhpur 273009, Uttar Pradesh, India

[3]Department of Biotechnology, Indian Institute of Technology, Kharagpur 721302, West Bengal, India

[4]Department of Mycology and Plant Pathology, Institute of Agriculture Science, Banaras Hindu University, Varanasi 221005, Uttar Pradesh, India

[5]School of Agriculture, Lovely Professional University, Phagwara 144411, Punjab, India

[6]Plant Pathology and Microbiology, College of Horticulture Under College of Agricultural Engineering and Post-harvest Technology, Central Agricultural University, Ranipool 737135, East Sikkim, India

[7]Department of Grassland Ecology, Development of Technology Science, International Development Education and Cooperation (IDEC), Hiroshima University, Hiroshima, Japan

[8]PHM-Division, National Institute of Plant Health Management, Ministry of Agriculture and Farmers Welfare, Government of India, Rajendranagar, 500030 Hyderabad, Telangana, India

[9]Department of Plant Pathology, Punjab Agriculture University, Ludhiana 141004, Punjab, India

[10]Department of Land Management, Universiti Putra Malaysia, 43400 UPM Serdang, Selangor Darul Ehsan, Malaysia

[11]Department of Microbiology, PSGVPM's, SIP Arts, GBP Science & STSKVS Commerce College, District Nandurbar, Shahada 425409, Maharashtra, India

[12]Department of Biology, Jackson State University, Jackson, MS 39217, USA

[13]Department of Biotechnology and Microbiology, Saaii College of Medical Science & Technology, Kanpur 209203, Uttar Pradesh, India

[14]Department of Biochemistry, College of Basic Sciences and Humanities, Punjab Agriculture University, Ludhiana 141004, Punjab, India

*Corresponding authors. E-mail: deepak.verma@agfe.iitkgp.ernet.in/ rajadkv@rediffmail.com; balarammohapatra09@gmail.com; vipulpathology@gmail.com; talukdardiganggana@gmail.com; huey-min.hwang@jsums.edu

3.1 INTRODUCTION

The information about the damage caused by plant diseases to mankind have been mentioned in some of the oldest books available (Old Testament, 750 B.C.; Homer, 1000 B.C.). In agricultural practice worldwide, plant diseases regularly cause severe crop losses that may devastate the staple of millions of people, thus causing famines, and collectively result in economic damage of billions of euros (Van Esse et al., 2008). Famous examples from the past are the Irish potato famine (1845–1847), caused by the oomycete pathogen *Phytophthora infestans*, and the great Bengal famine (1942–1943) when the rice pathogen *Helminthosporium oryzae* (named suggested by Brede de Haan, 1990) caused a food shortage that resulted in the death of 2 million of people (Padmanab, 1973). The fundamental knowledge of microbiology in modern agriculture is becoming proved as a gateway to solution for many practical problems especially for the microbial control of plant disease. The achievement of this success is only by the way of better understanding of the surrounding disease-retardant interactions between the microbes and plants with their knowledge in complex ecology. Microorganisms are the emerging and important tools of microbiology in the modern agriculture that is dealt by microbiologists for microbial control under the commercialization of agriculture and changing food habits (Manczinger et al., 2002) which afford effective means in increasing sophisticated measures to analyze the multichannel dialogue for microbial control. The multichannel dialogue for microbial control is consequence of the interactions between the plant, pathogens, microbial community, and biological agent (Handelsman, 2002) and is expected to attention the environmental security and to create the possibilities of increasing the agricultural crop's yields and food production by reduction of disease and it's pathogen suppression at a significant level to fulfill the growing food demands of the world's population, and also supplying the farm's produce to the consumer free from chemical residues (Narayanasamy, 2013) and to realize the practical potential of microorganism for microbial control to improve the sustainable environment and agriculture. In microbiology, the "biological control of plant disease" means dealing with the suppression of pathogenic populations by the actions of their native organism or introduction of their enemy organism to afford greater levels of protection and sustainability to the environments and agricultural yields. In the words of Eilenberg et al. (2001) "The use of living organisms to suppress the population density or impact of a specific pest organism, making it less abundant or less damaging than it would otherwise be." Another widely quoted and accepted definition of biological control of disease is "The reduction in the amount of inoculum or disease-producing activity of a pathogen accomplished by or through one or more organisms."

The unite knowledge of mechanistic interactions is very imperative with an appreciation of the complexity of the agro-ecosystem. In narrow sense, microbial control indicates to control one organism by another organism (Beirner, 1967) and has many definitions (Bull, 2001) proposed by various researchers time to time in different ways (Table 3.1). Various eco-friendly approaches are identified with research efforts in addition to uses of microbes for plant-diseases management (Narayanasamy, 2013).

The entitled chapter "Microbial Control: A Potential Solution for Plant Disease Management in Sustainable Environment and Agriculture" addresses to significant efforts on microbial control with the majority of new principles, community function, and new strategies of microorganism which are emerging and recognizing with many advantages. The advantages of these new tools are not limited (Table 3.2) for developing countries to sustain the agriculture and environment, whereas the other methods (like chemical treatment) are very costly, prohibitive, and spread the pollution through soil, air, and water.

3.2   MICROBIAL CONTROL: A POTENTIAL SOLUTION FOR PLANT DISEASE

Microbial control in agriculture is an alternative resource to the use of chemicals (Cook, 2000; Weller et al., 2002) to avoid the global crop

TABLE 3.1   Definition of Biological Control of Plant Disease.

S. N.	Proposed definition
1.	The action of parasites, predators, or pathogens in maintaining another organism's population density at a longer average than would occur in their absence (DeBach, 1964)
2.	The reduction of inoculum or disease producing activity of a pathogen accomplished by one or more organisms other than man (Cook and Baker, 1983)
3.	The use of natural organisms or genetically modified, genes or gene products the effects of undesirable organisms to favor organisms useful to human, such as crop, trees, animals, and beneficial microorganisms (Monte and Llobell, 2003)
4.	The use of natural organisms, or genetically modified genes or gene products, to reduce the effects of undesirable organisms to favor organisms useful to human, such as crops, trees, animals, and beneficial microorganisms (Cook and Chairman, 1987; Saba et al., 2012)
5.	The purposeful utilization of introduced or resident living organisms, other than disease resistant host plants, to suppress the activities and populations of one or more plant pathogens (Pal and Gardener, 2006; Ouda, 2014)
6.	Biological control involving microbial agents or biochemicals to control plant pathogens can be an eco-friendly and cost-effective component of an integrated disease management program (Mao et al., 1997)
7.	An applied field of research involving the human use of biological agents as microorganism to enhance crop productivity (Handelsman, 2002)

TABLE 3.2   Advantage of the Microbes as an Eco-friendly Approach for Control of Plant Disease.

S. N.	Advantage of the microbes
1.	To mitigate the ill effects of infection by microbial plant pathogens
2.	To reduce or replace the use of synthetic chemicals
3.	Integrate the compatible and synergistic strategies for enhancing the effectiveness of disease suppression
4.	Beneficial microorganism (including antagonists in soil) do not received any kind of adverse effects
5.	Safer to the environment and no chance for any kind of pollutions such as air, water, and soil
6.	Cost-effective because of less expensive and comparatively cheaper than any other approaches
7.	Avoid the development of resistant strains in pathogen and protect the crop throughout the crop period
8.	In addition of plant disease control, it also enhances the root and growth of plant by promoting the helpful soil microflora to facilitate the increases in crop yields

Source: Narayanasamy (2013).

production losses (i.e., about 13–20%) due to agricultural crop diseases (Anonymous, 1993) that have been concerned with the mankind since our agriculture began. The plant diseases occurrence and their counterattack for self-defense might have evolved simultaneously and also played a crucial role in the natural resources destruction. Thus, the steps toward use of microorganism in microbial control are depicted in Figure 3.1.

3.2.1 CHARACTERISTIC OF MICROORGANISMS IN MICROBIAL CONTROL

The reduction of the disease or amount of inoculum producing by the pathogenic activity is accomplished by or through microorganism (i.e., one or more) other than man is explained as biological control (Baker

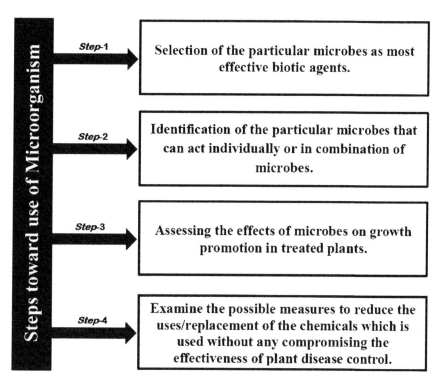

FIGURE 3.1 Steps toward use of microorganism in plant disease through microorganism. (*Source:* Concept adapted from Narayanasamy, 2013)

and Cook, 1974). There are various activity produced by disease which involves growth, infectivity, aggressiveness, virulence, and other qualities of pathogen or process that controls infections, symptom development, and reproduction. The potential of microorganism to control the pathogens of plant disease, if they have certain characteristic features, is described in Figure 3.2 (Ukey and Meshram, 2009).

3.2.2 FUNDAMENTAL CONCEPTS AND PRINCIPLE FOR MICROBES APPLICATION

The worldwide understanding on "plant diseases management" has enormously and positively influenced by the dynamics of plant–microbes interactions. General public considerations for microbes are widely regarded as natural and therefore nonthreatening products (Monte and Llobell, 2003). They are one of the most promising medium and new tools as microbial agents (MAs) to achieve the goal "microbial control of plant disease"

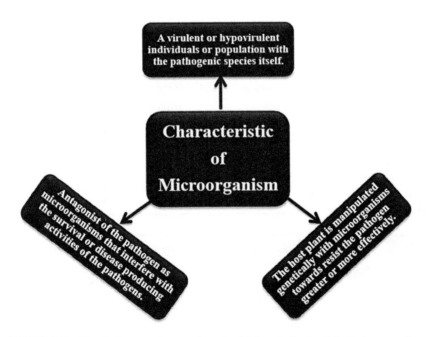

FIGURE 3.2   Certain features characterize microbial control potential of microorganism.

from their pathogen alone, or to integrate with reduced chemicals doses with minimal environmental impact (Chet and Inbar, 1994; Harman and Kubicek, 1998; Eilenberg, 2006) and although their effects on nontarget organisms and plants with risk assessments carrying out are subjected to be careful and very attention at the time of application (Monte and Llobell, 2003). The principle behind the microbial approach application for microbial control is implemented by Gnanamanickam et al. (2002) and is considered as shown in Figure 3.3.

3.2.2.1 INDUCED RESISTANCE

In the 21st century, microbial control confronts significant challenges (Table 3.12). The capacity of plants actively responds to a variety of biotic and abiotic stress. In addition, plants also have to have the capacity to respond to a variety of problems (such as to resist the infection, recover from diseases, and then avoid the future infections) with the help of chemical stimuli produced by different groups of MAs have been reported from the beginning of the 20th century in many botanical studies (Chester, 1933). How the plants are reacting with different types stresses like biotic as well as abiotic origin. These phenomena have been described (Fig. 3.4) by elicitation of defense reactions and intensively been studied at the cellular and molecular levels (Van Loon, 2000). Several evidences

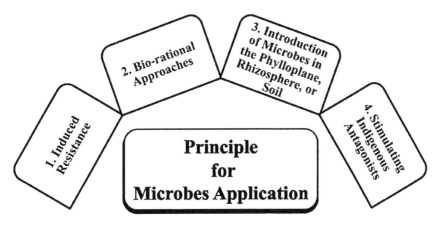

FIGURE 3.3   Principle for application of microorganisms in plant disease management.

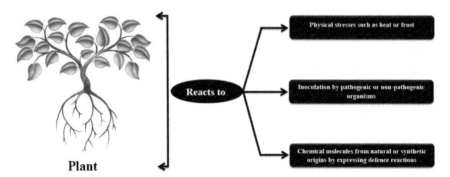

FIGURE 3.4   Plants reacting with stresses from biotic or abiotic origin. (*Source*: Concept adapted from Alabouvette et al., 2006)

on the natural phenomenon of induced resistance (IR) increased since the late 1950s and past many decades, its successful practical application has been accumulated and culminated (Kuc, 2001). The expanded exercise of traditional chemical inputs has led to environmental and economic dilemmas. Moreover, pathogenic microorganisms affecting plant health are a crucial and alarming threat to growing food demand and production, environmental security, and ecosystem sustainability throughout the world (Hammerschmidt et al., 2001; Anand et al., 2014).

IR is a plant response to challenge by harmful microbes other than beneficial such that following the inducing challenge de novo resistance to pathogens is shown in normally susceptible plants (De Wit, 1985). Ray (1901) and Beauverie (1901) were first recognized to the resistance in plants induced by pathogens. The phenomenon of inducing resistance in plants by pathogen confirmed by Chester (1933) with the help of his summarized field observations on such phenomenon and also supposed that this may help to the preserve and protect the crop plants from their enemies in nature.

IR can be defined as an increased expression of natural defense mechanisms of crop plants provoked by several biochemical changes and other external factors that enhance resistant against subsequent inoculation of a variety of pathogenic organisms (Hammerschmidt, 2007). Hence, inherent character of IR indicates low specificity and never effective against a broad range of pathogenic agents including bacteria, fungi, nematodes, viruses, parasitic plants, and even herbivores insect. To make it effective against the enemies of crop, plants need enhancement in resistance capacity of plants (Hammerschmidt and Kuc, 1995). IR can be local or systemic depending on the type, source, and amount of the MAs. Both the localized and systemically IR are known as nonspecific and employed against the wide range of pathogens, but many plants are found in the association of localized resistance, whereas systemic resistance is limited to some plants (De Wit, 1985). Plants during localized resistance react to the environmental stimulus by the activation of a variety of defense mechanisms that culminate in various biochemical and physiological changes in the plants including induced accumulation of chalcone synthase, chitinases, peroxidases, phenylalanine ammonia lyase, phytoalexins (production and alterations to plant cell walls such as increased production of hydroxyproline-rich glycoproteins, lignification, and suberin), polyphenol oxidase, and pathogenesis-related (PR) proteins (such as PR-1, PR-2), and correlations between resistance and lignin formation, and protease inhibitors and peroxidase activity have been found (Hammerschmidt et al., 1984; Dean and Kuc, 1987; Roby et al., 1987; Park and Kloepper, 2000; Ramamoorthy et al., 2001; Jeun et al., 2004). Many microbes have been reported to IR in plants for disease control. More and more studies are devoted to IR in the host plant from many past decades to present with the application of MAs against pathogens.

Some nonpathogenic MAs (organisms and avirulent pathogen) (Table 3.3) can also be used as inducers to IR in plants by a sustainable change resulting in an increased tolerance to subsequent infection by a pathogen. This phenomenon is described as induced systemic resistance (ISR). ISR-induced MAs do not cause visible symptoms on the host plants and

TABLE 3.3 Nonpathogenic MAs Involved in ISR in Plants.

Strains of MAs	Determinants of MAs	Plant species	References
Ba. mycoides strain Bac J and *Ba. pumilus*	Chitinase and beta-1,3 glucanase	Sugar beet	Bargabus et al. (2002)
Ba. subtilis	2,3-Butanediol	Arabidopsis	Ryu et al. (2004)
Psd. fluorescens CHA0	Antibiotics (DAPG)	Arabidopsis	Iavicoli et al. (2003)
Psd. fluorescens WCS374	Lipopolysaccharide	Radish	Leeman et al. (1995)
Psd. fluorescens WCS417	Lipopolysaccharide	Arabidopsis	Van Wees et al. (1997)
Psd. putida WCS358	Lipopolysaccharide	Arabidopsis	Meziane et al. (2005)
Ser. marcescens 90–166	Siderophore	Cucumber	Press et al. (2001)

differ from systemic acquired resistance (SAR), which refers to the host reaction in response to localized infection by pathogens, manifested as broad range of protection against other pathogens. ISR is mediated by jasmonate (JA) and ethylene (ET), which are produced following application of nonpathogenic MAs. *Bacillus* spp. *Burkholderia* and *Pseudomonas* are the examples of nonpathogenic MAs which have been shown to elicit ISR in different agricultural/horticultural crop plants, namely, arabidopsis (Mohammad et al., 2009), bean plants (Alstrom, 1991; De Meyer and Hofte, 1997; Audenaert et al., 2002; Ongena et al., 2002), betelvine (Singh et al., 2003), brinjal (Chakravarty and Kalita, 2011), carnation (Van Peer et al., 1991), chickpea (Saikia et al., 2006), chilies (Bharathi et al., 2004; Muthukumar et al., 2010), cucumber (Wie et al., 1991; Tomczyk, 2006), grapevine (Verhagen et al., 2010), hot pepper (Ramamoorthy et al., 2002), lentil (Siddiqui et al., 2007), mango (Vivekananthan et al., 2004), peppermint (Kamalakaman et al., 2003), radish (Leeman et al., 1996), rice (Nandakumar et al. 2001; Vidhyasekaran et al., 2001; Mathivanan et al., 2005), soybean (Khalimi and Suprapta, 2011), sugar beet (Dunne et al., 1998), sugarcane (Viswanathan and Samiyappan, 2001), tea (Saravanakumar et al., 2007a), sunflower (Srinivasan and Mathivanan, 2009), tobacco (Maurhofer et al., 1994), tomato (Duijff et al. 1998; Felix et al., 1999; Ramamoorthy et al., 2002), etc.

In ISR, the inoculation with nonpathogenic strains or weakly virulent strains of pathogenic form specially *Fusarium* and *Verticillium* species, or with other fungi or bacteria, all have shown different levels of crossprotection (Matta and Garibaldi, 1977; Ogawa and Komada, 1985; Sneh et al., 1985; Hillocks, 1986). The microbial control activity of the nonpathogenic strain Fo47 to systemic IR in tomato was attributed by Fuchs et al. (1997) which were correlated with an increased activity of chitinase, β-1–3–glucanase, and β-1–4-glucosidase. Smith and Metraux (1991) reported the hypersensitive necrotic reaction-causing bacterium *Psd. syringae* pv. Syringae induced SAR against the fungus *Pyric. oryzae* in rice. Importantly, stem inoculation of tomato by an avirulent strain of the bacteria *Clavibacter michiganensis* ssp. michiganensis induced long lasting, high-level protection against the virulent bacterial strain (Griesbach et al., 2000). Vegetable-like pea showed induce SAR by an avirulent strain of *Psd. syringae* pv. pisi against the fungus *Mycosphaerella pinodes* (Dann and Deverall, 2000). Conn et al. (2008) demonstrated the inoculation of

Arabidopsis thaliana with endophytic actinobacteria-induced SAR and JA/ET gene expression. The evidence of resistance induced by nonpathogenic/avirulent MAs against virulent pathogens is reported in literature by many workers (Brooks et al., 1994; Ryals et al., 1996; Yan et al., 2002; Claire et al., 2005; Abeysinghe, 2009; Feng et al., 2012).

Since, IR is a general phenomenon that can protect the plant against several pathogens and also help to sustained more attention than other mode of action of microbial control of plant disease. Taking into the account, there is an imperative need for additional approaches to microbial control of plant disease, and thus, IR renders the promise of durable, broad-spectrum disease control using the plant's own resistance, as a result, there has been extensive enthusiasm in the development of agents which can mimic natural inducers of resistance. There are certain criteria of IR (Table 3.4) that provide protection against crop disease and these criteria are different from conventional and chemical approaches.

TABLE 3.4   Criteria for Induced Resistance.

S. N.	Induced resistance criteria
1.	Absence of typical dose response correlation known for toxic compound
2.	IR is the genes for resistance or defense reaction exists in all plants. These genes are not expressed until after resistance inducing treatments activate or enhance their expression
3.	Lack of toxicity of the inducing agents toward the pathogens
4.	Local or systemic protection
5.	Necessity of time interval between applications of induce and the onset of protection of plants
6.	Nonspecificity of protection
7.	Protection of plant is based on the activation of plant defense mechanism or enhancement of their activities rather than elimination of the pathogens
8.	Suppression (concealment) of inducing resistance by previous application of specific inhibitor. However, increasing use of chemical inputs causes several negative effects, i.e., development of pathogen resistance to the applied agents and their nontarget environmental impacts. Furthermore, the growing cost of pesticides, particularly in less affluent regions of the world, and consumer demand for pesticide-free food has led to a search for substitutes for these products
9.	There are also a number of fastidious diseases for which chemical solutions are few, ineffective, or nonexistent. So IR is considered to be a biological plant protection procedure in which the plant is the target not the pathogen

3.2.2.2 BIORATIONAL APPROACHES

Biorational approaches (B-rAs) are important principle in modern agriculture to enhance the destructive potential of plant diseases. They are considered as recent approach with potent meaning *reduction in damage caused by plant pathogens*. Literally, B-rAs combines two major strategies, namely, host resistance and MAs to reduce or suppress disease incidence. Host resistance and MAs should complement each other in their activity against pathogens. Today, agricultural crops are threatened by a wide variety of plant's pathogens. To get the diagnostics from these problems, agricultural crops have now become heavily dependent on chemicals (De Waard et al., 1993). In such a way, the practices of B-rAs enhance disease susceptibility and dare to combat against wide ranges of pathogens of agricultural crops. The B-rAs for the microbial control of agricultural crop diseases are employed very few in literatures.

In B-rAs, a long-term crop rotation is used because it prevents the growth of inoculum (particularly, primary inoculum). The prevented primary inoculum is important in most soilborne diseases due to monocyclic. In B-rAs, monocropping is raised as an exceptions few because it increases disease and other than the monocropping, appropriate crop rotations are effective to suppress/kill the soilborne pathogens that induced most of the disease in agricultural crops. For example, if a crop rotation of long period (4–5 years) is applied on cultivated land, then the disease increases in the beginning, but after 4–5 years, the severity of the disease decreases to a level that cannot affect to agricultural crop yield or affect very few.

Under the above circumstances, the study on the effect of inoculum type, application rate, and time of application became very urgent and necessary to taking care in an order to ensure the efficacy of microbial control (Jones et al., 2004a,b) but sorry to say that the farmers of most developing countries are pressurized by economic constrains to grow repeatedly same agricultural crop on the same agricultural land that fails to the MAs.

3.2.2.3 INTRODUCTION OF MICROBES IN THE SOIL, RHIZOSPHERE, OR PHYLLOPLANE

Microbial protection against infection is skillful acts in which (1) the existing inocula is destroyed, (2) the formation of additional inocula is prevented, and/or (3) the existing virulent pathogen population is weakened and displaced. But, this skillful act can be achieved only by the protection of plant material and roots with microbially treated seeds or by the suppression of pathogens with the introduction of plant associated antagonists into the rhizosphere. The term rhizosphere proposed by Schmidt (1979) states the fact that the density of bacterial community is always higher than the nonrhizosphere soil comparatively (Foster and Rovira, 1978).

The introduction of microbes like as fungi and bacteria (actinomycetes) in the soil, rhizosphere, or phylloplane is easy to achieve successful control of plant disease because diverse habitats and substrates are colonized by them to play substantial role in yield and productivity of agricultural crop (Alabouvette et al., 2004). The phylloplane, rhizosphere, and/or soil are recognized as the best specialized ecological area for microbes to interact and grow with plants simultaneously either antagonistically, synergistically, and/or independently. These specialized ecological areas are also known for the natural resources for microbial metabolites such as amino acids, sugars, tannins, vitamins, etc. as carbon sources are exuded by the plant roots and used by microbes in their activity like growth and development (Loper and Schroth, 1986; Goddard et al., 2001). The ability of introduced bacteria (such as *Pseudomonas* strains, etc.) to suppress crop disease is concern relies mainly on ability of bacteria to colonize the roots (Chin-A-Woeng et al., 2000) and the density of bacterial population in rhizosphere (Raaijmakers and Weller, 1998) because rich colonization of roots must cause increased control activity of MAs against plant pathogens and also correlation to the efficiency of control activity MAs in many works has been reported with size of microbial population (Bull et al., 1991). For example, the population density of two strains of *Pseudomonas* spp., that is, Q2-87 and Q8r1-96 needed an average 1.2×105 CFU g^{-1} and 4.6×105 CFU g^{-1} root, respectively, for significant suppression of take-all of wheat (Raaijmakers and Weller, 1998).

The successful controls by microbial means in the phylloplane that have been reported in the literature involve mainly rusts, powdery mildews, and diseases caused by the pathogens of following genera: *Alternaria, Drechslera, Epicoccum, Erwinia, Plasmopara, Pseudomonas, Sclerotinia, Septoria*, and *Venturia*. The successful microbial control in soil has been reported for species of *Fusarium, Pythium, Rhizoctonia, Sclerotinia*, and *Sclerotium*. Table 3.5 shows that the countries have already been registered to various microbial control agents effective against various crop disease and their pathogens.

TABLE 3.5　Country-wise Registered/Commercialized MAs Used against Crop Diseases and Their Pathogens.

Country	Microbial agents	Disease/pathogens
Australia	*Ag. radiobacter*	Crown gall
	Pseudomonas fluorescens	Bacterial blotch
Brazil	*Ba. pumilus*	Foliar blight
Canada	*Ba. subtilis*	Rot (*Rzt. solani*, *Pythium* spp., *Fusarium* spp.)
	St. lydicus	Root rot
Colombia	*Ba. cepacia*	Soilborne pathogens
Europe	*T. viride*	Timber pathogens
Finland	*St. lydicus*	Root rot
Germany	*Ba. subtilis*	Root rot (*Rzt. solani*)
	Ba. amyloliquefaciens	Soilborne pathogens
	St. lydicus	Root rot
India	*Ba. subtilis*	Fungal diseases of cereals (*Fusarium* spp., *Pythium* spp.)
	Psd. fluorescens	Rice (blast, brown spot, root rot, seedling rot, sheath blight, sheath rot, stem rot, and wilt), wheat (root rot, stem rot, and wilt), and maize diseases (root rot, stem rot, and wilt)
	Pseudomonas chlororaphis	Wheat, rye, and triticale disease (*Fusarium* spp.)
Italy	*Bacillus* spp.	Root rot
Japan	*Ba. subtilis*	Ear rots minor (*Botryt. cinerea*)
	F. oxysporum	*F. oxysporum*
Mexico	*Ba. subtilis*	Stillborn and foliar pathogens
NZ	*Ag. radiobacter*	Crown gall
UK	*Peniophora gigantea*	*Fommes annosus*
USA	*Ag. radiobacter*	Crown gall
	Ba. subtilis	Leaf spot (*Er. carotovora*), rice bacterial spot (*Xanthomonas* spp.), rice blast (*Pyric. oryzae*), rice leaf spot (*Bi. maydis*), rice root rot (*Rzt. oryzae*), root rot (*F. oxysporum*, *Pythium* spp.), rot (*Rzt. solani*, *Pythium* spp., *Fusarium* spp.), seedling blight (*Rzt. solani*), seedling pathogens, stillborn, and foliar pathogens
	Burkholderia cepacia	Barley and wheat diseases (*Rhizoctonia* spp., *Pythium* spp., *Fusarium* spp.)

TABLE 3.5 (Continued)

Country	Microbial agents	Disease/pathogens
	Gliocadium virens	Seedling diseases
	Psd. aureofaciens	Rot
	Py. oligandrum	Seedling diseases
	St. lydicus	Root rot
	T. harzianum	Root diseases
	T. harzianum/ polysporum	Wood decay
USSR	*Py. oligandrum*	*Pythium* spp.
	Trichoderma spp.	Root diseases

Sources: Elad and Chet (1995), Nakkeeran et al. (2006), Cawoy et al. (2011), Tapadar and Jha (2013), Khan (2013).

3.2.2.4 STIMULATING INDIGENOUS ANTAGONISTS

Several species of MAs have been documented due to their antagonism effect/behavior (Table 3.6) toward the microbial control of diseases in various agricultural crops. However, major impediments to the commercial exploitation of microbes for disease treatments by microbial approaches have been subjected to the mediocre efficacy, effective, and environmental eco-friendly as compared with synthetic chemical fungicides. The responses to the microbes application act as antagonists for pathogenic organism caused disease to agricultural crops and have also to be found very effective on their control. The antagonistic microbes have employed by various methods to protect agricultural crops from their pathogens.

There are numerous studies that examined microbial control of plant pathogen using microbial antagonists such as *As. gigentus*, *As. tamarii*, *Bacillus* spp., *Botryo. theobromae*, *Pcl. purpurescens*, and *Psd. fluorescens* with success through direct antagonism, mixed-path antagonism, and indirect antagonism types (Table 3.6). The activity of microbial control can be manipulated with the addition of some exotic antagonists to the soil, or by stimulating the endogenous antagonist's activity by the addition of mulches or composts (Erwin and Ribeiro, 1996). For example, the use of composted pine bark, mulches, etc. types of organic media that have higher microbiological activity and lower pH (Hoitink and Fahy, 1986; You et al., 1996) which lead and offer the promising options to control

TABLE 3.6 Types of Antagonisms Leading to the Pathogenic Control in Agricultural Crops.

Type	Mechanism	Examples
Direct antagonism	Hyperparasitism/ predation	*Am. quisqualis*, *Ly. enzymogenes*, lytic/some nonlytic mycoviruses, *Pasteuria penetrans*, *T. virens*
Mixed-path antagonism	Antibiotics	Phenazines, cyclic lipopeptides, 2,4-diacetylphloroglucinol
	Lytic enzymes	Proteases, glucanases, chitinases
	Unregulated waste products	Ammonia, carbon dioxide, hydrogen cyanide
	Physical/chemical interference	Blockage of soil pores, germination signals consumption, molecular crosstalk confused
Indirect antagonism	Competition	Exudates/leachates consumption, physical niche occupation, siderophore scavenging
	Induction of host resistance	Contact with fungal cell walls, detection of pathogen-associated, molecular patterns, phytohor-mone-mediated induction

Source: Pal and Gardener (2006).

Psd. cinnamomi in container-grown plants in nurseries. On the other hand, organic amendments have also been successfully extrapolated in the field to control of apple replant disease (Utkhede and Smith, 1994). Mycorrhizae may also provide a powerful tool as microbial control against *Psd. cinnamomi* and identified in crops, namely, pines (Marais and Kotzé, 1976) and pineapple (Guillemin et al., 1994).

3.2.3 FUNCTIONAL MECHANISM WITH ANTAGONISTS BEHAVIOR OF MICROORGANISM

The plant pathogens are an organism capable of causing disease in its host. There are various common examples of these organisms including bacteria, fungi, virus, etc. with their specific strains. The pathogenic organisms of agricultural crops are suppressed and controlled through different mechanisms/modes of action of microbial sources as well as chemicals (Cook and Baker, 1983). The mechanisms/modes of action of microbial sources to control the fungal pathogen of agricultural plant have been described by some examples. Such as, Figure 3.5A indicates hyphae of the fungus

Arthrobotrys coiled around a hypha of a pathogenic fungus *Rhizoctonia* resulting in the death of the latter, whereas the hypha of *Sclerotium* parasitized (revealed by penetration hole) by a parasitic fungus, *Trichoderma* (Fig. 3.5B); Figure 3.5C depicts the hypha of a plant parasitic fungus colonized by bacteria, and only a few fragments of the wall are left (Fig. 3.5D).

In the past few decades, microbes have been concerned as a potential control strategy for pathogenic organisms, and search for such MAs is increasing day by day which causes and helps in reduction of agricultural yields loss (Kumar and Mukerji, 1996). Moreover, the knowledge

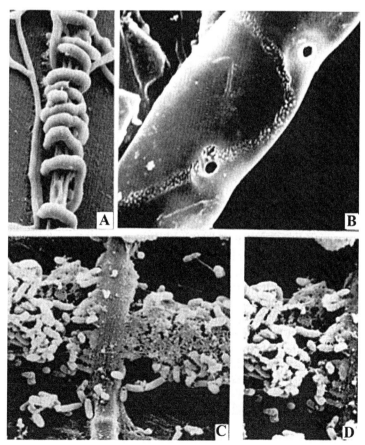

FIGURE 3.5   Mechanisms/modes of action of microbial sources to control the pathogenic fungi of agricultural crops. (*Source:* Adapted from Campbell, 1989)

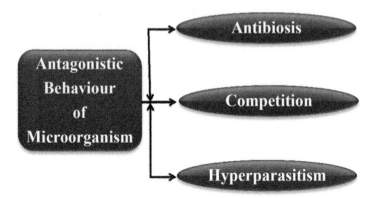

FIGURE 3.6   Antagonistic behavior responsible to lead microbial control of plant disease.

concerning the behavior of microorganism is essential for their effective use of such antagonists in agriculture to reduce the crop yields loss and increase agricultural production (Monte and Llobell, 2003).

An antagonist is an organism that exerts a damaging effect on another organism by the production of antibiotics, lytic enzymes, or by competition. Antagonism may be general or specific. The antagonistic behavior of microorganism is interactions which can lead and play a crucial role in microbial control (Cook and Baker, 1983) which is shown in Figure 3.6. General antagonism is the reduction of disease or pathogen inocula due to general microbial activity. In some cases, this antagonism can be achieved by adding organic matter in the soil. However, the specific antagonism is the reduction of disease or pathogen inocula due to the activity of a particular species or strain of microorganism (Fig. 3.7).

3.2.3.1   ANTIBIOSIS

A microbial interaction between two or more organisms in which at least one of them is detrimental and have direct inhibitory effect due to production of antibiotics or toxic metabolites. *Antibiotics* are a substance (toxin or bacteriocin) produced by MAs that is damaging another organism (i.e., pathogenic) of low concentration as reported by many authors (Table 3.7). *Bacteriocins* are chemicals, usually proteins, which are produced by

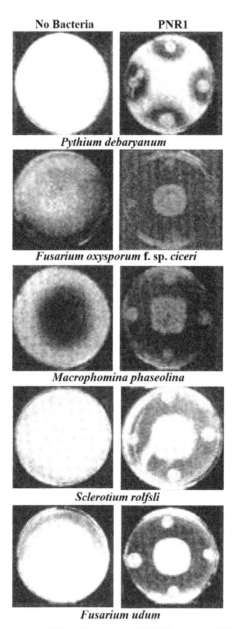

FIGURE 3.7 Antagonism of *Psd. fluorescens* (PNR1) against different soilborne plant pathogens. (Reprinted with permission from Desai, S.; Reddy, M. S.; Kloepper, J. W., © 2002 Elsevier.)

TABLE 3.7 Antibiotics Produced by MAs Used Against Target Pathogens Caused Diseases in Crops.

Antibiotics	Producer MAs	Target pathogens	References
2,4-Diacetylphloroglucinol (DAPG)	*Pseudomonas* spp.	*Xanthomonas oryzae* pv. *oryzae*	Velusamy et al. (2006)
Amphisin	*Psd. fluorescens*	*Py. ultimum, Rzt. solani*	Andersen et al. (2003)
Bacillomycin	*Bacillus* spp.	*As. flavus*	Moyne et al. (2001)
Geldanomycin	*St. hygroscopicus* var. *geldonus*	*Rzt. solani*	Rothrock and Gottlieb (1984)
Iturin	*Ba. subtilis*	*F. oxysporum, M. phaseoli, Py. ultimum, Rzt. solani, Sclt. sclerotiorum*	Constantinescu (2001)
Iturin A and surfactin	*Ba. subtilis*	*Rzt. solani*	Asaka and Shoda (1996)
Kanosamine	*Ba. cereus*	*Phy. medicaginis*	Milner et al. (1996)
Kasugamycin	*St. kasugaensis*	*Psd. oryzae*	Umezawa et al. (1965)
Oligomycin A	*S. libani*	*Botryt. cinerea*	Kim et al. (1999)
Phenazine-1-caboxylate	*Psd. fluorescens*	*Ga. graminis* var. *tritici*	Gurusiddaiah et al. (1986), Thomashow et al. (1990)
	Psd. aureofaciens	*Pythium* spp.	Gurusiddaiah et al. (1986)
		Sclt. homeocarpa	Powell et al. (2000)
Phenazine-1-carboxamide	*Psd. chlororaphis*	*F. oxysporum* f. sp. *radicis-lycopersici*	Chin-A-Woeng et al. (1998), Bolwerk et al. (2003)
Phenazines	*Psd. fluorescens*	*Ga. graminis* var. *tritici*	Weller and Cook (1983), Brisbane and Rovira (1988)
Polyoxin D	*St. cacaoi*	*Rzt. solani*	Isono et al. (1965)
Pyoluteorin	*Psd. fluorescens*	*Py. ultimum*	Howell and Stipanovic (1980)
		Ga. graminis var *tritici*	Tazawa et al. (2000)

TABLE 3.7 (Continued)

Antibiotics	Producer MAs	Target pathogens	References
		Rzt. solani	Howell and Stipanovic (1979)
	Bu. cepacia	*Rzt. solani*	El-Banna and Winkelmann (1988)
	Ent. agglomerans	*Ag. tumefaciens, Clavibacterium michiganense, Psd. syringae* pv. *syringae, X. campestris*	Chernin et al. (1996)
Rhamnolipids	*Psd. aeruginosa, Psd. chlororaphis*	*Pythium* spp.	Ligon et al. (2000)
Viscosinamide	*Psd. fluorescens*	*Rzt. solani*	Nielsen et al. (2002)
		Py. ultimum	Thrane et al. (2000)
Zwittermycin A	*Bacillus* spp.	*Sclt. sclerotiorum*	Zhang and Fernando (2004)
	Bsd. cereus	*Phy. parasitica* var. *nicotianae*	He et al. (1994)
	Ba. cereus, Ba. thuringiensis	*Phytophthora* spp.	Silo-Suh et al. (1998)

microorganisms, and act as antibiotics against closely related species or strains of microorganisms.

In simple words, *antibiosis* mean against life (Fravel, 1988). The French bacteriologist, Jean Paul Vuillemin, had introduced this term, and in 1877, Louis Pasteur and Robert Koch had first time described this in bacteria when they found the growth of *B. anthracis* could be inhibited by an air-borne *Bacillus* (Landsberg, 1949; Foster and Raoult, 1974). The example of antibiosis includes relationship between antibiotics and bacteria and the relationship between plants and disease causal organism (Fravel, 1988; Howell, 1998). In nature, there are various microorganisms (viz., *Bacillus* spp., fluorescent *Pseudomonas* spp., *Streptomyces* spp., and *Trichoderma* spp.) and have requirement of antibiosis as a very common phenomenon for their activities (Fravel, 1988; Loper and Lindow, 1993). The production of antibiotics by microbe like fungi, bacteria, and acti-nomycetes is very simple to demonstrate in vivo. The species of *Glioca-dium*, *Trichoderma*, and several species of the genus *Pseudomonas* are well-known MAs that produce a wide range of antibiotics that are active against pathogens in vitro (Dennis and Webster, 1971a,b; Claydon et al., 1987; Ghisalberti and Sivasithamparam, 1991) and involved due to their ability to control pathogenic agents of agricultural crops (Fravel, 1988).

3.2.3.2   COMPETITION

Competition usually occurs when two or more microorganisms, for example, MAs and the pathogen causing disease which require the same resources in excess of their supply. These resources can include space, nutrients, that is, carbon (C), nitrogen (N), iron (Fe), and/or oxygen (O). Implicit from the definition is to understand that how combative interac-tions, such as antibiotic production or mycoparasitism, or the occurrence of IR in the host are not included, even though these mechanisms may form an important part of the overall processes occurring in the interac-tion. If, there is competition for nutrients then it occurs on root surfaces and leaf surface where it reduces the infection due to the prevention of fungal spore's germination (Alabouvette et al., 2006).

In microbial control of plant disease, pathogens as out-competes are the less efficient competitor while in-competes the more efficient one, that is, the MAs (Bandyopadhyay and Cardwel, 2003), for example, the competition for nutrients "C" between pathogenic and nonpathogenic *Fusarium oxysporum* (Alabouvette et al., 2006). Another example is in the rhizosphere, competition for space as well as nutrients are of the major importance. Thus, an important attribute of a successful rhizosphere, microbial control agent would have the ability to remain at high population density on the root surface, providing protection to the root for whole duration of its life.

When competition is well-thought-out as the main mode of action, the population of the MAs used in control must be at least as large. If not larger, than that of the pathogen population in an order to achieve control, can be done easily in laboratory experiments but much more difficult to reach under the field conditions where the inoculum density of the pathogen is not known (Alabouvette et al., 2006). The role of competition in the microbial control of pathogens causing disease in agricultural crops has been studied with special emphasis on bacterial MAs for many years (Weller, 1988).

3.2.3.3   HYPERPARASITISM

Hyperparasitism of a plant pathogen by other microorganisms (including viruses) is a well-known phenomenon (Alabouvette et al., 2006). This phenomenon is also known as predation, that is, the results of biotrophic or necrotrophic interactions which lead to parasitism of the plant pathogen by the MAs. Some microorganisms particularly those in soil that can reduce to the diseases damage by promoting plant growth or by inducing host resistance against the variety of pathogenic agents (Hutchinson, 1998; Cook, 2000; Kerry, 2000).

In addition of antibiosis, competition and hyperparasitism, efficient MAs often express more than one mode of action for suppressing the plant pathogens in disease management.

3.3   IDENTIFIED MICROBIAL COMMUNITY USED IN PLANT DISEASE MANAGEMENT

Several types of biotic stresses such as fungal, bacterial, or viral that causes infections are challenging to plants and leading to a great loss to yield of agricultural crops. Farmers are using various options (like biological control, chemical pesticides, crop rotation, development of resistant cultivars and tillage, etc.) for protection of agricultural crop from their pathogens. In such way, MAs are one of the best options that come under the biological control and are always a good player to provide the protection against the pathogen of agricultural crops since its first report in history. The strategy of biological control was proposed in 1965 that was more than five decades ago during a symposium held in Berkeley was entitled "Ecology of Soil-borne Plant Pathogens; Prelude to Biological Control" (Baker and Snyder, 1965). Therefore, considerable efforts have been accomplished to devise environmental-friendly strategies for the check of plant diseases and thus to save mankind from health hazard (El-Gamal et al., 2007).

The suppression of disease is the consequence of the interactions between the plant, pathogens, and the microbial community which is mediated by beneficial MAs. The success of these approaches depends on the complex interactions between these beneficial MAs and with patho-genic organism and plants should be established in the ecosystem (Vinale et al., 2008). Indeed, they cannot perform their function with the lack of any one of the components among them. In microbial control, they are receiving enormous attention and are globally acclaiming as ideal candi-dates for obvious reasons. However, MAs are typically have a relatively very complex spectrum of their activities if compared with synthetic chemicals and exhibit very limited commercial use in agriculture for pathogenic control of agricultural crops (Mathre et al., 1999; Chaube et al., 2002). Many researches are conducted to overcome this limitation and to increase the commercialization of MAs and their product formulations with significant increasing effect spectrum.

The worldwide interest on microbial control of plant diseases for sustainable production of crop yield and productivity has been signifi-cantly increased. There are several species of MAs are identified as bacte-rial as fungal (Table 3.8) that have been used in plant disease control. These microorganisms are considered as the beneficial MAs have been affected by either directly contributing to the growth of plants or indi-rectly by reducing plants disease incidence (Jetiyanon et al., 2003; Gray and Smith, 2005). Recent studies shown that these MAs are very oppor-tunistic, avirulent plant symbionts as well as parasites on other microbes which responsible for disease in agricultural/horticultural crops. The uses of MAs against pathogen of agricultural crop plants are the fundamental matter of eco-friendly management of the organism's community. In the

TABLE 3.8   Microbial Community Used Against Pathogens of Agricultural/Horticultural Crops.

Bacterial community	Fungal community
[a]*Agrobacterium* spp.	[a]*Ampelomyces* spp.
[a]*Bacillus* spp.	*Aspergillus* spp.
Burkholderia spp.	[a]*Candida* spp.
Enterobacter spp.	*Chaetomium* spp.
Erwinia spp.	[a]*Coniothyrium* spp.
Pantoea spp.	*Cryptococcus* spp.
Pasteuria spp.	*Dicyma* spp.
[a]*Pseudomonas* spp.	*Fusarium* spp.
Rhizobacteria spp.	[a]*Gliocladium* spp.
Rhizobium spp.	*Glomus* spp.
Serratia spp.	*Paecilomyces* spp.
[a]*Streptomyces* spp.	*Penicillium* spp.
	Sporidesmium spp. (*Teratosperma* spp.)
	Talaromyces spp.
	[a]*Trichoderma* spp.
	Verticillium spp.

[a] Species registered are available as commercial products.

Sources: Vinale et al. (2008) and Verma et al. (2016).

nature, diverse groups of microbes such as bacteria, fungi, viruses, etc. are existing; among them, the MAs are not limited to any specific group but very few majority of them like bacterial as well as fungal have been identified since its first report in literature, namely, *Agrobacterium* spp., *Ampelomyces* spp., *Aspergillus* spp., *Bacillus* spp., *Burkholderia* spp., *Candida* spp., *Chaetomium* spp., *Coniothyrium* spp., *Cryptococcus* spp., *Dicyma* spp., *Enterobacter* spp., *Erwinia* spp., *Fusarium* spp., *Gliocladium* spp., *Glomus* spp., *Paecilomyces* spp., *Pasteuria* spp., *Penicillium* spp., *Pseudomonas* spp., *Rhizobacteria* spp., *Rhizobium* spp., *Serratia* spp., *Sporidesmium* spp. (*Teratosperma* spp.), *Streptomyces* spp., *Talaromyces* spp., *Trichoderma* spp. *Verticillium* spp., etc. other than the pathogenic agents or damaged hosts which causing several types of disease to agricultural/horticultural crops. Among them, *Agrobacterium* spp., *Ampelomyces* spp., *Bacillus* spp., *Candida* spp., *Coniothyrium* spp., *Gliocladium* spp., *Pseudomonas* spp., *Streptomyces* spp., and *Trichoderma* spp. identified MAs are registered and available as commercial products in the market (Vinale et al., 2008). These MAs include use of beneficial microbes and their strains, genes, and/or formulated products such as metabolites, etc. which can be employed against the disease of various agricultural/horticultural crops by reducing the negative effects of pathogens on agricultural/horticultural crops and to promote positive responses in the terms of crop yields received from the crop plants (Vinale et al., 2008; Verma et al., 2016). In Europe, Alabouvette et al. (2006) review on biological control of plant diseases and reported some registered microbes, namely, *Am. quisqualis*, *Coniothyrium minitans*, *Pae. fumosoroseus*, *Psd. chlororaphis*, and their formulated products are used in the European Union.

Many researches concluded that the microbes were always helpful to provide the protection from pathogens of agricultural/horticultural crops (Tables 3.9 and 3.10). In addition, they are also effective in plants to mobilize and acquire nutrients (Postma et al., 2003; Khan et al., 2004; Perner et al., 2006; Borriss et al., 2011; Gopalakrishnan et al., 2011a,b). The potential of these MAs was found at high incidence and has been reported in many crops, namely, rice, wheat (Kumar et al., 1993; Hashemi et al., 2013), barley (Arabi et al., 2013), maize (Babu et al., 1998; Pal et al., 2001; Veerabhadrasamy and Garampalli, 2011), sorghum (Gopalakrishnan et al., 2011a), pea (Bodker et al., 1998; Tokala et al., 2002), chickpea (Jayaraman and Kumar, 1995; Singh and Singh, 1995; Akhtar and Siddiqui, 2007), cowpea (Sadeghi et al., 2006), common bean/French bean (Muchovej et al., 1991; Nassar et al., 2003; Afsharmanesh et al., 2006), tomato (Al-Raddad, 1995; Berta et al., 2005; Akkopru and Demir, 2005; Bolwerk et al., 2005; Dababat and Sikora, 2007; El-Tarabily, 2008; Fiorilli et al., 2011), groundnut (Doley and Jite, 2012), tobacco (Giovannetti et al., 1991), cucumber (Huang et al., 2012), potato (Iqbal et al., 1990; Tariq et al., 2010), cotton (Egamberdieva and Jabborova, 2013), lettuce (Erlacher et al., 2014), etc.

TABLE 3.9 Major Diseases of Agricultural Crops and Their Microbial Control.

Crops	Disease	Pathogens	Employed MAs	References
Rice	Blast	*Pyric. grisea* (Cooke) Sacc.	*Psd. fluorescens*	Gnanamanickam and Mew (1992), Vasudevan et al. (2002)
		Pyric. oryzae	*St. sindeneusis*	Zarandi et al. (2009)
		Magnaporthe oryzae	*Ba. coagulans, Ba. polymyxa, Ba. pumilus*	Tapadar and Jha (2013)
			En. agglomerans	Tapadar and Jha (2013)
			Psd. fluorescens	Gnanamanickam and Mew (1992), Valasubramanian (1994)
	Brown spot	*Bi. oryzae* (Breda de Haan) shoemaker	*Psd. aeruginosa, Ba. subtilis*	Vasudevan et al. (2002)
	Bacterial blight	*X. oryzae pv. oryzae*	*Bacillus* spp., *Ba. lentus, Ba. cereus, Ba. circulans*	Vasudevan et al. (2002), Vasudevan (2002), Velusamy and Gnanamanickam (2003)
			Ly. antibioticus	Ji et al. (2008)
			Psd. fluorescens	Velusamy et al. (2006)
	Sheath blight	*Rzt. solani* Kuhn	*Ba. coagulans, Ba. laterosporus, Ba. megaterium, Ba. polymyxa, Ba. pumilus, Ba. subtilis*	Vasudevan et al. (2002), Kanjanamaneesathian et al. (2007), Tapadar and Jha (2013)
			En. agglomerans	Tapadar and Jha (2013)
			Psd. aeruginosa, Psd. fluorescens, Psd. putida	Vasudevan et al. (2002), Sunish Kumar et al. (2005), Nagrajkumar et al. (2005), Tapadar and Jha (2013)
			Ser. marcescens	Vasudevan et al. (2002), Tapadar and Jha (2013)
	Sheath rot	*Sarocladium oryzae* (Sawada) W. Gams and D. Hawksworth	*Ba. subtilis, Ba. pumilus*	Sakthivel (1987), Vasudevan et al. (2002)
			Psd. aeruginosa, Psd. fluorescens	Sakthivel (1987), Sunish Kumar et al. (2005), Tapadar and Jha (2013), Vasudevan et al. (2002)

TABLE 3.9 (Continued)

Crops	Disease	Pathogens	Employed MAs	References
	Stem rot	Sclerotium oryzae Cattaneo	Ba. pumilus, Ba. subtilis	Vasudevan et al. (2002)
			Psd. aeruginosa, Psd. fluorescens	Vasudevan et al. (2002)
	Tungro	Rice tungro virus (RTV), vector—Nephotettix spp.	Psd. fluorescens (forvector)	Vasudevan et al. (2002)
Wheat	Foliar disease	Septoria tritici	Psd. aeruginosa	Flaishman et al. (1990), Baron et al. (1997)
	Root rot	Bi. sorokiniana	Gl. epigaeus, Gl. fasciculatum	Kumar et al. (1993), Hashemi et al. (2013)
			Psd. fluorescens	Hashemi et al. (2013)
			T. viride	Kumar et al. (1993)
		Rzt. solani	Psd. fluorescens	Huang et al. (2004)
		Py. ultimum	Psd. fluorescens	Meyer et al. (2010)
	Seedling blight/head blight	F. graminearum	Ba. cereus	Bello et al. (2002)
			Clonostachys rosea	Xue et al. (2014)
			Ly. Enzymogenes	Jochum et al. (2006)
			T. harzianum	Bello et al. (2002)
		F. culmorum	Chryseobacterium spp.	Khan et al. (2006)
			Psd. fluorescens, Psd. frederiksbergensis	Johansson et al. (2003), Khan et al. (2006), Khan and Doohan (2009)
			Pantoea spp.	Johansson et al. (2003)
		Microdochium nivale	Pantoea spp.	Johansson et al. (2003)
			Psd. fluorescens	Johansson et al. (2003), Amein et al. (2008)

TABLE 3.9 (Continued)

Crops	Disease	Pathogens	Employed MAs	References
	Septoria blotch	Mycosphaerella graminicola	Ba. megaterium	Kildea et al. (2008)
		Ga. graminis var. tritici	Psd. fluorescens	Thomashow et al. (1990), Keel et al. (1992), Raaijmakers and Weller (1998)
	Damping-off	Py. ultimum	En. cloacae	Kageyama and Nelson (2003)
Maize/corn	Bacterial stalk rot	Er. carotovora	Ba. thuringiensis	Dong et al. (2004)
	Black bundle disease	Cephalosporium acremonium	Aca. laevis	Veerabhadrasamy and Garampalli (2011)
			Gl. mosseae, Gl. fasciculatum	
	Banded leaf and sheath blight	Rzt. solani	Ba. subtilis	Muis and Quimiob (2006)
	Charcoal rot	Mac. phaseolina	Bacillus spp.	Tapadar and Jha (2013)
			Pseudomonas spp., Psd. fluorescens	Pal et al. (2001), Tapadar and Jha (2013)
	Damping-off	F. oxysporum	Psd. fluorescens	Georgakopoulos et al. (1994)
		F. graminearum	Ba. cepacia	Mao et al. (1998)
		Py. arrhenomanes	Ba. cepacia	Mao et al. (1998)
		Py. ultimum	Ba. cepacia	Mao et al. (1998)
			Psd. fluorescens	Callan et al. (1990)

TABLE 3.9 (Continued)

Crops	Disease	Pathogens	Employed MAs	References
	Downey mildew	Peronosclerospora sorghi	Ba. subtilis Psd. fluorescens	Sadoma et al. (2011)
	Ear and stalk rot	F. moniliforme F. graminearum	Pseudomonas spp.	Pal et al. (2001)
	Ear rot	Stenocarpella maydis	Ba. subtilis Psd. fluorescens Pan. agglomerans	Petatan-Sagahon et al. (2011)
	Foot rots and wilting	F. moniliforme	Bacillus spp. Pseudomonas spp.	Pal et al. (2001)
	Lesion (nematodes, parasitic)	Pratylenchus zea	Vesicular arbuscular mycorrhizae (VAM)	Babu et al. (1998)
	Southern leaf corn blight	Helminthosporium maydis	Acremonium strictum	Tapadar and Jha (2013)
			Bacillus strain B-TL2, Ba. cereus, Ba. subtilis	Zhang et al. (2008a), Huang et al. (2010), Yun-feng et al. (2012)
			Streptomyces spp.	Tapadar and Jha (2013)
Barley	Head blight	F. culmorum	Psd. fluorescens, Psd. frederiksbergensis	Khan et al. (2006), Khan and Doohan (2009), Petti et al. (2010)
	Net blotch	Pyrenophora teres	Psd. fluorescens, Psd. frederiksbergensis	Khan and Doohan (2009), Khan et al. (2010)
	Root rot	Cochliobolus sativus	Vesicular arbuscular mycorrhizae (VAM)	Arabi et al. (2013)

TABLE 3.9 (Continued)

Crops	Disease	Pathogens	Employed MAs	References
Sorghum	Anthracnose	*Col. graminicola*	*Psd. fluorescens*	Singh (2008)
			T. harzianum	
Pearl millet	Downy mildew	*Sclerospora graminicola*	*Psd. fluorescens*	Umesha et al. (1998), Raj et al. (2003)
Alfalfa	Damping-off	*Phy. megasperma* f. sp. *medicaginis*	*Ba. cereus*	Handelsman et al. (1990), Silo-Suh et al. (1994)
	Root rot	*Phy. medicaginis*	*Streptomyces* spp.	Xiao et al. (2002)
Chickpea	Root rot	*Phy. medicaginis*	*Mesorhizobium ciceri*	Misk and Franco (2011)
			Streptomyces spp.	
Common bean/ French bean (*Phaseolus vulgaris*)	Damping-off	*Rzt. solani*	*Pseudomonas* spp.	Afsharmanesh et al. (2006)
	Fusarium root rot	*F. solani*	*Gl. macrocarpum*	Muchovej et al. (1991)
Groundnut	Stem rot (southern blight)	*Sclerotium rolfsii*	*Streptomyces* spp.	Adhilakshmi et al. (2014a)
	Collar rot	*As. niger*	*Pseudomonas* spp.	Kishore et al. (2005)
	Late leaf spot	*Phaeoisariopsis personata*	*Ba. circulans*	Kishore et al. (2005)
			Ser. marcescens	
	Root rot	*Mac. phaseolina*	*Gl. fasciculatum*	Doley and Jite (2012)
		Cercospora arachidicola	*Aca. lacunospora*	Hemavati and Thippeswamy (2014)

TABLE 3.9 (Continued)

Crops	Disease	Pathogens	Employed MAs	References
Mung bean	Root rot	*Mac. phaseolina*	*Burkholderia* spp. strain	Sathya et al. (2011)
			Psd. fluorescens	Karthikeyan et al. (2005), Saravanakumar et al. (2007b), Thilagavathi et al. (2007)
			Streptomyces spp.	Adhilakshmi et al. (2014b)
			Trichoderma spp.	Raguchander et al. (1993), Indira and Gayatri (2003), Raj et al. (2003)
	Charcoal rot	*Mac. phaseolina*	*Gliocladium virens*	Hussain et al. (1990)
			Pae. lilacinus	
			Rhizobium meliloti	
			Streptomyces spp.	
			T. harzianum	
Pea	Aphanomyces root rot	*Aphanomyces euteiches* f. sp. *pisi*	*Ba. megaterium*	Wakelin et al. (2002)
			Gl. intraradices	Bodker et al. (1998)
			Psd. cepacia	King and Parke (1993)
	Damping-off/ seed rot	*Pythium* spp., *Py. ultimum*	*Psd. cepacia,*	Paulitz and Loper (1991), King and Parke (1993), Wang et al. (2003), Bainton et al. (2004)
			Psd. fluorescens,	
			Psd. putida	
	Fusarium root rot	*F. solani* f. sp. *pisi*	*Psd. fluorescens*	Negi et al. (2005)

TABLE 3.9 (Continued)

Crops	Disease	Pathogens	Employed MAs	References
Soybean	Damping-off	*Py. ultimum*	*Psd. putida*	Paulitz and Loper (1991)
	Root rot	*Phy. sojae, Phy. medicaginis*	*Ba. cereus* *Hypochytrium catenoids* *Streptomyces* spp.	Osburn et al. (1995) Filnow and Lockwood (1985) Xiao et al. (2002)
Rape	Verticillium wilt	*Verticillium longisporum*	*Ser. plymuthica*	Berg et al. (1999)
Sunflower	Charcoal rot	*Mac. phaseolina*	*Gliocladium virens* *Pae. lilacinus* *Streptomyces* spp. *T. harzianum*	Hussain et al. (1990)
	Downy mildew	*Pl. halstedii*	*Ba. pumilus*	Nandeeshkumar et al. (2008)
	Head and Stem rot	*Sclt. sclerotiorum*	*Streptomyces* spp.	Baniasadi et al. (2009)
Cotton	Damping-off/seedling damping-off	*Rzt. solani*	*Psd. extremorientalis,* *Psd. fluorescens,* *Psd. cepacia*	Howell and Stipanovic (1979, 1980), Cartwright et al. (1995), Ligon et al. (2000), Egamberdieva and Jabborova (2013)
		Py. ultimum	*En. cloacae*	van Dijk and Nelson (2000), Kageyama and Nelson (2003)

TABLE 3.9 (Continued)

Crops	Disease	Pathogens	Employed MAs	References
Tobacco	Black root rot	Py. ulimum	Psd. fluorescens	Howell and Stipanovic (1979), Kageyama and Nelson (2003)
		Thielaviopsis basicola	Gl. monosporum	Giovannetti et al. (1991)
	Blue mold	Peronopora tabacina	Ba. pasteurii, Ba. pumilus	Zhang et al. (2001)
			Psd. fluorescens	
			Ser. marcescens	
Sugarcane	Red rot	Col. falcatum	Psd. putida	Malathi et al. (2002)

Aca., Acaulospora; Ag., Agrobacterium; Al., Alternaria; Am., Ampelomyces; As., Aspergillus; Au., Aureobasidium; Ba., Bacillus; Bi., Bipolaris; Botryo., Botryodiplodia; Botryt., Botrytis; Ca., Candida; Col., Colletotrichum; Cr., Cryptococcus; De., Debaryomyces; En., Enterobacter; Er., Erwinia; F., Fusarium; Ga., Gauemannomyces; Gl., Glomus; Kl., Kloeckera; La., Lasiobasidium; Ly., Lysobacter; Mac., Macrophomina; Pae., Paecilomyces; Pan., Pantoea; Pcl., Penicillium; Pho., Phomopsis; Phy., Phytophthora; Pi., Pichia; Pl., Plasmopora; Psd., Pseudomonas; Py., Pythium; Pyric., Pyricularia; Rhdc., Rhodococcus; Rhdt., Rhodotorula; Rzp., Rhizopus; Rzt., Rhizoctonia; Sclt., Sclerotinia; Ser., Serratia; St., Streptomyces; T., Trichoderma; X., Xanthomonas.

TABLE 3.10 Major Diseases of Horticultural Crops and Their Microbial Control.

Crops	Disease	Pathogens	Employed MAs	References
Apple	Blue mold	*Pcl. expansum*	*Ca. sake* (CPA-1)	Usall et al. (2001), Morales et al. (2008)
			Cr. albidus (Saito) Skinner	Fan and Tian (2001)
			Met. pulcherrima	Spadaro et al. (2002, 2004)
			Pan. agglomerans	Nunes et al. (2001b), Morales et al. (2008)
			Psd. syringae	Zhou et al. (2002)
			Rahnella aquatilis	Calvo et al. (2007)
			Rhdt. glutinis	Zhang et al. (2009)
	Gray mold	*Botryt. cinerea*	*Cr. albidus* (Saito) Skinner	Fan and Tian (2001)
			Met. pulcherrima	Spadaro et al. (2002, 2004)
			Psd. syringae	Zhou et al. (2001)
			Ra. aquatilis	Calvo et al. (2003, 2007)
			Rhdt. glutinis	Zhang et al. (2009)
			T. harzianum	Tronsmo (1991)
		Botryt. mali Ruehle	*Psd. fluorescens* Migula	Mikani et al. (2008)
	Seedling disease	*Pythium* spp.	*Psd. fluorescens*	Mazzola et al. (2007)
Apricot	Anthracnose	*Col. gloeosporioides*	*Pestalotiopsis neglecta* (Thuemen) Steyaert	Adikaram and Karunaratne (1998)
	Brown rot	*Lasiodiplodia theobromae*	*Ba. subtilis*	Pusey and Wilson (1984)
Avocado	Stem end rot	*Botryo. theobromae* Pat.	*Ba. subtilis*	Demoz and Korsten (2006)
Banana	Anthracnose	*Col. musae*	*Bu. cepacia*	Costa and de Erabadupitiya (2005)
			T. harzianum	Devi and Arumugam (2005)
	Blossom end rot	*Col. musae*	*Bu. cepacia*	Costa and de Erabadupitiya (2005)

TABLE 3.10 (Continued)

Crops	Disease	Pathogens	Employed MAs	References
	Crown rot	Col. musae	Ca. oleophila	Lassois et al. (2008)
			Pi. anomala (Hansen) Kurtzman	
			Pseudomonas spp.	Costa and de Subasinghe (1998)
	Panama disease (Fusarium wilt)	F. oxysporum f. sp. cubense	F. oxysporum	Forsyth et al. (2006)
Cherry	Botrytis rot	Botryt. cinerea	Ba. subtilis	Utkhede and Sholberg (1986)
			Kl. apiculata (Rees) Janke	Karabulut et al. (2005)
	Gray mold	Botryt. cinerea	T. pullulans (Lindner) Didlens and Lodder	Qin et al. (2004)
	Alternaria rot	Al. alternata	En. aerogenes Hormaeche and Edwards	Utkhede and Sholberg (1986)
	Brown rot	Monilinia fructicola	Cr. laurentii	Tian et al. (2004), Qin et al. (2006)
Citrus	Green mold	Pcl. digitatum	Ba. subtilis	Singh and Deverall (1984)
			De. hansenii	Singh (2002)
			Kl. apiculata (Rees) Janke	Long et al. (2006, 2007)
			Pan. agglomerans	Teixido et al. (2001), Torres et al. (2007)
			Pi. guilliermondii	Wilson and Chalutz (1989), Chalutz and Wilson (1990)
			Psd. glathei, Psd. syringae	Wilson and Chalutz (1989), Iqbal et al. (1990), Huang et al. (1995)
			T. viride	De-Matos (1983)

TABLE 3.10 (Continued)

Crops	Disease	Pathogens	Employed MAs	References
	Blue mold	*Pcl. italicum*	*De. hansenii*	Chalutz and Wilson (1990), Singh (2002)
			Kl. apiculata (Rees) Janke	Long et al. (2006, 2007)
			Pan. agglomerans	Teixido et al. (2001), Torres et al. (2007)
			Psd. syringae	Wilson and Chalutz (1989)
	Sour rot	*Geotrichum candidum*	*Ba. subtilis*	Singh and Deverall (1984)
			De. hansenii	Chalutz and Wilson (1990)
			Trichoderma spp.	De-Matos (1983)
	Root rot	*Phy. parasitica*	*Psd. fluorescens*	Yang et al. (1994)
	Stem end rot	*Botryo. theobromae, Pho. citri* Fawc., *Al. citri* Ell. and Pierce	*Ba. subtilis*	Singh and Deverall (1984)
	Penicillium rot	*Penicillium* spp. (*Pcl. digitatum* and *Pcl. italicum*)	*Au. pullulans*	Wilson and Chalutz (1989)
			Ca. oleophila	El-Neshawy and El-Sheikh (1998), Lahlali et al. (2004), Lahlali et al. (2005)
			Kl. apiculata (Rees) Janke	Long et al. (2006, 2007)
			Pan. agglomerans	Plaza et al. (2001)
			Pi. anomala (Hansen) Kurtzman	Lahlali et al. (2004)

TABLE 3.10 (Continued)

Crops	Disease	Pathogens	Employed MAs	References
Cucumber	Damping-off	*Py. mamillatum*	*Ba. mycoides*	Bernard et al. (1995)
		Py. aphanidermatum	*Psd. fluorescens*	Ongena et al. (1999)
			Ser. plymuthica	Pang et al. (2009)
		Py. ultimum	*En. cloacae*	Kageyama and Nelson (2003)
			Psd. fluorescens, Psd. corrugate, Psd. putida	Howell and Stipanovic (1980), Paulitz and Loper (1991), Georgakopoulos et al. (2002)
			Ba. subtilis	Georgakopoulos et al. (2002)
	Seed infection	*Py. ultimum*	*En. cloacae*	Windstam and Nelson (2008a,b)
	Cottony leak	*Phytophthora* spp.	*Ba. cereus*	Smith et al. (1993)
		F. oxysporum f. sp. *cucumerinum*	*Streptomyces* spp. *Paenibacillus* spp.	Singh et al. (1999)
	Root and crown rot	*Py. aphanidermatum*	*Ly. enzymogenes*	Folman et al. (2003), Folman et al. (2004), Postma et al. (2009)
Grape	Botrytis (grey rot or noble rot)	*Botryt. cinerea*	*Au. pullulans*	Schena et al. (2003)
			Metschnikowia fructicola	Karabulut et al. (2003); Karabulut and Baykal (2003)
			Pi. guilliermondii	Chalutz et al. (1988)
			T. harzianum	Batta (2007)
	Soft rot	*Monilinia laxa*	*Au. pullulans*	Barkai-Golan (2001)
	Rhizopus rot	*Rzp. stolonifer*	*Pi. guilliermondii*	Chalutz et al. (1988)

TABLE 3.10 (Continued)

Crops	Disease	Pathogens	Employed MAs	References
Grapevine	Dieback	*Eutypa lata*	*Ba. subtilis*	Schmidt et al. (2001)
			Er. herbicola	
			Ser. plymuthica	
	Downy mildew	*Pl. viticola*	*Ba. Subtilis*	Furuya et al. (2011)
			Streptomyces spp.	Abdalla et al. (2011), Islam et al. (2011), Zinada et al. (2011)
Guava	Fruit rot	*La. theobromae, Pho. psidi, Rhizopus* spp.	*Trichoderma* spp.	Majumdar and Pathak (1995)
Jujube	Alternaria rot	*Al. alternata*	*Rhdt. glutinis*	Tian et al. (2005)
			Cr. laurentii	Qin and Tian (2004), Tian et al. (2005)
	Penicillium rot	*Pcl. expansum*	*Rhdt. glutinis*	Tian et al. (2005)
			Cr. laurentii	Qin and Tian (2004), Tian et al. (2005)
Kiwifruit	Gray mold	*Botryt. cinerea*	*T. harzianum*	Batta (2007)
Litchi	Alternaria rot	*Al. alternata* (Fr.) Keissler	*Ba. subtilis*	Jiang et al. (1997, 2001)
Mango	Anthracnose	*Col. gloeosporioides*	*Ba. licheniformis* (Weigmann) Verhoeven	Govender et al. (2005)
			Brevundimonas diminuta (Leifson and Hugh) Segers	Kefialew and Ayalew (2008)
			Ca. membranifaciens Hansen	
	Stem end rot	*Dothiorella gregaria* Sacc.	*Ba. licheniformis* (Weigmann) Verhoeven	Govender et al. (2005)
	Stem end rot	*Botryo. theobromae*	*T. viride*	Kota et al. (2006)

TABLE 3.10 (Continued)

Crops	Disease	Pathogens	Employed MAs	References
	Fruit rot	La. theobromae, Rhizopus spp.	Trichoderma spp.	Pathak (1997)
Muskmelon	Alternaria rot	Al. alternate	Ba. subtilis	Yang et al. (2006)
Orange	Green mold	Pcl. digitatum	Psd. cepacia	Huang et al. (1993)
Papaya	Anthracnose	Col. gloeosporioides	Ca. oleophila	Gamagae et al. (2003)
Peach	Gray mold	Botryt. cinerea	Ca. guilliermondii, Ca. oleophila	Tian et al. (2002), Karabulut and Baykal (2004)
			Cr. laurentii	Zhang et al. (2007c)
	Blue mold	Pcl. expansum	Cr. laurentii	Zhang et al. (2007c)
	Rhizopus rot	Rzp. stolonifer	Cr. laurentii	Zhang et al. (2007c)
			De. hansenii	Singh (2004, 2005), Mandal et al. (2007)
			En. cloacae	Wilson et al. (1987)
	Brown rot	Lasiodiplodia theobromae	Ba. subtilis	Pusey and Wilson (1984)
		Monilinia spp.	Cr. laurentii	Yao and Tian (2005)
			Pcl. frequentans Westling	Guijarro et al. (2007)
			Psd. cepacia, Psd. corrugata Roberts and Scarlett, Psd. syringae	Smilanik et al. (1993), Zhou et al. (1999)
Pear	Gray mold	Botryt. cinerea	Ba. pumilus	Mari et al. (1996)
			Cr. laurentii	Zhang et al. (2005)
			Psd. cepacia	Janisiewicz and Roitman (1988)
			Rhdt. glutinis	Zhang et al. (2008b)
			T. harzianum	Batta (2007)

TABLE 3.10 (Continued)

Crops	Disease	Pathogens	Employed MAs	References
	Blue mold	Pcl. expansum	Ca. sake (CPA-1)	Torres et al. (2006)
			Cr. laurentii	Zhang et al. (2003)
			Psd. cepacia	Janisiewicz and Roitman (1988)
	Blue rot	Pcl. expansum	Rhdt. glutinis	Zhang et al. (2008b)
	Rhizopus rot	Rzp. stolonifer	Pan. agglomerans	Nunes et al. (2001a,b)
	Mucor rot	Mucor piriformis	Cr. albidus (Saito) Skinner, Cr. flavus, Cr. laurentii	Roberts (1990)
Pineapple	Penicillium rot	Penicillium spp.	Penicillium spp. (attenuated strains)	Tong-Kwee and Rohrbock (1980)
Raspberry	Root rot	Phy. fragariae	Streptomyces spp.	Valois et al. (1996)
Strawberry	Gray mold	Botryt. cinerea	Ba. subtilis	Zhao et al. (2007)
			Rhdt. glutinis	Zhang et al. (2007a)
			T. harzianum, T. viride	Batta (2007), Tronsmo and Denis (1977)
	Rhizopus rot	Rzp. stolonifer	Cr. laurentii	Zhang et al. (2007b)
Cabbage	Bacterial soft rot	Er. carotovora sub sp. carotovora	Psd. aeruginosa (Schroter) Migula	Adeline and Sijam (1999)
	Bottom-rot	Rzt. solani	Coprinellus curtus	Nakasaki et al. (2007)
Carrot	Damping-off	Py. ultimum	En. cloacae	Kageyama and Nelson (2003)
Chilies	Anthracnose	Col. capsici (Syd.) Butler and Bisby	Pi. guilliermondii	Chanchaichaovivat et al. (2007)
		Col. gleosporioides	St. ambofaciens	Heng et al. (2015)
	Damping-off	Py. aphanidermatum	Psd. fluorescens	Muthukumar et al. (2011)

TABLE 3.10 (Continued)

Crops	Disease	Pathogens	Employed MAs	References
Lettuce	Damping-off	*Py. ultimum*	*En. cloacae*	Kageyama and Nelson (2003)
Onion	Botrytis rot	*Botryt. allii*	*Ba. licheniformis*	Lee et al. (2001)
	Fusarium rot	*F. oxysporum*	*Ba. amyloliquefaciens*	
Pepper	Fruit rot/blight/crown blight	*Phy. capsici*	*Ba. megnaterium, Ba. subtilis*	Sid Ahmed et al. (2003), Akgül and Mirik (2008)
			Ly. antibioticus	Ko et al. (2009)
			Psd. aeruginosa	Kim et al. (2000); Rettinassababady et al. (2002)
	Root rot	*Py. aphanidermatum*	*Psd. chlororaphis*	Khan et al. (2003), Chatterton et al. (2004)
		Py. dissotocum	*Psd. chlororaphis*	Chatterton et al. (2004)
Potato	Black scurf	*Rzt. solani*	*Pseudomonas* spp.	Tariq et al. (2010)
	Common scab	*Streptomyces* spp.	*Pseudomonas* spp.	Singhai et al. (2011)
	Drop (sclerotinia rot)	*Rzt. solani*	*As. niger*	Iqbal et al. (1990), Huang et al. (1995)
			Pseudomonas spp.	Tariq et al. (2010)
	Late blight	*Phy. infestans*	*Ba. subtilis*	Stephan et al. (2005)
	Pink rot	*Phy. erythroseptica*	*Psd. fluorescens*	Schisler et al. (2009)
	Soft rot	*Er. carotovora* sub sp. *carotovora, Er. carotovora* sub sp. *atroseptica*	*Psd. putida* (Trevisan) Migula, *Psd. fluorescens*	Cronin et al. (1997)
	Fusarium rot	*F. roseum* var. *sambucinum*	*Bacillus* spp.	Sadfi et al. (2002)

TABLE 3.10 (Continued)

Crops	Disease	Pathogens	Employed MAs	References
	Dry rot	*Gibberella pulicans*	*Pan. agglomerans*	Schisler et al. (2000)
			Psd. fluorescens	
		F. sambucinum, F. oxys-porum, F. culmorum	*Bu. Cepacia*	Al-Mughrabi (2010)
	Silver scurf	*Helminthosporium solani*	*Ba. Cereus*	Martinez et al. (2002)
			Cellulomonas fimi	
			Kocuria varians	
			Psd. putida	
			Rhdc. erythropolis, Rhdc. globerulus	
Radish	Fusarium wilt	*F. oxysporum*	*Psd. putida*	Scher and Baker (1982)
Sugar beet	Damping-off	*Py. ultimum*	*En. cloacae*	Kageyama and Nelson (2003)
	Damping-off	*Sclecrotium rolfsii*	*Streptomyces* spp.	Errakhi et al. (2007)
		Ap. cochlioides	*Pseudomonas* spp., *Psd. jessenii, Psd. fluorescens*	Deora et al. (2005), Islam and Fukushi (2010)
		Pythium spp., *Py. ultimum*	*Ba. subtilis*	Georgakopoulos et al. (2002)
			Ly. enzymogenes	Palumbo et al. (2005)
			Psd. corrugate, Psd. fluorescens	Georgakopoulos et al. (2002), Schmidt et al. (2004)
			Stenotrophomonas maltophilia	Dunne et al. (1997, 1998)
	Bipolaris leaf spot of tall fescue	*Pythium* spp., *Py. ultimum*	*Ly. enzymogenes*	Palumbo et al. (2005)
		Cercospora spp.	*Ba. subtilis*	Collins and Jacobsen (2003)

TABLE 3.10 (Continued)

Crops	Disease	Pathogens	Employed MAs	References
Spinach	Damping-off	*Ap. cochlioides*	*Pseudomonas* spp.	Deora et al. (2005)
Squash	Phytopthora blight	*Phy. capsici*	*Ba. pumilus, Ba. safensis, Ba. macauensis, Ba. subtilis* subsp. *Subtilis*	Zhang et al. (2010)
			Lysinibacillus boronitolerans	Zhang et al. (2010)
Tomato	Alternaria rot	*Alternata alternata*	*Pi. guilliermondii*	Chalutz et al. (1988)
	Crown and root rot	*F. oxysporum* f. sp. *radicis-lycopersici*	*F. oxysporum* Fo47	Bolwerk et al. (2005)
			Psd. chlororaphis	Chin-A-Woeng et al. (1998)
	Damping-off	*Psd. splendens*	*Psd. fluorescens, Psd. aeruginosa*	Buysens et al. (1994), Buysens et al. (1996)
		Py. aphanidermatum	*Ba. subtilis*	Ongena et al. (2005), Leclere et al. (2005)
			Psd. corrugate Psd. fluorescens, Psd. marginalis, Psd. putida, Psd. syringae, Psd. viridiflava	Ramamoorthy et al. (2002), Gravel et al. (2005)
		Py. ultimum	*En. cloacae*	Kageyama and Nelson (2003)
			Psd. corrugate, Psd. fluorescens, Psd. marginalis, Psd. putida, Psd. syringae, Psd. viridiflava	Gravel et al. (2005)
	Fusarium wilt	*F. oxysporum* f. sp. *lycopersici*	*Fusarium* spp.	Larkin and Fravel (2002)
			Gl. intraradices	Akkopru and Demir (2005)

TABLE 3.10 (Continued)

Crops	Disease	Pathogens	Employed MAs	References
	Gray mold	*Botryt. cinerea*	*Ca. oleophila, Ca. guilliermondii*	Saligkarias et al. (2002)
			Cr. laurentii	Xi and Tian (2005)
			Gl. mosseae	Fiorilli et al. (2011)
			Pi. guilliermondii	Chalutz et al. (1988)
	Late blight	*Phy. infestans*	*Ba. pumilus*	Yan et al. (2002)
			Psd. fluorescens	Yan et al. (2002), Tran et al. (2007)
	Rhizopus rot	*Rzp. nigricans*	*Pi. guilliermondii*	Zhao et al. (2008)
	Root-knot	*Meloidogyne incognita*	*F. oxysporum*	Dababat and Sikora (2007)
			Gl. mosseae	
			Pae. Lilacinus	Al-Raddad (1995)
		Rzt. solani	*Gl. mosseae*	Berta et al. (2005)
			Psd. fluorescens	Berta et al. (2005)
Yams	Rot	*Botryo. theobromae*	*Ba. subtilis*	Okigbo (2002), Swain et al. (2008)
		F. moniliforme	*Ba. subtilis*	Okigbo (2002)
	Blue and green rot	*Pcl. sclerotigenum*	*Pseudomonas* spp.	Okigbo (2002)

Aca., Acaulospora; Ag., Agrobacterium; Al., Alternaria; Am., Ampelomyces; As., Aspergillus; Au., Aureobasidium; Ba., Bacillus; Bi., Bipolaris; Botryo., Botryodiplodia; Botryt., Botrytis; Ca., Candida; Col., Colletotrichum; Cr., Cryptococcus; De., Debaryomyces; En., Enterobacter; Er., Erwinia; F., Fusarium; Ga., Gauemannomyces; Gl., Glomus; Kl., Kloeckera; La., Lasiobasidium; Ly., Lysobacter; Mac., Macrophomina; Pae., Paecilomyces; Pan., Pantoea; Pcl., Penicillium; Pho., Phomopsis; Phy., Phytophthora; Pi., Pichia; Pl., Plasmopora; Psd., Pseudomonas; Py., Pythium; Pyric., Pyricularia; Rhdc., Rhodococcus; Rhdt., Rhodotorula; Rzp., Rhizopus; Rzt., Rhizoctonia; Sclt., Sclerotinia; Ser., Serratia; St., Streptomyces; T., Trichoderma; X., Xanthomonas.

3.4   MICROORGANISMS IN SUSTAINABLE ENVIRONMENT AND AGRICULTURE

3.4.1   MICROORGANISMS AND SUSTAINABLE ENVIRONMENT

The value of nature's services to production of agriculture and horticulture and also the importance of maintaining their viable environment have long been considered not only by ecologists and environmental scientists but also plant pathologists have stressed on environmental influences by emphasizing on disease triangle (Fig. 3.8) in their study of plant disease management (PDM) through microbial control (Krupinsky et al., 2002; Garrett, 2008; Grulke, 2011). In the context of sustainable environment, if MAs are used for plant disease control successful is become very much essential and urgently needed to their biological and ecological study due to continue ensue of safety problems and ecological disruptions (Verma et al., 2014).

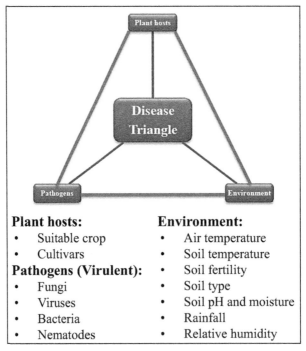

Plant hosts:
- Suitable crop
- Cultivars

Pathogens (Virulent):
- Fungi
- Viruses
- Bacteria
- Nematodes

Environment:
- Air temperature
- Soil temperature
- Soil fertility
- Soil type
- Soil pH and moisture
- Rainfall
- Relative humidity

FIGURE 3.8   Interactions between plant hosts, pathogens, and environment cause disease. (*Source*: Concept adapted from Krupinsky et al., 2002)

Toward the sustainable environment and environmental security, considerable effort has been directed by microbial control with the help of different discipline of sciences such as microbiology, modern chemistry, molecular biology, biotechnology, genetic engineering, etc. as the new technology which has been implemented to replace hazardous chemicals and or nontoxic to environments and others organism including humans but still emerging. Microbes are used as MAs to combat the pathogenic strain of agricultural/horticultural crops is a practical option and also known for promising environmental approach with high potential against pathogenic organisms without effect on the nontarget and beneficial organisms, and environment. There presently exist numerous reports on the potential use of MAs as replacements of agrochemicals (Shimizu et al., 2000; Yang et al., 2008).

3.4.2 MICROORGANISMS AND SUSTAINABLE AGRICULTURE

Today, many microbes including fungi and bacteria are well known useful potential solution in crop disease control (Romero et al., 2007; Suarez-Estrella et al., 2007; Whipps and McQuilken, 2009) and have been found environmental friendly, which do not negatively affect human health (Kaewchai et al., 2009). Microbial control traced to environmental concerns interest in the agricultural sustainability. Various different views have come in existence to be used to imply greater sustainability over prevailing at both levels preindustrial and industrialized in agriculture. Sustainable agriculture is the act of farming which can be understood an ecosystem approach to agriculture by using ecology principles and the study of relationships between organisms and their environment (Altieri, 1995; Verma et al., 2014). It means keep in existence, keep up, maintain or prolong, etc. many goals encompasses in literally meaning. As a more sustainable agriculture seeks to make the best use of nature's goods and services, so technologies and practices must be locally adapted and fitted to place. There are various terms like biodynamic, eco-agriculture, ecological, environmentally-sensitive, extensive, farm-fresh, free-range, low-input, organic, permaculture, sustainable, wise-use, etc. have been used time to time used for sustainability in agriculture (Pretty, 1995; Conway,

1997; NRC, 2000; McNeely and Scherr, 2003; Clements and Shrestha, 2004; Cox et al, 2004; Gliessman, 2005). These terms are qualified as sustainable agriculture after continuing and intense debate on such agricultural practices (Balfour, 1943; Lampkin and Padel, 1994; Altieri, 1995; Trewevas, 2001). Simply defined, sustainable agriculture is a gateway to agriculture that contributes to the livelihood of communities and does not degrade to the environment during producing food. In words of Lewandowski et al. (1999), sustainable agriculture can fulfill to the significant ecological, economic, and social functions at the different levels, namely, local, national, and global, and also never harm to other ecosystems in addition to the agriculture because it focused on to maintains agricultural productivity, biological diversity, regeneration capacity, vitality and ability to function, etc. through well-recognized management and utilization of the agricultural ecosystem (Rodrigues et al., 2003). Agricultural sustainability indicates the need to fit these factors to the specific circumstances of different agricultural systems.

Today, more food and income are needed for growing population especially from agriculture and agricultural-related activities that have been the most significant challenges faced by our sustainable agriculture (Dordas, 2009). According to the United Nations and Food and Agricultural Organization, often asserted in the next few decades about 9 billion people will approach from the global population where is a need for 70–100% more food (Godfray et al., 2010). In such a way, significant losses are gradually being affected by various factors, namely, diseases, pests, droughts, decreased soil fertility due to use of hazardous chemical pesticides, pollution and global warming in agricultural/horticultural yield, and quality of crops. It is due to the traditional agricultural management practices worldwide. In addition of all, plant diseases continue to stand/raise as a major limiting factor in agricultural/horticultural production which cannot fulfill to that particular global demand. Thus, there is a need for some eco-friendly management of crop disease through MAs for sound plant health because it is key to sustainable agriculture and may help to find out the appropriate solutions for some of these problems. In recent years, the significance and meaning of microbial control have been subjected to the scientist, researchers, and academician because of one of the most important and concerning issues for sustainable agriculture. Due to nature-friendly and ecological approach, the use of specific microorganisms has dictated the need for alternative PDM techniques that interfere with plant pathogens and also made to solve and overcome to the serious concerns about food safety and health issues, environmental quality, and standard chemical methods of PDM (Weller et al., 2002; Harman et al., 2004). To achieve the goal of double food production from global agriculture by 2050 in order to feed the worldwide growing population and at the same time reduce its reliance on synthetic chemicals, there is an urgent need of microbial control to harness the multiple beneficial interactions that occur between plants and microorganisms to solve PDM and global food demands.

3.5 OPPORTUNITIES FOR FUTURE WORK AND CHALLENGES IN RESEARCH

3.5.1 OPPORTUNITIES

In the plant disease control, the MAs are fundamentally very much different from synthetic chemical used against crop pathogen and as they grow and proliferate effectively. Therefore, the effective antagonist of MAs under favorable conditions in crop ecosystem is to establish the active action against targeted pathogens (Caldwell, 1958; Lewis and Papavizas, 1984). The microbial control is emerging tools in which the agents are receiving much more attention due to the importance of their controlling strength of plant pathogens and for environmental protection of natural ecosystems. They hold promise with the future to provide solutions for both, disease problems affecting worldwide agricultural/horticultural production and to needs for environmental security in natural ecosystems.

The use of MAs and their formulated products to employ in management of pathogens caused crop diseases is very common in many developed countries, especially the Australia, Europe, New Zealand, and USA. Despite the many research efforts dealing with MAs, formulated products, and its application is limited. In the opinion of Alabouvette et al. (2006) in order to make microbial control more trustworthy, much more research should be devoted to the conditions required for the expression of the beneficial effects of microorganism under field conditions. The field-oriented research related to microbes in microbial control for plant disease is a potential solution which taking care under sustainable environment and agriculture are very limited and very few are supported by research institutions (Table 3.11). There is great opportunities which have very perspectives future to microbial control the plant disease through the application of microbe and many potentially interesting because microorganisms have fewer been tested under field conditions.

TABLE 3.11   Prospective Area for Future Research and Some Raised Question Related to MAs and Their Role in Crop Disease and Pathogens Control.

S. N.	Prospective area for research	Raised question
1.	The ecology of plant-associated microbes	How are pathogens and their antagonists distributed in the environment?
		Under what conditions do microbial control agents exert their suppressive capacities?
		How do native and introduced populations respond to different management practices?
		What determines successful colonization and expression of microbial control traits?
		What are the components and dynamics of plant host defense induction?
2.	Application of current strains/ inoculant strategies	Can more effective strains or strain variants be found for current applications?
		Will genetic engineering of microbes and plants be useful for enhancing microbial control?
		How can formulations be used to enhance activities of known microbial control agents?
3.	Discovering novel strains and mechanisms of action	Can previously uncharacterized microbes act as microbial control agents?
		What other genes and gene products are involved in pathogen suppression?
		Which novel strain combinations work more effectively than individual agents?
		Which signal molecules of plant and microbial origin regulate the expression of microbial control traits by different agents?
4.	Practical integration into agricultural systems	Which production systems can most benefit from microbial control for disease management?
		Which microbial control strategies best fit with other IPM system components?
		Can effective microbial control-cultivar combinations be developed by plant breeders?

Source: Pal and Gardener (2006) with some modification.

3.5.2 CHALLENGES

In addition of prospective future and opportunities, there are several challenges rising and facing, and they are not allaying to the general public concerns, not satisfying to the regulatory agencies and also not promoting to the commercial acceptance in microbial control of plant disease. Some of them are illustrated in Table 3.12.

TABLE 3.12 Challenges Toward Application of Microbiological Control Agent in PDM.

S. N.	Challenges
1.	To select suitable naturally occurring strains of microbiological control agents or create them through genetic engineering by incorporating useful genes to made them ecologically sustainable as well as socially acceptable (Kerr, 1987)
2.	Modifications in farm equipment and practices to accommodate microbiological treatment as one component in PDM
3.	The quality control of commercial microbiological control agents should be strictly enforced which ensure that their bioformulation products are not treating as well as hazards chemicals
4.	For considering the environmental, social, economic as well as sustainable values, there are no any easier, earlier, and fast techniques for plant pathologists have been developed to detect the pathogens and also to postulate appropriate management practices
5.	Microbiological control moves beyond the suppression of plant disease pathogen which include the impacts on human health and disasters because of no safety of biological control agents, as well as their efficacy

3.6 SUMMARY AND CONCLUSIONS

The new era in the agriculture has begun with the implementation of microbial control for plant disease. Owing to unpredictable performance of biological agents in the field developed thus far has plagued researchers and their efforts to exploit them for commercial application. Microbial control is rapidly developing as an important component of integrated disease management (IDM) because of its cost-effectiveness and the focus on environmental security. Microbial control for plant disease involves the participation of biological agents against crop pathogen. Commonly used agents for microbial control of plant disease are like bacteria, fungi, virus, etc. or their genes or their products. In the last 10 years, our understanding regarding microbial control mechanisms and plant responses has increased tremendously. The choice of the right microbial candidate is indeed as one of the most important factors governing the success of microbial control programs on the commercial basis. *Bacillus* spp., *Pseudomonas* spp., and *Trichoderma* spp. are more reliable candidates in addition of *Agrobacterium* spp., *Streptomyces* spp., *Ampelomyces* spp., *Candida* spp., *Coniothyrium* spp., *Gliocladium* spp. as they are highly effective against crop pathogen to limit the yield loss. This emerging tool is currently receiving inadequate attention because of their significant role in sustainable environment and agriculture. As it is illustrated by numerous research studies that microbial control has given new dimensions in the field of agricultural microbiology as it is eco-friendly and complementary to environmental sustainability. The research toward microbial control improvement and implementation is gaining momentum. There is a need of enthusiastic attempt of disseminating knowledge about microbial control to increase production efficiency. It is assumed that cost-effective formulations of MAs that perform consistently in their fields when made available, either by themselves or as part of an IDM package, will be revolutionary innovation in the field of agricultural microbiology than never before. Further research is needed into the basic biology of MAs and plant pathogens in order to understand their ecology, physiology, biochemistry, and genetics which are necessary to predict the behavior of these candidates under the different environmental conditions and to develop reliable microbial control systems as the potential solution for plant disease.

KEYWORDS

- *Phytophthora infestans*
- microbial agents
- pathogen
- biorational approaches
- antibiosis
- biological control

REFERENCES

Abdalla, M. A.; Win, H. Y.; Islam, M. T.; von Tiedemann, A.; Schuffler, A.; Laatsch, H. Khatmiamycin, a Motility Inhibitor and Zoosporicide against the Grapevine Downy

Mildew Pathogen Plasmopara Viticola from *Streptomyces* sp. ANK313. *J. Antibiot.* **2011**, *64*, 655–659.

Abeysinghe, S. Use of Nonpathogenic *Fusarium Oxysporum* and Rhizobacteria for Suppression of *Fusarium* Root and Stem Rot of *Cucumis sativus* Caused by *Fusarium oxysporum* f sp Radicis-cucumerinum. *Arch. Phytopathol. Plant Prot.* **2009**, *42*, 73–82.

Adeline, T. S. Y.; Sijam, K. In *Biological Control of Bacterial Soft Rot of Cabbage*, Biological Control in the Tropics: Towards Efficient Biodiversity and Bioresource Management for Effective Biological Control: Proceedings of the Symposium on Biological Control in the Tropics; Hong, L. W., Sastroutomo, S. S., Caunter, I. G., Ali, J., Yeang, L. K., Vijaysegaran, S., Sen, Y. H., Eds.; CABI Publishing: Wallingford, UK, 1999; Alabouvett, pp 133–134.

Adhilakshmi, M.; Latha, P.; Paranidharan, V.; Balachandar, D.; Ganesamurthy, K.; Velazhahan, R. Biological Control of Stem Rot of Groundnut (*Arachis hypogaea* L.) Caused by *Sclerotium rolfsii* Sacc. with Actinomycetes. *Arch. Phytopathol. Plant Prot.* **2014a**, *47*(3), 298–311.

Adhilakshmi, M.; Paranidharana, V.; Balachandarb, D.; Ganesamurthyc, K.; Velazhahana, R. Suppression of Root Rot of Mung Bean (*Vigna radiata* L.) by *Streptomyces* sp. is Associated with Induction of Peroxidase and Polyphenol Oxidase. *Arch. Phytopathol. Plant Prot.* **2014b**, *47*(5), 571–583.

Adikaram, N. K. B.; Karunaratne, A. Suppression of Avocado Anthracnose and Stem-end Rot Pathogens by Endogenous Antifungal Substances and a Surface Inhabiting Pestalotiopsis sp. *ACIAR Proc. Ser.* **1998**, *80*, 72–77.

Afsharmanesh, H.; Ahmadzadeh, M.; Sharifi-Tehrani, A. Biocontrol of *Rhizoctonia solani*, the Causal Agent of Bean Damping-Off by Fluorescent Pseudomonads. *Commun. Agric. Appl. Biol. Sci.* **2006**, *71*(3 Pt B), 1021–1029.

Akgül, D. S.; Mirik, M, Biocontrol of *Phytophthora capsici* on Pepper Plants by *Bacillus Megnaterium* strains. *J. Plant Pathol.* **2008**, *90*(S1), 29–34.

Akhtar, M. S.; Siddiqui, Z. A. Effects of *Glomus fasciculatum* and *Rhizobium* sp. on the Growth and Root-Rot Disease Complex of Chickpea. *Arch. Phytopathol. Plant Prot.* **2007**, *40*, 37–43.

Akkopru, A.; Demir, S. Biocontrol of Fusarium wilt in Tomato Caused by *Fusarium oxysporum* f. sp. *lycopersici* by AMF *Glomus intraradices* and some *Rhizobacteria. J. Phytopathol.* **2005**, *153*, 544–550.

Alabouvette, C.; Olivain, C.; Steinberg, C. Biological Control of Plant Diseases: The European Situation. *Eur. J. Plant Pathol.* **2006**, *114*, 329–341.

Alabouvette, C.; Backhouse, D.; Steinberg, C.; Donovan, N. J.; Edel-Hermann, V.; Burgess, L. W. Microbial Diversity in Soil—Effects on Crop Health. In *Managing Soil Quality: Challenges in Modern Agriculture*; Schjønning, P., Elmholt, S., Christensen, B. T., Eds.; CABI Publishing: Wallingford, UK, 2004; pp 121–138.

Al-Mughrabi, K. I. Biological Control of the Potato Dry Rot Caused by *Fusarium* Species Using PGPR Strains. *Biol. Control.* **2010**, *53*, 280–284.

Al-Raddad, A. M. Interaction of *Glomus mosseae* and *Paecilomyces lilacinus* on *Meloidogyne incognita* of Tomato. *Mycorrhiza* **1995**, *5*(3), 233–236.

Alstrom, S. Induction of Disease Resistance in Common Bean Susceptible to Halo Blight Bacterial Pathogen after Seed Bacterization with Rhizosphere Pseudomonads. *J. Gen. Appl. Microbiol.* **1991**, *37*, 495–501.

Altieri, M. A. *Agroecology: The Science of Sustainable Agriculture*; Westview Press: Boulder, CO, 1995.

Amein, T.; Omer, Z.; Welch, C. Application and Evaluation of *Pseudomonas* Strains for Biocontrol of Wheat Seedling Blight. *Crop Prot.* **2008**, *27*, 532–536.

Anand, Y. R.; Singh, S. J.; Verma, D. K.; Panyam, K. R.; Sumitra, P.; Meitei, K. M.; Gurumurthy, S.; Asthir, B. Recent Advances in Induced Resistance for Plant Disease Management: An Overview. In *Innovations in Plant Science and Biotechnology*; Wani, S. H., Malik, C. P., Hora, A., Kaur, R., Eds.; Pub. Agrobios, India, 2014; pp 107–143.

Andersen, J. B.; Koch, B.; Nielsen, T. H.; Sørensen, D.; Hansen, M.; Nybroe, O.; Christophersen, C.; Sørensen, J.; Molin, S.; Givskov, M. Surface Motility in *Pseudomonas* sp. DSS73 is Required for Efficient Biological Containment of the Root-pathogenic Microfungi *Rhizoctonia solani* and *Pythium ultimum*. *Microbiology* **2003**, *149*, 1147–1156.

Anonymous. *Production Year-Book*; Food and Agriculture Organization (FAO): Rome, 1993.

Arabi, M. I. E.; Kanacri, S.; Ayoubi, Z.; Jawhar, M. Mycorrhizal Application as a Biocontrol Agent against Common Root Rot of Barley. *Res. Biotechnol.* **2013**, *4*(4), 7–12.

Asaka, O.; Shoda, M. Biocontrol of *Rhizoctonia solani* Damping-off of Tomato with *Bacillus subtillis* RB14. *Appl. Environ. Microbiol.* **1996**, *62*, 4081–4085.

Audenaert, K.; Pattery, T.; Cornelis, P.; Hofte, M. Induction of Systemic Resistance to *Botrytis cinerea* in Tomato by *Pseudomonas aeruginosa* 7NSK2: Role of Salicylic Acid, Pyochelin, and Pyocyanin. *Mol. Plant-Microbe Interact.* **2002**, *15*, 1147–1156.

Babu, R. S.; Sankaranarayan, C.; Joshi, G. Management of *Pratylenchus zeae* on Maize by Biofertilizers and VAM. *Indian J. Nematol.* **1998**, *28*(1), 77–80.

Bainton, N. J.; Lynch, J. M.; Naseby, D.; Way, J. A. Survival and Ecological Fitness of *Pseudomonas fluorescens* Genetically Engineered with Dual Biocontrol Mechanisms. *Microbiol. Ecol.* **2004**, *48*, 349–357.

Baker, K. F.; Cook R. J. *Biological Control of Plant Pathogen*; Willey Freeman: San Francisco, 1974, Reprinted *Edn. Am. Phytopathol. Soc.*; St. Paul, MN, USA, pp 433.

Baker, K. F.; Snyder, W. C. *Ecology of Soil-Borne Plant Pathogens, Prelude to Biological Control*; University of California Press: Berkeley, Los Angeles, 1965; pp 571.

Balfour, E. B. *The Living Soil*; Faber and Faber: London, 1943.

Bandyopadhyay, R.; Cardwel, K. F. Species of *Trichoderma* and *Aspergillus* as Biological Control Agents Against Plant Diseases in Africa. In *Biological Control in IPM Systems in Africa*; Neuenschwander, P., Borgemeister, C., Langewald, J., Eds.; CABI: Oxfordshire, United Kingdom 2003; pp 193–206.

Baniasadi, F.; Shahidi Bonjar, G. H.; Baghizadeh, A.; Karimi Nik, A.; Jorjandi, M.; Aghighi, S.; Rashid, F. P. Biological Control of *Sclerotinia sclerotiorum*, Causal Agent of Sunflower Head and Stem Rot Disease, by Use of Soil Borne Actinomycetes Isolates. *Am. J. Agric. Biol. Sci.* **2009**, *4*, 146–151.

Bargabus, R. L.; Zidack, N. K.; Sherwood, J. W.; Jacobsen, B. J. Characterization of Systemic Resistance in Sugar Beet Elicited by a Non-pathogenic, Phyllosphere

Colonizing *Bacillus mycoides*, Biological Control Agent. *Physiol. Mol. Plant Pathol.* **2002**, *61*, 289–298.

Darkai-Golan, R. *Postharvest Diseases of Fruit and Vegetables: Development and Control*; Elsevier Sciences: Amsterdam, The Netherlands; 2001.

Baron, S. S.; Teranova, G.; Rowe, J. J. Molecular Mechanism of the Antimicrobial Action of Pyocyanin. *Curr. Microbiol.* **1997**, *18*, 223–230.

Batta, Y. A. Control of Postharvest Diseases of Fruit with an Invert Emulsion Formulation of *Trichoderma harzianum* Rifai. *Postharvest Biol. Technol.* **2007**, *43*(1), 143–150.

Beauverie, J. Essais d'immunisation des végétaux contre les maladies cryptogamiques. *CR. Acad. Sci. Paris*, **1901**, *133*, 107–110.

Beirner, B. P. Biological Control and Its Potential. *World Rev. Pest Control* **1967**, *6*, 7–20.

Bello, G. M. D.; Monaco, C. I.; Simon, M. R. Biological Control of Seedling Blight of Wheat Caused by *Fusarium graminearum* with Beneficial Rhizosphere Microorganisms. *World J. Microbiol. Biotechnol.* **2002**, *18*, 627–636.

Berg, G.; Frankowski, J.; Bahl, H. In *Biocontrol of Verticillium Wilt in Oilseed Rape by Chitinolytic Serratia plymuthica*, Proceeding of the 10th International Rapeseed Congress, Caberra, Australia, September 26–29, 1999; pp 369–375.

Bernard, P.; Romond, C.; Bhatnagar, T. Biological Control of *Pythium mamillatum* Causing Damping-off of Cucumber Seedlings by a Soil Bacterium, *Bacillus mycoides*. *Microbiol. Res.* **1995**, *150*, 71–75.

Berta, G.; Sampo, S.; Gamalero, E.; Musasa, N.; Lemanceau, P. Suppression of *Rhizoctonia* Root-rot of Tomato by *Glomus mosseae* BEG 12 and *Pseudomonas fluorescens* A6RI is Associated with Their Effect on the Pathogen Growth and on the Root Morphogenesis. *Eur. J. Plant Pathol.* **2005**, *111*, 279–288.

Bharathi, R.; Vivekananthan, R.; Harish, S.; Ramanathan, A.; Samiyappan, R. *Rhizobacteria* Based Bioformulations for the Management of Fruit Rot Infection in Chillies. *Crop Prot.* **2004**, *23*, 835–843.

Bodker, L.; Kjoller, R.; Rosendahl, S. Effect of Phosphate and the Arbuscular Mycorrhizal Fungus *Glomus intraradices* on Disease Severity of Root Rot of Peas (*Pisum sativum*) Caused by *Aphanomyces euteiches*. *Mycorrhiza.* **1998**, *8*(3), 169–174.

Bolwerk, A.; Lagopodi, A. L.; Wijfjes, A. H. M.; Lamers, G. E. M.; Chin-A-Woeng, T. F. C.; Lugtenberg, B. J. J.; Bloemberg, G. V. Interactions in the Tomato Rhizosphere of Two *Pseudomonas* Biocontrol Strains with the Phytophogenic Fungus *Fusarium oxysporum* f. sp. Radicis-lycopersici. *Mol. Plant Microbe. Interact.* **2003**, *11*, 983–993.

Bolwerk, A.; Lagopodi, A. L.; Lugtenberg, B. J. J.; Bloemberg, G. V. Visualization of Interactions Between a Pathogenic and a Beneficial *Fusarium* strain During Biocontrol of Tomato Food and Root rot. *Mol. Plant Microbe. Interact.* **2005**, *18*, 710–721.

Borriss, R.; Chen, X. H.; Rueckert, C.; Blom, J.; Becker, A.; Baumgarth, B.; Fan, B.; Pukall, R.; Schumann, P.; Sproer, C.; Junge, H.; Vater, J.; Puhler, A.; Klenk, H. P. Relationship of *Bacillus amyloliquefaciens* Clades Associated with Strains DSM7T and FZB42T: A Proposal for *Bacillus amyloliquefaciens* subsp. *Amyloliquefaciens* subsp. nov. and *Bacillus amyloliquefaciens* subsp. *Plantarum* subsp. nov. Based on Complete Genome Sequence Comparisons. *Int. J. Syst. Evol. Microbiol.* **2011**, *61*, 1786–1801.

Brede de Haan, J. V. Vorläufige beschreibung von pilzen bei tropischen kulturpflanzen beobachtet. *Bull. Inst. Bot. Buitenzorg.* **1900**, *6*, 11–13.

Brisbane, P. G.; Rovira, A. D. Mechanisms of Inhibition of *Gaeumannomyces graminis* var. *tritici* by Fluorescent Pseudomonads. *Plant Pathol.* **1988**, *37*,104–111.

Brooks, D. S.; Gonzales, C. F.; Apple, D. N.; Filer, T. H. Evaluation of Endophytic Bacteria as Potential Biological Control Agents for Oak Wilt. *Biol. Control.* **1994**, *4*, 373–381.

Bull, C. T. Biological Control. In *Encyclopedia of Plant Pathology*; Malloy, O. C., Murray, T. D., Eds.; J. Wiley and Sons: New York, 2001; Vol. 1, pp 128–135.

Bull, C. T.; Weller, D. M.; Thomashow, L. S. Relationship between Root Colonization and Suppression of *Gaeumannomyces graminis* var. *triciti* by *Pseudomonas fluorescens* strain 2–79. *Phytopathology.* **1991**, *81*, 954–959.

Buysens, S.; Heungens, K.; Poppe, J.; Hofte, M. Involvement of Pyochelin and Pyoverdin in Suppression of *Pythium* Induced Damping-off of Tomato by *Pseudomonas aeruginosa* 7NSK2. *Appl. Environ. Microbiol.* **1996**, *62*, 865–871.

Buysens, S.; Poppe, J.; Hofte, M. Role of Siderophores in Plant Growth Stimulation and Antagonism by *Pseudomonas aeruginosa* 7NSK2. In *Improving Plant Productivity with Rhizosphere Bacteria*; Ryder, M. H., Stephens, P. M., Bowen, G. D., Eds.; CSIRO: Adelaide, 1994; pp 139–141.

Caldwell, R. Fate of Spores of *Trichoderma viride* Pers. Ex. Ft. Introduced in the soil. *Nature* **1958**, *181*, 1144–1145.

Callan, N. W.; Mathre, D. E.; Miller, J. B. Bio-priming Seed Treatment for Biological Control of *Pythium ultimum* Preemergence Damping-off in sh2 Sweet Corn. *Plant Dis.* **1990**, *74*(5), 368–372.

Calvo, J.; Calvente, V.; de Orellano, M. E.; Benuzzi, D.; de Tosetti, M. I. S. Biological Control of Postharvest Spoilage Caused by *Penicillium expansum* and *Botrytis cinerea* in Apple by Using the Bacterium *Rahnella aquatilis. Int. J. Food Microbiol.* **2007**, *113*(3), 251–257.

Calvo, J.; Calvente, V.; Orellano, M.; Benuzzi, D.; Sanz-de-Tosetti, M. I. Improvement in the Biocontrol of Postharvest Diseases of Apples with the Use of Yeast Mixtures. *Biol. Control.* **2003**, *48*(5), 579–593.

Campbell, R. *Biological Control of Microbial Plant Pathogens*. Cambridge University Press, Cambridge, 1989; p 219.

Cartwright, D. K.; Chilton, W. S.; Benson, D. M. Pyrrolnitrin and Phenazine Production by *Pseudomonas cepacia*, Strain 5.5B, a Biocontrol Agent of *Rhizoctonia solani. Appl. Microbiol. Biotechnol.* **1995**, *43*, 211–216.

Cawoy, H.; Bettiol, W.; Fickers, P.; Ongena, M. Bacillus-Based Biological Control of Plant Diseases. In *Pesticides in the Modern World-pesticides and Management*; Stoytcheva, M., Ed.; Intech Open Access Publisher: Croatia, 2011; Vol. I., pp 273–302.

Chakravarty, G.; Kalita, M. C. Comparative Evaluation of Organic Formulations of *Pseudomonas fluorescens* Based Biopesticides and Their Application in the Management of Bacterial Wilt of Brinjal (*Solanum melongena* L.). *Afr. J. Biotechnol.* **2011**, *10*, 7174–7182.

Chalutz, E.; Wilson, C. L. Postharvest Biocontrol of Green and Blue Mold and Sour Rot of Citrus Fruit by *Debaryomyces hansenii. Plant Dis.* **1990**, *74*, 134–137.

Chalutz, E.; Ben-Arie, R.; Droby, S.; Cohen, L.; Weiss, B.; Wilson, C. L. Yeasts as Biocontrol Agents of Postharvest Diseases of Fruit. *Phytoparasitica* **1988**, *16*, 69–75.

Chanchaichaovivat, A.; Ruenwongsa, P.; Panijpan, B. Screening and Identification of Yeast Strains from Fruit and Vegetables: Potential for Biological Control of Postharvest Chilli Anthracnose (*Colletotrichum capscii*). *Biol. Control.* **2007**, *42*(3), 326–335.

Chatterton, S.; Sutton, J. C.; Boland, G. J. Timing *Pseudomonas chlororaphis* Applications to Control *Pythium aphanidermatum*, *Pythium dissotocum*, and Root Rot in Hydroponic Peppers. *Biol. Control.* **2004**, *30*, 360–373.

Chaube, H. S.; Mishra, D. S.; Varshney, S.; Singh, U. S. Biological Control of Plant Pathogens by Fungal Antagonistic: Historical Background, Present Status and Future Prospects. *Annu. Rev. Plant Pathol.* **2002**, *2*, 1–42.

Chernin, L.; Brandis, A.; Ismailov, Z.; Chet, I. Pyrrolnitrin Production by an Enterobacter Agglomerans Strain with a Broad Spectrum of Antagonistic Activity Towards Fungal and Bacterial Phytopathogens. *Curr. Microbiol.* **1996**, *32*, 208–212.

Chester, K. S. The Problem of Acquired Physiological Immunity in Plants. *Q. Rev. Biol.* **1933**, *8*, 275–324.

Chet, I.; Inbar, J. Biological Control of Fungal Pathogens. *Appl. Biochem. Biotechnol.* **1994**, *48*, 37–43.

Chin-A-Woeng, T. F. C.; Bloemberg, G. V.; Mulders, I. H. M.; Dekkers, L. C.; Lugtenberg, B. J. J. Root Colonization by Phenazine-1-carboxamide-producing Bacterium *Pseudomonas chlororaphis* PCL1391 Is Essential for Biocontrol of Tomato Root Rot. *Mol. Plant-Microbe Interact.* **2000**, *12*, 1340–1345.

Chin-A-Woeng, T. F. C.; Bloemberg, G. V.; van der Bij, A. J.; van der Drift, K. M. G. M.; Schripsema, J.; Kroon, B.; Scheffer, R. J.; Keel, C.; Bakker, P. A. H. M.; Tichy, H. V.; de Bruijn, F. J.; Thomas-Oates, J. E.; Lugtenberg, B. J. J. Biocontrol by Phenazine-1-carboxamide-producing *Pseudomonas chlororaphis* PCL1391 of Tomato Root Rot Caused by *Fusarium oxysporum f. sp. radicis-lycopersici. Mol. Plant-Microbe Interact.* **1998**, *11*, 1069–1077.

Claire, V. F.; Laugé, R.; Langin, T. Nonpathogenic Strains of Colletotrichum Lindemuthianum Trigger Progressive Bean Defense Responses During Appressorium-Mediated Penetration. *Appl. Environ. Microbiol.* **2005**, *71*(8), 4761–4770.

Claydon, N.; Allan, M.; Hanson, J. R.; Avont, A. G. Antifungal Alkyl Pyrones of *Trichoderma harzianum. Trans. Br. Mycol. Soc.* **1987**, *88*, 503–513.

Clements, D.; Shrestha, A. *New Dimensions in Agroecology*; Food Products Press: Binghamton, NY; 2004.

Collins, D. P.; Jacobsen, B. J. Optimizing a *Bacillus subtilis* Isolate for Biological Control of Sugar Beet Cercospora Leaf Spot. *Biol. Control.* **2003**, *26*, 153–161.

Conn, V. M.; Walker, A. R.; Franco, C. M. M. Endophytic Actinobacteria Induce Defense Pathways in *Arabidopsis thaliana. Mol. Plant–Microbe Interact.* **2008**, *21*, 208–218.

Constantinescu, F. Extraction and Identification of Antifungal Metabolites Produced by Some *B. subtilis* Strains. *Analele Institutului de Cercetari Pentru Cereale Protectia Plantelor.* **2001**, *31*, 17–23.

Conway, G. R. *The Doubly Green Revolution*; Penguin: London; 1997.

Cook, R. J. Advances in Plant Health Management in the 20th Century. *Annu. Rev. Phytopathol.* **2000**, *38*, 95–116.

Cook, R. J.; Baker, K. F. *The Nature and Practice of Biological Control of Plant Pathogens*; APS Press: St. Paul, Minnesota, USA, 1983.

Cook, R. J.; In *Research Briefing Panel on Biological Control in Managed Ecosystems.* Committee on Science, Engineering, and Public Policy, National Academy of Sciences,

National Academy of Engineering, and Institute of Medicine; National Academy Press: Washington, DC, 1987.

Costa, D. M.; de Erabadupitiya, H. R. U. T. An Integrated Method to Control Postharvest Diseases of Banana Using a Member of the *Burkholderia cepacia* Complex. *Postharvest Biol. Technol.* **2005**, *36*(1), 31–39.

Costa, D. M.; de Subasinghe, S. S. N. S. Antagonistic Bacteria Associated with the Fruit Skin of Banana in Controlling its Postharvest Diseases. *Trop. Sci.* **1998**, *38*(4), 206–212.

Cox, T. S.; Picone, C.; Jackson, W. Research Priorities in Natural Systems Agriculture. In *New Dimensions in Agroecology*; Clements, D., Shrestha, A., Eds.; Food Products Press: Binghamton, NY, 2004.

Cronin, D.; Moënne-Loccoz, Y.; Fenton, A.; Dunne, C.; Dowling, D. N.; Gara, F. O. Ecological Interaction of a Biocontrol *Pseudomonas fluorescens* Strain Producing 2,4-diacetylphloroglucinol with the Soft Rot Potato Pathogen *Erwinia carotovora* subsp. *atroseptica*. *FEMS Microbiol. Ecol.* **1997**, *23*, 95–106.

Dababat, A. E. F. A.; Sikora, R. A. Induced Resistance by the Mutualistic Endophyte, *Fusarium oxysporum* Strain 162, toward *Meloidogyne incognita* on Tomato. Biocontrol Sci. Technol. **2007**, *17*, 969–975.

Dann, E. K.; Deverall, B. J. Activation of Systemic Disease Resistance in Pea by an Avirulent Bacterium or a Benzothiadiazole, but not by a Fungal Leaf Spot Pathogen. *Plant Pathol.* **2000**, *49*, 324–332.

De Meyer, G.; Hofte, M. Salicylic acid Produced by the Rhizobacterium *Pseudomonas aeruginosa* 7NSK2 Induces Resistance to Leaf Infection by *Botrytis cinerea* on Bean. *Phytopathology* **1997**, *87*, 588–593.

De Waard, M. A.; Georgopoulos, S. G.; Hollomon, D. W.; Ishii, H.; Leroux, P.; Ragsdale, N. N.; Schwinn, F. J. Chemical Control of Plant Diseases: Problems and Prospects. *Annu. Rev. Phytopathol.* **1993**, *31*, 403–421.

De Wit, P. J. G. M. Induced Resistance to Fungal and Bacterial Diseases. In *Mechanisms of Resistance to Plant Diseases*; Fraser, R. S. S., Ed.; Nijhoff/Junk: Dordrecht, 1985; pp 405–424.

Dean, R. A.; Kuc, J. Rapid Lignification in Response to Wounding and Infection as a Mechanism for Induced Systemic Protection in Cucumber. *Physiol. Mol. Plant Pathol.* **1987**, *31*, 69–81.

DeBach, P. *Biological Control of Insect Pests and Weeds*; DeBach, P., Ed.; Reinhold: New York, New York, 1964; pp 844.

De-Matos, A. P. Chemical and Microbiological Factors Influencing the Infection of Lemons by *Geotrichum candidum* and *Penicillium digitatum*. Ph.D. Thesis, University of California, Riverside, 1983.

Demoz, B. T.; Korsten, L. *Bacillus subtilis* Attachment, Colonization, and Survival on Avocado Flowers and Its Mode of Action on Stem-End Rot Pathogens. *Biol. Control.* **2006**, *37*(1), 68–74.

Dennis, C.; Webster, J. Antagonistic Properties of Species-groups of *Trichoderma*. I. Production of Non-Volatile Antibiotics, *Trans. Br. Mycol. Soc.* **1971a**, *57*, 25–39.

Dennis, C.; Webster, J. Antagonistic Properties of Species-groups of *Trichoderma*. II. Production of Volatile Antibiotics. *Trans. Br. Mycol. Soc.* **1971b**, *57*, 41–48.

Deora, A.; Hashidoko, Y.; Islam, M. T.; Tahara, S. Antagonistic Rhizoplane Bacteria Induce Diverse Morphological Alterations in Peronosporomycete Hyphae During in vitro Interaction. *Eur. J. Plant Pathol.* **2005**, *112*, 311–322.

Desai, S.; Reddy, M. S.; Kloepper, J. W. Comprehensive Testing of Biocontrol Agents. In *Biological Control of Crop Diseases*; Gnanamanickman, S. S. E., Ed.; Marcel Dekker Inc: New York, 2002; pp 387–420.

Devi, A. N.; Arumugam, T. Studies on the Shelf Life and Quality of Rasthali Banana as Affected by Postharvest Treatments. *Orissa J. Hortic.* **2005**, *33*(2), 3–6.

Doley, K.; Jite, P. K. Effect of Arbuscular Mycorrhizal Fungi on Growth of Groundnut and Disease Caused by *Macrophomina phaseolina. J. Exp. Sci.* **2012**, *3*(9), 46–50.

Dong, Y-H.; Zhang, X-F.; Xu, J-L.; Zhang, L-H. Insecticidal *Bacillus* Thuringiensis Silences *Erwinia carotovora* Virulence by a New Form of Microbial Antagonism, Signal Interference. *Appl. Environ. Microbiol.* **2004**, *70*(2), 954–960.

Dordas C. Role of Nutrients in Controlling Plant Diseases in Sustainable Agriculture. In *Sustainable Agriculture*; Lichtfouse, E., Navarrete, M., Debaeke, P., Véronique, S., Alberola, C., Eds.; Springer Science + Business Media: New York, 2009; pp 443–460.

Duijff, B. J.; Pouhair, D.; Olivain, C.; Alabouvette, C.; Lemanceau, P. Implication of Systemic Induced Resistance in the Suppression of Fusarium Wilt of Tomato by *Pseudomonas fluorescens* WCS417r and by Nonpathogenic *Fusarium oxysporum* Fo47. *Eur. J. Plant Pathol.* **1998**, *104*, 903–910.

Dunne, C.; Crowley, J. J.; Moenne-Loccoz, Y.; Dowling, D. N.; de Bruijn, F. J.; O'Gara, F. Biological Control of Pythium Ultimum by *Stenotrophomonas maltophilia* W81 Is Mediated by an Extracellular Proteolytic Activity. *Microbiology* **1997**, *143*, 3921–3391.

Dunne, C.; Moenne-Loccoz, Y.; McCarthy, J.; Higgins, P.; Powell, J.; Dowling, D. N.; O'Gara, F. Combining Proteolytic and Phloroglucinol-Producing Bacteria for Improved Biocontrol of Pythium Mediated Damping-Off of Sugar Beet. *Plant Pathol.* **1998**, *47*, 299–307.

Egamberdieva, D.; Jabborova, D. Biocontrol of Cotton Damping-Off Caused by *Rhizoctonia solani* in Salinated Soil with Rhizosphere Bacteria. *Asian Aust. J. Plant Sci. Biotechnol.* **2013**, *7*(2), 31–38.

Eilenberg, J. Concepts and Visions of Biological Control. In *An Ecological and Societal Approach to Biological Control*; Eilenberg, J., Hokkanen, H. M. T., Eds.; Springer: Dordrecht, the Netherlands, 2006; pp 1–11.

Eilenberg, J.; Hajek, A.; Lomer, C. Suggestions for Unifying the Terminology in Biological Control. *Biol. Control.* **2001**, *46*, 387–400.

Elad, Y.; Chet, I. Practical Approaches for Biocontrol Implementation. In *Novel Approaches to Integrated Pest Management*; Reuveni, R., Ed.; CRC Press: Boca Raton, FL, 1995; pp 369.

El-Banna, N.; Winkelmann, G. Pyrrolnitrin from *Burkholderia cepacia*: Antibiotic Activity against Fungi and Novel Activities Against Streptomycetes. *J. Appl. Microbiol.* **1988**, *85*, 69–78.

El-Gamal, N. G.; Abd-El-Kareem, F.; Fotouh Y. O.; El-Mougy, N. S. Induction of Systemic Resistance in Potato Plants against Late and Early Blight Diseases Using Chemical Inducers Under Greenhouse and Field Conditions. *Res. J. Agric. Biol. Sci.* **2007**, *3*(2), 73–81.

El-Neshawy, S. M.; El-Sheikh, M. M. Control of Green Mold on Oranges by *Candida oleophila* and Calcium Treatments. *Ann. Agric. Sci. Cairo.* **1998**, *3*(Special Issue), 881–890.

El-Tarabily, K. A. Promotion of Tomato (*Lycopersicon esculentun* mill.) Plant Growth by Rhizosphere Competent 1-aminocyclopropane-1-carboxylic Acid Deaminase-producing *Streptomycete actinomycetes*. *Plant Soil* **2008**, *308*, 161–174.

Erlacher, A.; Cardinale, M.; Grosch, R.; Grube, M.; Berg, G. The Impact of the Pathogen *Rhizoctonia solani* and Its Beneficial Counterpart *Bacillus amyloliquefaciens* on the Indigenous Lettuce Microbiome. *Front. Microbiol.* **2014**, *5*(175), 1–8.

Errakhi, R.; Bouteau, F.; Lebrihi, A.; Barakate, M. Evidences of Biological Control Capacities of *Streptomyces* spp. Against *Sclecrotium rolfsii* Responsible for Daming-Off Disease in Sugar Beet (*Beta vulgaris* L.). *World J. Microbiol. Biotechnol.* **2007**, *23*, 1503–1509.

Erwin, D. C.; Ribeiro, O. K. *Phytophthora Diseases Worldwide*; APS Press: St Paul, Minnesota, USA, 1996.

Fan, Q.; Tian, S. P. Postharvest Biological Control of Grey Mold and Blue Mold on Apple by *Cryptococcus albidus* (Saito) Skinner. *Postharvest Biol. Technol.* **2001**, *21*(3), 341–350.

Felix, G.; Duran, J. D.; Volko, S.; Boller, T. Plants have a Sensitive Perception System for the Most Conserved Domain of Bacterial Flagellin. *Plant J.* **1999**, *18*, 265–276.

Feng, D. X.; Tasset, C.; Hanemian, M.; Barlet, X.; Hu, J.; Trémousaygue, D.; Deslandes, L.; Marco, Y. Biological Control of Bacterial Wilt in *Arabidopsis thaliana* Involves Abscisic Acid Signalling. *New Phytol.* **2012**, *194*, 1035–1045.

Filnow, A. B.; Lockwood, J. L. Evaluation of Several Actinomycetes and the Fungus *Hypochytrium catenoids* as Biocontrol Agents of *Phytophthora* Root Rot of Soybean. *Plant Dis.* **1985**, *69*, 1033–1036.

Fiorilli, V. Catoni, M. Francia, D. Cardinale, F.; Lanfranco, L. The Arbuscular Mycorrhizal Symbiosis Reduces Disease Severity in Tomato Plants Infected by *Botrytis cinerea. J. Plant Pathol.* **2011**, *93*(1), 237–242.

Flaishman, M.; Eyal, Z.; Voisard, C.; Haas, D. Suppression of Septoria triticii by Phenazine or Siderophore-deficient Mutants of *Pseudomonas. Curr. Microbiol.* **1990**, *20*, 121–124.

Folman, L. B.; De Klein, M. J. E. M.; Postma, J.; van Veen, J. A. Production of Antifungal Compounds by Lysobacter Enzymogenes Isolate 3.1 T8 Under Different Conditions in Relation to its Efficacy as a Biocontrol Agent of *Pythium Aphanidermatum* in Cucumber. *Biol. Control.* **2004**, *31*, 145–154.

Folman, L. B.; Postma, J.; Van Veen, J. A. Characterization of Lysobacter Enzymogenes (Chtistensen and Cook 1978) Strain 3.1 T8, a Powerful Antagonist of Fungal Diseases of Cucumber. *Microbiol. Res.* **2003**, *158*, 107–115.

Forsyth, L. M.; Smith, L. J.; Aitken, E. A. B. Identification and Characterisation of Non-Pathogenic *Fusarium oxysporum* Capable of Increasing and Decreasing Fusarium Wilt Severity. *Mycol. Res.* **2006**, *110*, 929–935.

Foster, R. C.; Rovira, A. D. The Ultrastructure of the Rhizosphere of *Trifolium subterraneum* L. In *Microbial Ecology*; Loutit, M. W., Miles, J. A. R., Eds.; Springer: Berlin, 1978; pp 278–290.

Foster, W.; Raoult, A. Early Descriptions of Antibiosis. *J. R. Coll. Gen. Pract. Occas Pap.* **1974**, *24*(149), 889–894.

Fravel, D. Role of Antibiosis in the Biocontrol of Plant Diseases. *Annu. Rev. Phytopathol.* **1988**, *26*, 75–91.

Fuchs, J. G.; Moënne-Loccoz, Y.; Défago, G. Non-pathogenic *Fusarium oxysporum* Strain Fo47 Induces Resistance to Fusarium wilt in Tomato. *Plant Dis.* **1997,** *81,* 492–496.

Furuya, S.; Mochizuki, M.; Aoki, Y.; Kobayashi, H.; Takayanagi, T.; Shimizu, M.; Suzuki, S. Isolation and Characterization of *Bacillus subtilis* KS1 for the Biocontrol of Grapevine Fungal Diseases. *Biol. Sci. Technol.* **2011,** *21,* 705–720.

Gamagae, S. U.; Sivakumar, D.; Wilson-Wijeratnam, R. S.; Wijesundra, R. L. C. Use of Sodium Bicarbonate and Candida Oleophila to Control Anthracnose in Papaya During Storage. *Crop Prote.* **2003,** *22*(5), 775–779.

Garrett, K. A. Climate Change and Plant Disease Risk. In *Global Climate Change and Extreme Weather Events: Understanding the Contributions to Infectious Disease Emergence*; Relman, D. A., Hamburg, M. A., Choffnes, E. R., Mack, A., Eds.; National Acade. Press: Washington, DC, 2008; pp 143–155.

Georgakopoulos, D. G.; Fiddaman, P.; Leifert, C.; Malathrakis, N. E. Biological Control of Cucumber and Sugar Beet Damping-Off Caused by Pythium Ultimum with Bacterial and Fungal Antagonists. *J. Appl. Microbiol.* **2002,** *92,* 1078–1086.

Georgakopoulos, D. G.; Hendson, M.; Panopoulos, N. J.; Schroth, M. N. Analysis and Expression of a Phenazine Biosynthesis locus of *Pseudomonas aureofaciens* PGS12 on Seeds with a Mutant Carrying a Phenazine Biosynthesis Locus-Ice Nucleation Reporter Gene Fusion. *Appl. Environ. Microbiol.* **1994,** *60,* 4573–4579.

Ghisalberti, E. L.; Sivasithamparam, K. Antifungal Antibiotics Produced by *Trichoderma* spp. *Soil Biol. Biochem.* **1991,** *23,* 1011–1020.

Giovannetti, M.; Tosi, L.; Tore, G. D.; Zazzerini, A. Histological, Physiological and Biochemical Interactions Between Vesicular-arbuscular Mycorrhizae and *Thielaviopsis basicola* in Tobacco Plants. *J. Phytopathol.* **1991,** *131*(4), 265–274.

Gliessman, S. R. Agroecology and agroecosystems. In *The Earthscan Reader in Sustainable Agriculture*; Pretty, J., Ed.; Earthscan: London, 2005.

Gnanamanickam, S. S.; Mew, T. W. Biological Control of Blast Disease of Rice (*Oryza sativa* L.) with Antagonistic Bacteria and Its Mediation by a *Pseudomonas* Antibiotic. *Ann. Phytopathol. Soc. Jpn.* **1992,** *58,* 380–385.

Gnanamanickam, S. S.; Vasudevan, P.; Reddy, M. S.; Kloepper, J. W.; Defago, G. Principles of Biological Control. In *Biological Control of Crop Diseases*; Gnanamanickam, S. S., Ed.; Marcel Dekker, Inc.: New York, NY, 2002.

Goddard, V. J.; Bailey, M. J.; Darrah, P.; Lilley, A. K.; Thompson, I. P. Monitoring Temporal and Spatial Variation in Rhizosphere Bacterial Population Diversity: A Community Approach for the Improved Selection of Rhizosphere Competent Bacteria. *Plant Soil* **2001,** *232,* 181–193.

Godfray, H. C. J.; Beddington, J. R.; Crute, J. I.; Haddad, L.; Lawrence, D.; Muir, J. F.; Pretty, J.; Robinson, S.; Thomas, S.; Toulmin, C. Food Security: The Challenge of Feeding 9 Billion People. *Science* **2010,** *327,* 812–818.

Gopalakrishnan, S.; Humayun, P.; Kiran, B. K.; Kannan, I. G. K.; Vidya, M. S.; Deepthi, K.; Rupela, O. Evaluation of Bacteria Isolated from Rice Rhizosphere for Biological Control of Charcoal Rot of Sorghum Caused by *Macrophomina phaseolina* (Tassi) Goid. *World J. Microbiol. Biotechnol.* **2011a,** *27,* 1313–1321.

Gopalakrishnan, S.; Pande, S.; Sharma, M.; Humayun, P.; Kiran, B. K.; Sandeep, D.; Vidya, M. S.; Deepthi, K.; Rupela, O. Evaluation of Actinomycete Isolates Obtained from Herbal Vermicompost for Biological Control of Fusarium wilt of Chickpea. *Crop Pro.* **2011b**, *30*, 1070–1078.

Govender, V.; Korsten, L.; Sivakumar, D. Semi-commercial Evaluation of *Bacillus licheniformis* to Control Mango Postharvest Diseases in South Africa. *Postharvest Biol. Technol.* **2005**, *38*(1), 57–65.

Gravel, V.; Martinez, C.; Antoun, H.; Tweddell, R. J. Antagonist Microorganisms with the Ability to Control Pythium Damping-off Tomato Seeds in Rockwool. *Biol. Control.* **2005**, *50*, 771–786.

Gray, E. J.; Smith, D. L. Intracellular and Extracellular PGPR: Commensalities and Distinctions in the Plant-bacterium Signaling Processes. *Soil Biol. Biochem.* **2005**, *37*, 395–412.

Griesbach, E.; Eisbein, K.; Kramer, I.; Muller, J.; Volksh, B. Induction of Resistance to Bacterial Pathogens in the Pathosystem Tomato/*Clavibacter michiganensis* sub sp. Michiganensis. I. Characterization of the Resistance Induction. *J. Plant Prot.* **2000**, *107*, 449–463.

Grulke, N. E. The Nexus of Host and Pathogen Phenology: Understanding the Disease Triangle with Climate Change. *New Phytol.* **2011**, *189*, 8–11.

Guijarro, B.; Melgarejo, P.; Torres, R.; Lamarca, N.; Usall, J.; de Cal, A. Effects of Different Biological Formulations of Penicillium Frequentans on Brown Rot of Peaches. *Biol. Control.* **2007**, *42*(1), 86–96.

Guillemin, J. P.; Gianinazzi, S.; Gianinazzipearson, V.; Marchal, J. Contribution of Arbuscular Mycorrhizas to Biological Protection of Micropropagated Pineapple (*Ananas comosus* (L) Merr) against *Phytophthora cinnamomi* Rands. *Agric. Sci. Finland* **1994**, *3*, 241–251.

Gurusiddaiah, S.; Weller, D. M.; Sarkar, A.; Cook, R. J. Characterization of an Antibiotic Produced by a Strain of *Pseudomonas fluorescens* Inhibitory to *Gaeumannomyces Graminis* var. tritici and *Pythium* spp. *Antimicrob. Agents Chemother.* **1986**, *29*, 488–495.

Hammerschmidt, R. Introduction: Definitions and Some History. In *Induced Resistance for Plant Disease Control: A Sustainable Approach to Crop Protection*; Walters, D., Newton, A., Lyon, G., Eds.; Blackwell Publishing: Oxford, 2007; pp 1–8.

Hammerschmidt, R.; Kuc, J. *Induced Resistance to Disease in Plants*; Kluwer Academic Publishers: Dordrecht, 1995.

Hammerschmidt, R.; Lamport, D. T. A.; Muldoon, E. Cell Wall Hydroxyproline Enhancement and Lignin Deposition as an Early Event in the Resistance of Cucumber to *Cladosporium cucumerinum. Physiol. Plant Pathol.* **1984**, *24*, 43–47.

Hammerschmidt, R.; Metraux, J. P.; Van Loon, L. C. Inducing Resistance: A Summary of Papers Presented at the First International Symposium on Induced Resistance to Plant Diseases, Corfu. *Eur. J. Plant Pathol.* **2001**, *107*, 1–6.

Handelsman, J. Future Trends in Biocontrol. In *Biological Control of Crop Diseases*; Gnanamanickam, S. S., Ed.; Marcel Dekker, Inc.: New York, NY, 2002.

Handelsman, J.; Raffel, S. J.; Mester, E. H.; Wunderlich, L.; Grau, C. R. Biological Control of Damping-off of Alfalfa Seedlings with *Bacillus cereus* UW85. *Appl. Environ. Microbiol.* **1990**, *56*, 713–718.

Harman, G. E.; Kubicek, C. P. *Trichoderma and Gliocladium*; Taylor and Francis: London, 1998; pp 278.

Harman, G. E.; Howell, C. R.; Viterbo, A.; Chet, I.; Lorito, M. *Trichoderma* Species—Opportunistic, Avirulent Plant Symbionts. *Nat. Rev. Microbiol.* **2004**, *2*, 43–56.

Hashemi, S. G.; Rouhani, H.; Tarighi, S. Study of Interaction Between Mycorrhiza *Glomus fasciculatum* and *Pseudomonas fluorescens* on Control of Common Root Rot of Wheat Caused by *Bipolaris sorokiniana. J. Plant Prot.* **2013**, *27*(2), 13.

He, H.; Silo-Suh, L. A.; Handelsman, J.; Clardy, J. Zwittermicin A, an Antifungal and Plant Protection Agent from *Bacillus cereus. Tetrahedron Lett.* **1994**, *35*, 2499–2502.

Hemavati, C.; Thippeswamy, B. Effect of Arbuscular Mycorrhizal Fungi, *Acaluspora lacunospora* on Growth of Groundnut Disease Caused by *Cercospora arachidicola. Int. J. Res. Appl. Nat. Soc. Sci.* **2014**, *2*(4), 57–60.

Heng, J. L. S.; Shah, U. K. M.; Rahman, N. A. A.; Shaari, K.; Hamzah, H. *Streptomyces ambofaciens* S2—A Potential Biological Control Agent for *Colletotrichum gleosporioides* the Causal Agent for Anthracnose in Red Chilli Fruits. *J. Plant Pathol. Microbiol.* **2015**, *1*(Special Issue), 1–6.

Hillocks, R. J. Cross Protection between Strains of *Fusarium oxysporum* f. sp. *vasinfectum* and its Effect on Vascular Resistance Mechanisms. *J. Phytopathol.* **1986**, *117*, 712–715.

Hoitink, H. A.; Fahy, P. C. Basis for the Control of Soilborne Plant Pathogens with Composts. *Annu. Rev. Phytopathol.* **1986**, *24*, 93–114.

Howell, C. R. The Role of Antibiosis in Biocontrol. In *Trichoderma and Gliocladium. Enzymes, Biological Control and Commercial Application*; Harman, G. E., Kubicek, C. P., Eds.; Taylor and Francis Ltd.: London, 1998; Vol. 2, pp 173–183.

Howell, C. R.; Stipanovic, R. D. Control of *Rhizoctonia solani* on Cotton Seedlings with *Pseudomonas fluorescens* and with an Antibiotic Produced by the Bacterium. *Phytopathology* **1979**, *69*, 480–482.

Howell, C. R.; Stipanovic, R. D. Suppression of *Pythium ultimum*-induced Damping off of Cotton Seedlings by *Pseudomonas fluorescens* and its Antibiotic, Pyoluteorin. *Phytopathology* **1980**, *70*, 712–715.

Huang, C. H.; Vallad, G. E.; Zhang, S.; Wen, A.; Balogh, B.; Figueiredo, J. F. L.; Behlau, F.; Jones, J. B.; Momol, M. T.; Olson, S. M. Effect of Application Frequency and Reduced Rates of Acibenzolar-S-methyl on the Field Efficacy of Induced Resistance Against Bacterial Spot on Tomato. *Plant Dis.* **2012**, *96*, 221–227.

Huang, C.-J.; Yang, K.-H.; Liu, Y.-H.; Lin, Y.-J.; Chen, C.-Y. Suppression of Southern Corn Leaf Blight by a Plant Growth-promoting Rhizobacterium *Bacillus cereus* C1L. *Ann. Appl. Biol.* **2010**, *157*, 45–53.

Huang, Y.; Daverall, B. J.; Morris, S. C.; Wild, B. L. Biocontrol of Postharvest Orange Diseases by a Strain of *Pseudomonas cepacia* Under Semi Commercial Conditions. *Postharvest Biol. Technol.* **1993**, *3*(4), 293–304.

Huang, Y.; Deverall, B. J.; Morris, S. C. Postharvest Control of Green Mold on Oranges by a Strain of *Pseudomonas glathei* and Enhancement of Its Biocontrol by Heat Treatment. *Postharvest Biol. Technol.* **1995**, 13, 129–137.

Huang, Z. Y.; Bonsall, R F.; Mavrodi, D. V.; Weller D. M, Thomashow L. S. Transformation of *Pseudomonas fluorescens* with Genes for Biosynthesis of Phenazine-1-carboxylic Acid Improves Biocontrol of *Rhizoctonia* Root Rot. *FEMS Microbiol. Ecol.* **2004**, *49*, 243–251.

Hussain, S.; Ghaffar, A.; Aslam, M. Biological Control of *Macrophomina phaseolina* Charcoal Rot of Sunflower and Mung Bean. *J. Phytopathol.* **1990**, *130*(2), 157–160.

Hutchinson, S. W. Current Concepts of Act We Defence in Plants. *Annu. Rev. Phytopathol.* **1998,** *36,* 59–90.

Iavicoli, A.; Boutet, E.; Buchala, A.; Metraux, J. P. Induced Systemic Resistance in *Arabidopsis thaliana* in Response to Root Inoculation with *Pseudomonas fluorescens* CHA0. *Mol. Plant Microbe. Interact.* **2003,** *16,* 851–858.

Indira, N.; Gayatri, S. Management of Blackgram Root Rot Caused by *Macrophomina phaseolina* by Antagonistic Microorganisms. *Madras Agric. J.* **2003,** *90,* 490–494.

Iqbal, S. H.; Shahbaz, R.; Khalid, A. N.; Khan, M. The Influence of a Vesicular-arbuscular Mycorrhiza (VAM) and *Aspergillus niger* as Deterrents Against *Rhizoctonia solani* in Potatoes. *Sarhad J. Agric.* **1990,** *6*(5), 481–484.

Islam, M. T.; Fukushi, Y. Growth Inhibition and Excessive Branching in *Aphanomyces cochlioides* Induced by 2,4-diacetylphloroglucinol Is Linked to Disruption of Filamentous Actin Cytoskeleton in the Hyphae. *World J. Microbiol. Biotechnol.* **2010,** *26,* 1163–1170.

Islam, M. T.; von Tiedemann, A.; Laatsch, H. Protein Kinase C Is Likely to be Involved in Zoosporogenesis and Maintenance of Flagellar Motility in the Peronosporomycete Zoospores. *Mol. Plant–Microbe Interact.* **2011,** *24,* 938–947.

Isono, K.; Nagatsu, J.; Kawashima, Y.; Suzuki, S. Studies on Polyoxins, Antifungal Antibiotics. Part I. Isolation and Characterization of Polyoxins A and B. *Agric. Biol. Chem.* **1965,** *29,* 848–854.

Janisiewicz, W. J.; Roitman, J. Biological Control of Blue Mold and Gray Mold on Apple and Pear with *Pseudomonas cepacia. Phytopathology* **1988,** *78,* 1697–1700.

Jayaraman, J.; Kumar, D. VAM Fungi-pathogen-fungicide Interactions in Gram. *Indian Phytopathol.* **1995,** *48*(3), 294–299.

Jetiyanon, K.; Fowler, W. D.; Kloepper, J. W. Broad Spectrum Protection against Several Pathogens by PGPR Mixtures under Field Conditions. *Plant Dis.* **2003,** *87,* 1390–1394.

Jeun, Y. C.; Park, K. S.; Kim, C. H.; Fowler, W. D.; Kloepper, J. W. Cytological Observations of Cucumber Plants During Induced Resistance Elicited by *Rhizobacteria. Biol. Control.* **2004,** *29,* 34–42.

Ji, G.-H.; Wei, L.-F.; He, Y.-Q; Wu, Y.-P.; Bai, X.-H. Biological Control of Rice Bacterial Blight by *Lysobacter antibioticus* Strain 13–1. *Biol. Control.* **2008,** *45,* 288–296.

Jiang, Y. M.; Chen, F.; Li, Y. B.; Liu, S. X. A Preliminary Study on the Biological Control of Postharvest Diseases of Litchi Fruit. *J. Fruit Sci.* **1997,** *14*(3), 185–186.

Jiang, Y. M.; Zhu, X. R.; Li, Y. B. Postharvest Control of Litchi Fruit Rot by *Bacillus subtilis. Lebensm. Wiss. Technol.* **2001,** *34*(7), 430–436.

Jochum, C. C.; Osborne, L. E.; Yuen, G. Y. *Fusarium* Head Blight Biological Control with *Lysobacter enzymogenes* Strain C3. *Biol. Control.* **2006,** *39,* 336–344.

Johansson, P. M.; Johnsson, L.; Gerhardson, B. Suppression of Wheat-seedling Diseases Caused by *Fusarium culmorum* and *Microdochium nivale* Using Bacterial Seed Treatment. *Plant Pathol.* **2003,** *52,* 219–227.

Jones, E. E.; Mead, A.; Whipps, J. M. Effect of Inoculum Type and Timing of Application of *Coniothyrium minitans* on *Sclerotinia sclerotiorum*: Control of Sclerotinia Disease in Glasshouse Lettuce. *Plant Pathol.* **2004a,** *53,* 611–620.

Jones, E. E.; Clarkson, J. P.; Mead, A.; Whipps, J. M. Effect of Inoculum Type and Timing of Application of *Coniothyrium minitans* on *Sclerotinia sclerotiorum*: Influence on Apothecial Production. *Plant Pathol.* **2004b,** *53,* 621–628.

Kaewchai, S.; Soytong, K.; Hyde, K. D. Mycofungicides and Fungal Biofertilizers, *Fungal Divers* **2009**, *38*, 25–50.

Kageyama, K.; Nelson, E. B. Differential Inactivation of Seed Exudate Stimulation of *Pythium ultimum* Sporangium Germination by *Enterobacter cloacae* Influences Biological Control Efficacy on Different Plant Species. *Appl. Environ. Microbiol.* **2003**, *69*, 1114–1120.

Kamalakaman, A.; Mohan, L.; Kavitha, K.; Harish, S.; Radjacommare, R.; Nakkeeran, S.; Parthiban, V. K.; Karuppiah, R.; Angayarkanni, T. Enhancing Resistance to Stem and Stolen Rot of Peppermint (*Mentha piperita* Lin.) Using Biocontrol Agents. *Acta Phytopathol. Ent. Hung.* **2003**, *38*, 293–305.

Kanjanamaneesathian, M.; Wiwattanapatapee, R.; Pengnoo, A.; Oungbho, K.; Chumthong, A. Efficacy of Novel Formulations of *Bacillus megaterium* in Suppressing Sheath Blight of Rice Caused by *Rhizoctonia solani*. *Plant Pathol. J.* **2007**, *6*(2), 195–201.

Karabulut, O. A.; Baykal, N. Biological Control of Postharvest Diseases of Peaches and Nectarines by Yeasts. *J. Phytopathol.* **2003**, *151*(3), 130–134.

Karabulut, O. A.; Baykal, N. Integrated Control of Postharvest Diseases of Peaches with a Yeast Antagonist, Hot Water and Modified Atmosphere Packaging. *Crop Prot.* **2004**, *23*(5), 431–435.

Karabulut, O. A.; Arslan, U.; Kadir, I.; Gul, K. Integrated Control of Post Harvest Diseases of Sweet Cherry with Yeast Antagonist and Sodium Bicarbonate Applications within a Hydrocooler. *Postharvest Biol. Technol.* **2005**, *37*, 135–141.

Karabulut, O. A.; Smilanick, J. L.; Gabler, F. M.; Mansour, M.; Droby, S. Nearharvest Applications of *Metschnikowia fructicola*, Ethanol, and Sodium Bicarbonate to Control Postharvest Diseases of Grape in Central California. *Plant Dis.* **2003**, *87*(11), 1384–1389.

Karthikeyan, M.; Bhaskaran, R.; Radhika, K.; Mathiyazhahan, S.; Jayakumar, V.; Sandoss-kumar, R.; Velazhahan, R. Endophytic *Pseudomonas fluorescens* Endo2 and Endo35 Induce Resistance in Black Gram [*Vigna mungo* (L.) Hepper] to the Pathogen *Macrophomina phaseolina*. *J. Plant Interact.* **2005**, *1*, 135–143.

Keel, C.; Schnider, U.; Maurhofer, M.; Voisard, C.; Laville, J.; Burger, U.; Wirthner, P.; Haas D.; Defago, G. Suppression of Root Diseases by *Pseudomonas fluorescens* CHAO: Importance of the Bacterial Secondary Metabolite 2,4-diacetylphloroglucinol. *Mol. Plant-Microbe Interact.* **1992**, *5*, 4–13.

Kefialew, Y.; Ayalew, A. Postharvest Biological Control of Anthracnose (*Colletotrichum gloeosporioides*) on Mango (*Mangifera indica*). *Postharvest Biol. Technol.* **2008**, *50*(1), 8–11.

Kerr, A. The Impact of Molecular Genetics on Plant Pathology. *Annu. Rev. Phytopathol.* **1987**, *25*, 87–110.

Kerry, B. R. Rhizosphere Interactions and the Exploitation of Microbial Agents for the Biological Control of Plant-Parasitic Nematodes. *Annu. Rev. Phytopathol.* **2000**, *38*, 423–441.

Khalimi, K.; Suprapta, D. N. Induction of Plant Resistance Against Soybean Stunt Virus Using Some Formulations of *Pseudomonas aeruginosa*. *J. ISSAAS*. **2011**, *17*, 98–105.

Khan, A.; Sutton, J. C.; Grodzinski, B. Effects of *Pseudomonas chlororaphis* on *Pythium aphanidermatum* and Root Rot in Peppers Grown in Small-Scale Hydroponic Troughs. *Biocontrol Sci. Technol.* **2003**, *13*, 615–630.

Khan, M. R. Beneficial Bacteria for Biological Control of Fungal Pathogens of Cereals. In *Bacteria in Agrobiology: Disease Management*; Maheshwari, D. K., Ed.; Springer: Heidelberg, New York, Dordrecht, London, 2013; pp 153–165.

Khan, M. R.; Doohan, F. M. Bacterium-mediated Control of *Fusarium* Head Blight Disease of Wheat and Barley and Associated Mycotoxin Contamination of Grain. *Biol. Control.* **2009**, *48*(1), 42–47.

Khan, M. R.; Fischer, S.; Egan, D.; Doohan, F. M. Biological Control of *Fusarium* Seedling Blight Disease of Wheat and Barley. *Phytopathology* **2006**, *96*, 386–394.

Khan, M. R.; Khan, S. M.; Mohiddin, F. A. Biological Control of Fusarium wilt of Chickpea Through Seed Treatment with the Commercial Formulation of *Trichoderma harzianum* and/or *Pseudomonas fluorescens*. *Phytopathol. Mediterr.* **2004**, *43*, 20–25.

Khan, M. R.; O'Brien, E.; Carney, B. F.; Doohan, F. M. A Fluorescent Pseudomonad Shows Potential for the Control of Net Blotch Disease of Barley. *Biol. Control.* **2010**, *54*, 41–45.

Kildea, S.; Ransbotyn, V.; Khan, M. R.; Fagan, B.; Leonard, G.; Mullins, E.; Doohan, F. M. *Bacillus megaterium* Shows Potential for the Biocontrol of Septoria Tritici Blotch of Wheat. *Biol. Control.* **2008**, *47*, 34–45.

Kim, B. S.; Lee, J. Y.; Hwang, B. K. In vivo Control and in vitro Antifungal Activity of Rhamnolipid B, a Glycolipid Antibiotic, Against *Phytophthora capsici* and *Colletotrichum orbiculare*. *Pest Manage. Sci.* **2000**, *56*, 1029–1035.

Kim, B. S.; Moon, S. S.; Hwang, B. K. Isolation, Identification and Antifungal Activity of a Macrolide Antibiotic, Oligomycin A, Produced by *Streptomyces libani*. *Can. J. Bot.* **1999**, *77*, 850–858.

King, E. B.; Parke, J. L. Biocontrol of Aphanomyces Root Rot and *Pythium* Damping-Off by *Pseudomonas cepacia* AMMD on Four Pea Cultivars. *Plant Dis.* **1993**, *77*, 1185–1188.

Kishore, G. K.; Pande, S.; Podile, A. R. Biological Control of Late Leaf Spot of Peanut (*Arachis hypogaea*) with Chitinolytic Bacteria. *Phytopathology* **2005**, *95*, 1157–1165.

Ko, H. S.; Jin, R. D.; Krishnan, H. B.; Lee, S. B.; Kim, K. Y. Biocontrol Ability of *Lysobacter antibioticus* HS124 Against *Phytophthora* Blight Is Mediated by the Production of 4-hydroxyphenylacetic Acid and Several Lytic Enzymes. *Curr. Microbiol.* **2009**, *59*, 608–615.

Kota, V. R.; Kulkarni, S.; Hegde, Y. R. Postharvest Diseases of Mango and Their Biological Management. *J. Plant Dis. Sci.* **2006**, *1*(2), 186–188.

Krupinsky, J. M.; Bailey K. L.; McMullen, M. P.; Gossen, B. D.; Turkington, T. K. Managing Plant Disease Risk in Diversified Cropping Systems. *Agron. J.* **2002**, *94*, 198–209.

Kuc, J. Concepts and Direction of Induced Systemic Resistance in Plants and Its Application. *Eur. J. Plant Pathol.* **2001**, *107*, 7–12.

Kumar, C. P. C.; Garibova, I. V.; Vellickanov, L. L.; Durinina, E. P. Biocontrol of Wheat Root Rots Using Mixed Cultures of *Trichoderma viride* and *Glomus epigaeus*. *Indian J. Plant Prot.* **1993**, *21*(2), 145–148.

Kumar, R. N.; Mukerji, K. G. Integrated Disease Management Future Perspectives. In *Advances in Botany*; Mukerji, K. G., Mathur, B., Chamala, B. P., Chitralekha, C. Eds.; APH Publishing Corporation: New Delhi, 1996; pp 335–347.

Lahlali, R.; Serrhini, M. N.; Jijakli, M. H. Development of a Biological Control Method against Postharvest Diseases of Citrus Fruit. *Commun. Agric. Appl. Biol. Sci.* **2005**, *70*(3), 47–58.

Lahlali, R.; Serrhini, M. N.; Jijakli, M. H. Efficacy Assessment of *Candida oleophila* (Strain O) and *Pichia anomala* (Strain K) Against Major Postharvest Diseases of Citrus Fruit in Morocco. *Commun. Agric. Appl. Biol. Sci.* **2004,** 69(4), 601–609.

Lampkin, N. H.; Padel, S. The Economics of Organic Farming. In *An International Perspective*; Lampkin, N. H., Padel S., Eds.; CAB International: Wallingford, 1994.

Landsberg, H. Prelude to the Discovery of Penicillin. *Isis* **1949,** *40*(3), 225–227.

Larkin, R. P.; Fravel, D. R. Effects of Varying Environmental Conditions on Biological Control of Fusarium wilt of Tomato by Nonpathogenic *Fusarium* spp. *Phytopathology* **2002,** *92*, 1160–1166.

Lassois, L.; de Bellaire, L.; Jijakli, M. H. Biological Control of Crown Rot of Bananas with *Pichia anomala* Strain K and *Candida oleophila* Strain O. *Biol. Control.* **2008,** *45*(3), 410–418.

Leclere, V.; Bechet, M.; Adam, A.; Guez, J. S.; Wathelet, B.; Ongena, M.; Thonart, P.; Gancel, F.; Chollet-Imbert, M.; Jacques, P. Mycosubtilin Overproduction by *Bacillus subtilis* BBG100 Enhances the Organism's Antagonistic and Biocontrol Activities. *Appl. Environ. Microbiol.* **2005,** *71*, 4577–4584.

Lee, J. T.; Bae, D. W.; Park, S. H.; Shim, C. K.; Kwak, Y. S.; Kim, H. K. Occurrence and Biological Control of Postharvest Decay in Onion Caused by Fungi. *Plant Pathol. J.* **2001,** *17*, 141–148.

Leeman, M.; den Ouden, F. M.; van Pelt, J. A.; Dirkx, F. P. M.; Steijl, H.; Bakker, P. A. H. M.; Schippers, B. Iron Availability Affects Induction of Systemic Resistance to Fusarium Wilt of Radish by *Pseudomonas fluorescens*. *Phytopathology* **1996,** *86*, 149–155.

Leeman, M.; Van Pelt, J. A.; Den Ouden, F. M.; Heinbroek, M.; Bakker, P. A. H. M. Induction of Systemic Resistance by *Pseudomonas fluorescens* in Radish Cultivars Differing in Susceptibility to Fusarium Wilt, Using Novel Bioassay. *Eur. J. Plant Pathol.* **1995,** *101*, 655–664.

Lewandowski, I.; Hardtlein, M.; Kaltschmitt, M. Sustainable Crop Production: Definition and Methodological Approach for Assessing and Implementing Sustainability. *Crop Sci.* **1999,** *39*, 184–193.

Lewis, J. A.; Papavizas, G. C. Chlamydospores Formation by *Trichoderma* spp. in Natural Substrates. *Can. J. Microbiol.* **1984,** *30*, 1–7.

Ligon, J. M.; Hill, D. S.; Hammer, P. E.; Torkewitz, N. R.; Hofmann, D.; Kempf, H. J.; van Pee, K. H. Natural Products with Antifungal Activity from *Pseudomonas* Biocontrol Bacteria. *Pest Manage. Sci.* **2000,** *56*, 688–695.

Long, C. A.; Deng, B. X.; Deng, X. X. Commercial Testing of *Kloeckera apiculata*, Isolate 34–9, for Biological Control of Postharvest Diseases of Citrus Fruit. *Ann. Microbiol.* **2007,** *57*(2), 203–207.

Long, C. A.; Deng, B. X.; Deng, X. X. Pilot Testing of *Kloeckera apiculata* for the Biological Control of Postharvest Diseases of Citrus. *Ann. Microbiol.* **2006,** *56*(1), 13–17.

Loper, J. E.; Lindow, S. E. Roles of Competition and Antibiosis in Suppression of Plant Diseases by Bacterial Biological Control Agents. In *Pest Management: Biologically Bases Technologies*; Lumsden, D., Vaugh, J. L., Eds.; American Chemical Society: Washington, DC, 1993; pp 144–155.

Loper, J. E.; Schroth, M. N. Importance of Siderophores in Microbial Interactions in the Rhizosphere. In *Iron Siderophores and Plant Disease*; Swinburne, T. R., Ed.; Plenum: New York, 1986; pp 85–98.

Majumdar, V. L.; Pathak, V. N. Biological Control of Postharvest Diseases of Guava Fruit by *Trichoderma* spp. *Acta Bot. Indica* **1995,** *23,* 263–267.

Malathi, P.; Viswanathan, R.; Padmanaban, P.; Mohanraj, D.; Sundar, A. R. Microbial Detoxification of *Colletotrichum falcatum* Toxin. *Curr. Sci.* **2002,** *83*(6), 745–749.

Manczinger, L.; Antal, Z.; Kredics, L. Ecophysiology and Breeding of Mycoparacitic *Trichoderma* Strains (a Review). *Acta Microbiol. Immunol. Hung.* **2002,** *49,* 1–14.

Mandal, G.; Singh, D.; Sharma, R. R. Effect of Hot Water Treatment and Biocontrol Agent (*Debaryomyces hansenii*) on Shelf Life of Peach. *Indian J. Hortic.* **2007,** *64*(1), 25–28.

Mao, W.; Lewis, J.; Hebber, P.; Lumsden, R. Seed Treatment with a Fungal or a Bacterial Antagonist for Reducing Corn Damping-Off Caused by Species of *Pythium* and *Fusarium*. *Plant Dis.* **1997,** *81,* 450–454.

Mao, W.; Lumsden, R. D.; Lewis, J. A.; Hebbar, P. K. Seed Treatment Using Pre-infiltration and Biocontrol Agents to Reduce Damping-Off of Corn Caused by Species of *Pythium* and *Fusarium*. *Plant Dis.* **1998,** *82,* 294–299.

Marais, L. J.; Kotzé, J. M. Ectomycorrhizae of Pinus Patula as Biological Deterrents to *Phytophthora cinnamomi*. *S. Afr. J. For.* **1976,** *99,* 35–39.

Mari, M.; Guizzardi, M.; Pratella, G. C. Biological Control of Gray Mold in Pears by Antagonistic Bacteria. *Biol. Control.* **1996,** *7*(1), 30–37.

Martinez, C.; Michaud, M.; Be´langer, R. R.; Tweddell, R. J. Identification of Soils Suppressive Against *Helminthosporium solani*, the Causal Agent of Potato Silver Scurf. *Soil Biol. Biochem.* **2002,** *34,* 1861–1868.

Mathivanan, N.; Prabavathy, V. R.; Vijayanandraj, V. R. Application of Talc Formulations of *Pseudomonas fluorescens* Migula and *Trichoderma viride* Pers. Ex S.F. Gray Decrease the Sheath Blight Disease and Enhance the Plant Growth and Yield in Rice. *J. Phytopathol.* **2005,** *153,* 697–701.

Mathre, D. E.; Cook, R. J.; Callan, N. W. From Discovery to Use: Traversing the World of Commercializing Biocontrol Agents for Plant Disease Control. *Phytopathology* **1999,** *83,* 972–983.

Matta, A.; Garibaldi. A. Control of *Verticillium* Wilt of Tomato by Preinoculation with Avirulent Fungi, *Neth. J. Plant Pathol.* **1977,** *83,* 457–462.

Maurhofer, M.; Hase, C.; Meuwly, P.; Metraux, J. P.; Defago, G. Induction of Systemic Resistance of Tobacco to Tobacco Necrosis Virus by the Root-Colonizing *Pseudomonas fluorescens* Strain CHAO: Influence of the gacA Gene and of Pyoverdine Production. *Phytopathology* **1994,** *84,* 139–146.

Mazzola, M.; Zhao, X.; Cohen, M. F.; Raaijmakers, J. M. Cyclic Lipopeptide Surfactant Production by *Pseudomonas fluorescens* SS101 Is Not Required for Suppression of Complex *Pythium* spp. Populations. *Phytopathology* **2007,** *97,* 1348–1355.

McNeely, J. A.; Scherr, S. J. *Ecoagriculture*; Island Press: Washington, DC, 2003.

Meyer, J. B.; Lutz, M. P.; Frapolli, M.; Pechy-Tarr, M.; Rochat, L.; Keel, C.; Defago, G.; Maurhofer, M. Interplay between Wheat Cultivars, Biocontrol Pseudomonads, and Soil. *Appl. Environ. Microbiol.* **2010,** *76,* 6196–6204.

Meziane, H.; Van der Sluis, I.; Van Loon, L. C.; Hofte, M.; Bakker, P. A. H. M. Determinants of *Pseudomonas putida* WCS358 Involved in Inducing Systemic Resistance in Plants. *Mol. Plant Pathol.* **2005,** *6*, 177–185.

Mikani, A.; Eteburiun, H. R.; Sholberg, P. L.; Gorman, D. T.; Stokes, S.; Alizadeh, A. Biological Control of Apple Gray Mold Caused by *Botrytis mali* with *Pseudomonas fluorescens* Strains. *Postharvest Biol. Technol.* **2008,** *48*(1), 107–112.

Milner, J. L.; Silo-Suh, L.; Lee, J. C.; He, H.; Clardy, J.; Handelsman, J. Production of Kanosamine by *Bacillus cereus* UW85. *Appl. Environ. Microbiol.* **1996,** *62*, 3061–3065.

Misk, A.; Franco, C. Biocontrol of Chickpea Root Rot Using Endophytic Actinobacteria. *Biol. Control.* **2011,** *56*, 811–822.

Mohammad, D.; Jesus, M. B.; Van Loon, L. C.; Bakker, P. A. H. M. Analysis of Determinants of *Pseudomonas fluorescens* WCS374r Involved in Induced Systemic Resistance in *Arabidopsis thaliana. Biological Control of Fungal and Bacterial Plant Pathogens. IOBC/WPRS Bull.* **2009,** *43*, 109–112.

Monte, E.; Llobell, A. *Trichoderma in Organic Agriculture.* In Proceedings V World Avocado Congress (Actas V Congreso Mundial del Aguacate), December 10–14, 2003, pp 725–733.

Morales, H.; Sanchis, V.; Usall, J.; Ramos, A. J.; Marín, S. Effect of Biocontrol Agents *Candida sake* and *Pantoea agglomerans* on *Penicillium expansum* Growth and Patulin Accumulation in Apples. *Int. J. Food Microbiol.* **2008,** *122*(1–2), 61–67.

Moyne, A. L.; Shalby, R.; Cleveland, T. E.; Tuzun, S. Bacillomycin, D, an Iturin with Antifungal Activity against *Aspergillus flavus. J. Appl. Microbiol.* **2001,** *90*, 622–629.

Muchovej, J. J.; Muchovej, R. M. C.; Gonçalves, E. J. Effect of Kind and Method of Fungicidal Treatment of Bean Seed on Infections by the VA-mycorrhizal Fungus *Glomus macrocarpum* and by the Pathogenic Fungus *Fusarium solani. Plant Soil,* **1991,** *132*(1), 47–51.

Muis, A.; Quimiob, A. J. Biological Control of Banded Leaf and Sheath Blight Disease (*Rhizoctonia solani* Kuhn) in Corn with Formulated *Bacillus subtilis* BR23. *Indian J. Agric. Sci.* **2006,** *7*(1), 1–7.

Muthukumar, A.; Bhaskaran, R.; Sanjeevkumar, K. Efficacy of Endophytic *Pseudomonas fluorescens* (Trevisan) Migula Against Chilli Damping-Off. *J. Biopest.* **2010,** *3*, 105–109.

Muthukumar, A.; Eswaran, A.; Sangeetha, G. Induction of Systemic Resistance by Mixtures of Fungal and Endophytic Bacterial Isolates against Pythium Aphanidermatum. *Acta Physiol. Plant.* **2011,** *33*(5), 1933–1944.

Nagrajkumar, M.; Jayaraj, J.; Muthukrishnan, S.; Bhaskaran, R. Detoxification of Oxalic Acid by *P. fluorescens* strain PfMDU2: Implication for the Biocontrol of Rice Sheath Blight Caused by *Rhizoctonia solani. Microbiol. Res.* **2005,** *160*, 291–298.

Nakasaki, K.; Saito, M.; Suzuki, N. *Coprinellus curtus* (Hitoyo-take) Prevents Diseases of Vegetables Caused by Pathogenic Fungi. *FEMS Microbiol. Lett.* **2007,** *275*, 286–291.

Nakkeeran, S.; Fernando, D. W. G.; Siddiqui, Z. A. Plant Growth Promoting *Rhizobacteria* Formulations and Its Scope in Commercialization for the Management of Pests and Diseases. In *PGPR: Biocontrol and Biofertilization*; Siddiqui, Z. A., Ed.; Springer, Dordrecht, Netherlands 2006; pp 257–296.

Nandakumar, R.; Babu, S.; Viswanathan, R.; Sheela, J.; Raguchander, T.; Samiyappan, R. A New Bioformulation Containing Plant Growth Promoting Rhizobacterial Mixture for the Management of Sheath Blight and Enhanced Grain Yield in Rice. *Biocontrol* **2001,** *46,* 493–510.

Nandeeshkumar, P.; Ramachandrakini, K.; Prakash, H. S.; Niranjana, S. R.; Shetty, H. S. Induction of Resistance Against Downy Mildew on Sunflower by *Rhizobacteria*. *J. Plant Interact.* **2008,** *3,* 256–262.

Narayanasamy, P. Introduction. In *Biological Management of Diseases of Crops*; Narayanasamy, P., Ed.; Integration of Biological Control Strategies with Crop Disease Management Systems. Springer Science + Business Media: New York, 2013; Vol. 2, pp 1–6.

Nassar, A. H.; El-Tarabily, K. A.; Sivasithamparam, K. Growth Promotion of Bean (*Phaseolus vulgaris* L.) by a Polyamine Producing Isolate of *Streptomyces griseoluteus*. *Plant Growth Regul.* **2003,** *40,* 97–106.

Negi, Y. K.; Garg, S. K.; Kumar, J. Cold-tolerant Fluorescent *Pseudomonas* Isolates from Garhwal Himalayas as Biocontrol in Pea. *Curr. Sci.* **2005,** *89,* 2151–2156.

Nielsen, T. H.; Sorensen, D.; Tobiasen, C.; Andersen, J. B.; Christophersen, C.; Givskov, M.; Sorensen, J. Antibiotic and Biosurfactant Properties of Cyclic Lipopeptides Produced by Fluorescent *Pseudomonas* spp from the Sugar Beet Rhizosphere. *Appl. Environ. Microbiol.* **2002,** *68,* 3416–3423.

NRC. *Our Common Journey: Transition towards Sustainability*; Board on Sustainable Development, Policy Division, National Research Council, National Academy Press: Washington, DC, 2000.

Nunes, C.; Teixido, N.; Usall, J.; Vinas, I. Biological Control of Major Postharvest Diseases on Pear Fruit with Antagonistic Bacterium *Pantoea agglomerans* (CPA-2). *Acta Hortic.* **2001a,** *553*(2), 403–404.

Nunes, C.; Usall, J.; Teixido, N.; Miro, M.; Vinas, I. Nutritional Enhancement of Biocontrol Activity of Candida Sake (CPA-1) Against *Penicillium expansum* on Apples and Pears. *Eur. J. Plant Pathol.* **2001b,** *107,* 543–551.

Ogawa, K.; Komada, H. Biological Control of Fusarium wilt of Sweet Potato with Crossprotection by Nonpathogenic *Fusarium oxysporum*, In *Ecology and Management of Soilborne Plant Pathogens;* Parker, C. A., Rovira, A. D., Moore, K. J., Wong, P. T. W., Kollmorgen, J. F., Eds.; American Phytopathological Society: St. Paul, 1985; pp 121–123.

Okigbo, R. N. Mycoflora of Tuber Surface of White Yam (*Dioscorea rotundata* Poir) and Postharvest Control of Pathogens with *Bacillus subtilis*. *Mycopathologia* **2002,** *156,* 81–85.

Ongena, M.; Daay, F.; Jacques, P.; Thonart, P.; Benhamou, N.; Paulitz, T. C.; Cornelis, P.; Koedam, N. M.; Be'langer, R. R. Protection of Cucumber Against Pythium Root Rot by Fluorescent Pseudomonads: Predominant Role of Induced Resistance Over Siderophores and Antibiosis. *Plant Pathol.* **1999,** *48,* 66–76.

Ongena, M.; Duby, F.; Jourdan, E.; Beaudry, T.; Jadin, V.; Dommes, J.; Thonart, P. *Bacillus subtilis* M4 Decreases Plant Susceptibility Towards Fungal Pathogens by Increasing Host Resistance Associated with Differential Gene Expression. *Appl. Microbiol. Biotechnol.* **2005,** *67,* 692–698.

Ongena, M.; Giger, A.; Jacques, P.; Dommes, J.; Thonart, P. Study of Bacterial Determinants Involved in the Induction of Systemic Resistance in Bean by *Pseudomonas putida* BTP1. *Eur. J. Plant Pathol.* **2002,** *108,* 187–196.

Osburn, R. M.; Milner, J. L.; Oplinger, E. S.; Smith, R. S.; Handelsman, J. Effect of *Bacillus cereus* UW85 on the Yield of Soybean at Two Field Sites in Wisconsin. *Plant Dis.* **1995**, *79*, 551–556.

Ouda, S. M. Biological Control by Microorganisms and Ionizing Radiation. *Int. J. Adv. Res.* **2014**, *2*, 314–356.

Padmanab, S. Y. The Great Bengal Famine. *Annu. Rev. Phytopathol.* **1973**, *11*, 11–26.

Pal, K. K.; Gardener, B. M. Biological Control of Plant Pathogens. *Plant Health Instrum.* **2006**, 1–25. DOI: 10.1094/PHI-A-2006-1117-02.

Pal, K. K.; Tilak, K. V.; Saxena, A. K.; Dey, R.; Singh, C. S. Suppression of Maize Root Diseases Caused by *Macrophomina phaseolina*, *Fusarium moniliforme* and *Fusarium graminearum* by Plant Growth Promoting *Rhizobacteria*. *Microbiol. Res.* **2001**, *156*(3), 209–223.

Palumbo, J. D.; Yuen, G. Y.; Jochum, C. C.; Tatum, K.; Kobayashi, D. Y. Mutagenesis of Beta-1,3-glucanase Genes in Lysobacter Enzymogenes Strain C3 Results in Reduced Biological Control Activity Toward Bipolaris Leaf Spot of Tall Fescue and Pythium Damping-Off of Sugar Beet. *Phytopathology.* **2005**, *95*, 701–707.

Pang, Y.; Liu, X.; Ma, Y.; Chernin, L.; Berg, G.; Gao, K. Induction of Systemic Resistance, Root Colonization and Biocontrol Activities of the Rhizospheric Strain of *Serratia plymuthica* are Dependent on *N*-acyl Homoserine Lactones. *Eur. J. Plant Pathol.* **2009**, *124*, 261–268.

Park, K. S.; Kloepper, J. W. Activation of PR-1a Promoter by *Rhizobacteria* which Induce Systemic Resistance in Tobacco Against *Pseudomonas syringae* pv. *tabaci*. *Biol. Control.* **2000**, *18*, 2–9.

Pathak, V. N. Postharvest Fruit Pathology: Present Status and Future Possibilities. *Indian Phytopathol.* **1997**, *50*, 161–185.

Paulitz, T. C.; Loper, J. E. Lack of a Role for Fluorescent Siderophore Production in the Biological Control of Pythium Damping off of Cucumber by a Strain of *Pseudomonas putida*. *Phytopathology* **1991**, *81*, 930–935.

Perner, H.; Schwarz, D.; George, E. Effect of Mycorrhizal Inoculation and Compost Supply on Growth and Nutrient Uptake of Young Leek Plants Grown on Peat-based Substrates. *Hortic. Sci.* **2006**, *41*, 628–632.

Petatan-Sagahon, I.; Anducho-Reyes, M. A.; Silvo-Rojas, H. V.; Arana-Cuenca, A.; Tellez-Jurado, A.; Cardenas-Alvarez, I. O. Isolation of Bacteria with Antifungal Activity Against the Phytopathogenic Fungi Stenocarpella Maydis and Stenocarpella Macrospora. *Int. J. Mol. Sci.* **2011**, *12*, 5522–5537.

Petti, C.; Khan, M.; Doohan, F. Lipid Transfer Proteins and Protease Inhibitors as Key Factors in the Priming of Barley Responses to *Fusarium* Head Blight Disease by a Biocontrol Strain of *Pseudomonas fluorescens*. *Funct. Integr. Genomics* **2010**, *10*, 619–627.

Plaza, P.; Torres, R.; Teixido, N.; Usall, J.; Abadias, M.; Vinas, I. Control of Green Mold by the Combination of *Pantoea agglomerans* (CPA-2) and Sodium Bicarbonate on Oranges. *Bull. OILB/SROP*, **2001**, *24*(3), 167–170.

Postma, J.; Montanari, M.; Van den, B. P. H. J. F. Microbial Enrichment to Enhance Disease Suppressive Activity of Compost. *Eur. J. Soil Biol.* **2003**, *39*, 157–163.

Postma, J.; Stvens, L. H.; Wiegors, G. L.; Davelaar, E.; Nijhuis, E. H. Biological Control of *Pythium aphanidermatum* in Cucumber with a Combined Application of *Lysobacter engymogenes* Strain 3.1T8 and Chitosan. *Biol. Control* **2009**, *48*, 301–309.

Powell, J. F.; Vargas, J. M.; Nair, M. G.; Detweiler, A. R.; Chandra, A. Management of Dollar Spot on Creeping Bentgrass with Metabolites of *Pseudomonas aureofaciens* (TX-1). *Plant Dis.* **2000**, *84*, 19–24.

Press, C. M.; Loper, J. E.; Kloepper, J. W. Role of Iron in *Rhizobacteria* Mediated Induced Systemic Resistance of Cucumber. *Phytopathology* **2001**, *91*, 593–598.

Pretty, J. *Regenerating Agriculture: Policies and Practice for Sustainability and Self-Reliance*; Earthscan: London; National Academy Press: Washington, 1995.

Pusey, P. L.; Wilson, C. L. Postharvest Biological Control of Stone Fruit Brown Rot by *Bacillus subtilis*. *Plant Dis.* **1984**, *68*, 753–756.

Qin, G. Z.; Tian, S. P. Biocontrol of Postharvest Diseases of Jujube Fruit by *Cryptococcus laurentii* Combined with a Low Doses of Fungicides Under Different Storage Conditions. *Plant Dis.* **2004**, *88*(5), 497–501.

Qin, G. Z.; Tian, S. P.; Xu, Y. Biocontrol of Postharvest Diseases on Sweet Cherries by Four Antagonistic Yeasts in Different Storage Conditions. *Postharvest Biol. Technol.* **2004**, *31*(1), 51–58.

Qin, G. Z.; Tian, S. P.; Xu, Y.; Chan, Z. L.; Li, B. Q. Combination of Antagonistic Yeasts with Two Food Additives for Control of Brown Rot Caused by *Monilinia fructicola* on Sweet Cherry Fruit. *J. Appl. Microbiol.* **2006**, *100*(3), 508–515.

Raaijmakers, J. M.; Weller, D. M. Natural Plant Protection by 2,4-Diacetylphloroglucinol Producing *Pseudomonas* spp. in Take-All Decline Soils. *Mol. Plant-Microbe Interact.* **1998**, *11*, 144–152.

Raguchander, T.; Samiyappan, R.; Arjunan, G. Biocontrol of Macrophomina Root Rot of Mungbean. *Indian Phytopathol.* **1993**, *46*, 379–382.

Raj, S. N.; Deepak, S. A.; Basavaraju, P.; Shetty, H. S.; Reddy, M. S.; Kloepper, J. W. Comparative Performance of Formulations of Plant Growth Promoting *Rhizobacteria* in Growth Promotion and Suppression of Downy Mildew in Pearl Millet. *Crop Prot.* **2003**, *22*, 579–588.

Ramamoorthy, V.; Raguchander, T.; Samiyappa, R. Enhancing Resistance of Tomato and Hot Pepper to Pythium Diseases by Seed Treatment with Fluorescent Pseudomonads. *Eur. J. Plant Pathol.* **2002**, *108*, 429–441.

Ramamoorthy, V.; Viswanathan, R.; Raguchander, T.; Prakasam, V.; Smaiyappan, R. Induction of Systemic Resistance by Plant Growth-promoting *Rhizobacteria* in Crop Plants Against Pests and Diseases. *Crop Prot.* **2001**, *20*, 1–11.

Ray, J. Les maladies cryptogamiques des végétaux. *Rev. Gen. Bot.* **1901**, *13*, 145–151.

Rettinassababady, C.; Ramadoss, N.; Thirumeni, S. Management of Root Rot [*Macrophomina phaseolina* (Tassi.) Goid] of Rice Fallow Blackgram by Resident Isolates of *Trichoderma viride*. *Indian J. Agric. Res.* **2002**, *36*, 118–122.

Roberts, R. G. Biological Control of Mucor Rot of Pears by *Cryptococcus laurentii*, C. Flaws and *C. albidus*. *Phytopathology* **1990**, *80*, 1051.

Roby, D.; Toppan, A.; Esquerre-Tugaye, M. T. Cell Surfaces in Plant Microorganisms Interactions. VIII. Increased Proteinase Inhibitor Activity in Melon Plants in Response to Infection by *Colletotrichum lagenarium* or to Treatment with an Elicitor Fraction from this Fungus. *Physiol. Mol. Plant Pathol.* **1987**, *30*, 453–460.

Rodrigues, G. S.; Campanhola, C.; Kitamura P. C. An Environmental Impact Assessment System for Agricultural R and D. *Environ. Impact Assess. Rev.* **2003**, *23*, 219–244.

Romero, D.; de Vicente, A.; Zeriouh, H.; Cazorla, F. M.; Fernandez-Ortuno, D.; Tores, J. A.; Perez-Garcia, A. Evaluation of Biological Control Agents for Managing Cucurbit Powdery Mildew on Greenhouse-grown Melon. *Plant Pathol.* **2007**, *56*, 976–986.

Rothrock, C. S.; Gottlieb, D. Role of Antibiosis in Antagonism of Streptomyces Hygroscopicus var. Geldanus to *Rhizoctonia solani* in Soil. *Can. J. Microbiol.* **1984**, *30*, 1440–1447.

Ryals, J. A.; Neuenschwander, U. H.; Willits, M. G.; Molina, A.; Steiner, H. Y.; Hunt, M. D. Systemic Acquired Resistance. *Plant Cell* **1996**, *8*, 1808–1819.

Ryu, C. M.; Farag, M. A.; Hu, C. H.; Reddy, M. S.; Kloepper, J. W.; Pare, P. W. Bacterial Volatiles Induce Systemic Resistance in Arabidopsis. *Plant Physiol.* **2004**, *134*, 1017–1026.

Saba, H.; Vibhash, D.; Manisha, M.; Prashant, K. S.; Farhan, H.; Tauseef, A. *Trichoderma*—A Promising Plant Growth Stimulator and Biocontrol Agent. *Mycosphere* **2012**, *3*, 524–531. http://mycosphere.org/pdfs/MC3_4_No14.pdf.

Sadeghi, A.; Van Damme, E. J. M.; Peumans, W. J.; Smagghe, G. Deterrent Activity of Plant Lectins on Cowpea Weevil *Callosobruchus maculatus. Phytochemistry* **2006**, *67*(18), 2078–2084.

Sadfi, N.; Cherif, M.; Hajlaoui, M. R.; Boudabbous, A. Biological Control of the Potato Tubers Dry Rot Caused by *Fusarium roseum* var. *sambucinum* Under Greenhouse, Field and Storage Conditions Using *Bacillus* spp. Isolates. *J. Phytopathol.* **2002**, *150*, 640–648.

Sadoma, M. T.; El-Sayed, A. B. B.; El-Moghazy, S. M. Biological Control Downy Mildew Diseases of Maize Caused by *Peronosclerospora sorghi* Using Certain Biocontrol Agents Alone or in Combination. *J. Agric. Res. Kafer Elsheikh Univ.* **2011**, *37*(1), 1–5.

Saikia, R.; Yadav, M.; Varghese, S.; Singh, B. P.; Gogoi, D. K.; Kumar, R.; Arora, D. K. Role of Riboflavin in Induced Resistance Against Fusarium wilt and Charcoal Rot Diseases of Chickpea. *J. Plant Pathol.* **2006**, *24*, 339–347.

Sakthivel, N. Biological Control of *Sarocladium oryzae* (Sawada) Gams and Hawksworth, Sheath Rot Pathogen of Rice by Bacterization with *Pseudomonas fluorescens* Migula. Ph.D. Dissertation, University of Madras, India, 1987.

Saligkarias, I. D.; Gravanis, F. T.; Epton, H. A. S. Biological Control of Botrytis Cinerea on Tomato Plants by the Use of Epiphytic Yeasts *Candida guilliermondii* Strains 101 and US 7 and *Candida oleophila* Strain I-182: In vivo Studies. *Biol. Control.* **2002**, *25*(2), 143–150.

Saravanakumar, D.; Harish, S.; Loganathan, M.; Vivekananthan, R.; Rajendran, L.; Raguchander, T.; Samiyappan, R. Rhizobacterial Formulation for the Effective Management of Macrophomina Root Rot in Mungbean. *Arch. Phytopathol. Plant Prot.* **2007b**, *40*, 323–337.

Saravanakumar, D.; Vijayakumar, C.; Kumar, N.; Samiyappan, R. PGPR-induced Defense Responses in the Tea Plant against Blister Blight Disease. *Crop Prot.* **2007a**, *26*, 556–565.

Sathya, V. K.; Vijayasamundeeswari, A.; Paranidharan, V.; Velazhahan, R. *Burkholderia* sp. Strain TNAU-1 for Biological Control of Root Rot in Mung Bean (*Vigna radiata* L.) Caused by *Macrophomina phaseolina. J. Plant Prot. Res.* **2011**, *51*, 273–278.

Schena, L.; Nigro, F.; Pentimone, I. A.; Ippolito, A. Control of Postharvest Rots of Sweet Cherries and Table Grapes with Endophytic Isolates of Aureobasidium pullulans. *Postharvest Biol. Technol.* **2003**, *30*(3), 209–220.

Scher, F. M.; Baker, R. Effect of *Pseudomonas putida* and a Synthetic Iron Chelator on Induction of Soil Suppressiveness to Fusarium wilt Pathogens. *Phytopathology* **1982**, *72*, 1567–1573.

Schisler, D. A.; Slininger, P. J.; Kleinkopf, G.; Bothast, R. J.; Ostrowski, R. C. Biological Control of *Fusarium* Dry Rot of Potato Tubers under Commercial Storage Conditions. *Am. J. Potato Res.* **2000**, *77*, 29–40.

Schisler, D. A.; Slininger, P. J.; Miller, J. S.; Woodell, L. K.; Clayson, S.; Olsen, N. Bacterial Antagonists, Zoospore Inoculum Retention Time and Potato Cultivar Influence Pink Rot Disease Development. *Am. J. Potato Res.* **2009**, *86*, 102–111.

Schmidt, C. S.; Agostini, F.; Leifert, C.; Killham, K.; Mullins, CE. Influence of Inoculum Density of the Antagonistic Bacteria *Pseudomonas fluorescens* and *Pseudomonas corrugata* on Sugar Beet Seedling Colonisation and Suppression of Pythium Damping off. *Plant Soil* **2004**, *265*, 111–122.

Schmidt, C. S.; Lorenz, D.; Wolf, G. A. Biological Control of the Grapevine Dieback Fungus *Eutypa lata* I: Screening of Bacterial Antagonists. *J. Phytopathol.* **2001**, *149*, 427–435.

Schmidt, E. L. Initiation of Plant Root Microbe Interactions. *Annu. Rev. Microbiol.* **1979**, *33*, 355–376.

Shimizu, M.; Nakagawa, Y.; Sato, Y.; Furumai, T.; Igarashi, Y.; Onaka, H.; Yoshida, R.; Kunoh H. Studies on Endophytic Actinomycetes (I) *Streptomyces* sp. Isolated from *Rhododendron* and Its Antifungal Activity. *J. Gen. Plant Pathol.* **2000**, *66*(1), 360–366.

Sid Ahmed, A.; Ezziyyani, M.; Perez-Sanchez, C.; Candela, M. E. Effect of Chitin on Biological Control Activity of *Bacillus* spp. and *Trichoderma harzianum* against Root Rot Disease in Pepper (*Capsicum annuum*) Plants. *Eur. J. Plant Pathol.* **2003**, *109*, 418–426.

Siddiqui, Z. A.; Baghel, G.; Akhtar, M. S. Biocontrol of *Meloidogyne javanica* by Rhizobium and Plant Growth-promoting *Rhizobacteria* on Lentil. *World J. Microbiol. Biotechnol.* **2007**, *23*, 435–441.

Silo-Suh, L. A.; Lethbridge, B. J.; Raffel, S. J.; He, H.; Clardy, J.; Handelsman, J. Biological Activities of Two Fungistatic Antibiotics Produced by *Bacillus cereus* UW85. *Appl. Environ. Microbiol.* **1994**, *60*, 2023–2030.

Silo-Suh, L. A.; Stab, V. E.; Raffel, S. R.; Handelsman, J. Target Range of Zwittermicin A, an Aminopolyol Antibiotic from *Bacillus cereus*. *Curr. Microbiol.* **1998**, *37*, 6–11.

Singh, A.; Mehta, S.; Singh, H. B.; Nautiyal, C. S. Biocontrol of Collar rot Disease of Betelvine (*Piper betel* L.) Caused by *Sclerotium rolfsii* by Using Rhizosphere Competent *Pseudomonas fluorescens* NBRI-N6 and *Pseudomonas fluorescens* NBRI-N. *Curr. Microbiol.* **2003**, *47*, 153–158.

Singh, D. Bioefficacy of *Debaryomyces hansenii* on the Incidence and Growth of *Penicillium italicum* on Kinnow Fruit in Combination with Oil and Wax Emulsions. *Ann. Plant Prot. Sci.* **2002**, *10*(2), 272–276.

Singh, D. Effect of Debaryomyces Hansenii and Calcium Salt on Fruit Rot of Peach (*Rhizopus macrosporus*). *Ann. Plant Prot. Sci.* **2004**, *12*(2), 310–313.

Singh, D. Interactive Effect of *Debaryomyces hansenii* and Calcium Chloride to Reduce Rhizopus Rot of Peaches. *J. Mycol. Plant Pathol.* **2005,** *35*(1), 118–121.

Singh, P. P.; Shin, Y. C.; Park, C. S.; Chung, Y. R. Biological Control of Fusarium wilt of Cucumber by Chitinolytic Bacteria. *Phytopathology* **1999,** *89*, 92–99.

Singh, R. S.; Singh, D. Effect of *Glomus aggregatum* Inoculation and P-amendments on *Fusarium oxysporum* f. sp. *ciceri* Causing Chickpea wilt. In *Mycorrhizae: Biofertilizers for the Future*; Adholeya, A., Singh, S., Eds.; TERI: New Delhi, 1995; pp 119–123.

Singh, V.; Deverall, B. J. *Bacillus subtilis* as a Control Agent against Fungal Pathogens of Citrus Fruit. *Trans. Br. Mycol. Soc.* **1984,** *83*, 487–490.

Singh, Y. Effect of Soil Solarization and Biocontrol Agents on Plant Growth and Management of Anthracnose of Sorghum. *Int. J. Agric. Sci.* **2008,** *4*, 188–191.

Singhai, P. K.; Sarma, B. K.; Srivastava, J. S. Biological Management of Common Scab of Potato through *Pseudomonas* Species and Vermicompost. *Biol. Control.* **2011,** *57*, 150–157.

Smilanik, J. L.; Denis-Arrue, R.; Bosch, J. R.; Gonjalez, A. R.; Henson, D.; Janisiwicz, W. J. Control of Postharvest Brown Rot of Nectarines and Peaches by *Pseudomonas* Species. *Crop Prot.* **1993,** *12*(7), 513–520.

Smith, J. A.; Metraux, J. P. *Pseudomonas syringae* pv.68 Aglika *Edreva syringae* Induces Systemic Resistance to *Pyricularia oryzae* in Rice. *Physiol. Mol. Plant Pathol.* **1991,** *39*, 451–461.

Smith, K. P.; Havey, M. J.; Handelsman, J. Suppression of Cottony Leak of Cucumber with *Bacillus cereus* Strain UW85. *Plant Dis.* **1993,** *77*, 139–142.

Sneh, B.; Agami, O.; Baker, R. Biological Control of *Fusarium*-wilt in Carnation with *Serratia liquefaciens* and *Hafnia alvei* Isolated from Rhizosphere of Carnation. *Phytopathology* **1985,** *113*, 271–276.

Spadaro, D.; Garibaldi, A.; Gullino, M. L. Control of *Penicillium expansum* and *Botrytis cinerea* on Apple Combining a Biocontrol Agent with Hot Water Dipping and Acibenzolar-*S*-methyl, Baking Soda, or Ethanol Application. *Postharvest Biol. Technol.* **2004,** *33*(2), 141–151.

Spadaro, D.; Vola, R.; Piano, S.; Gullino, M. L. Mechanisms of Action and Efficacy of Four Isolates of the Yeast *Metschnikowia pulcherrima* Active Against Postharvest Pathogens on Apples. *Postharvest Biol. Technol.* **2002,** *24*(2), 123–134.

Srinivasan, K.; Mathivanan, N. Biological Control of Sunflower Necrosis Virus Disease with Powder and Liquid Formulations of Plant Growth Promoting Microbial Consortia Under Field Conditions. *Biol. Control.* **2009,** *51*, 395–402.

Stephan, D.; Schmitt, A.; Carvalho, S. M.; Seddon, B.; Koch, E. Evaluation of Biocontrol Preparation and Plant Extracts for the Control of *Phytophthora infestans* on Potato Leaves. *Eur. J. Plant Pathol.* **2005,** *112*, 235–246.

Suarez-Estrella, F.; Vargas-Garcia, C.; Lopez, M. J.; Capel, C.; Moreno, J. Antagonistic Activity of Bacteria and Fungi from Horticultural Compost against *Fusarium oxysporum* f. sp. *melonis. Crop Prote.* **2007,** *26*, 46–53.

Sunish Kumar, R.; Ayyadurai, N.; Pandiaraja, P.; Reddy, A. V.; Venkateswarlu, Y.; Prakash, O.; Sakthivel, N. Characterization of Antifungal Metabolite Produced by a New Strain *Pseudomonas aeruginosa* PUPa3 that Exhibits Broad-spectrum Antifungal Activity and Biofertilizing Traits. *J. Appl. Microbiol.* **2005,** *98*, 145–154.

Swain, M. R.; Ray, R. C.; Nautiyal, C. S. Biocontrol Efficacy of *Bacillus subtilis* Strains Isolated from Cow Dung against Post-harvest Yam (*Dioscorea rotundata* L.) Pathogens. *Curr. Microbiol.* **2008**, *55*(5), 407–411.

Tapadar, S. A.; Jha, D. K. Disease Management in Staple Crops: A Bacteriological Approach. In *Bacteria in Agrobiology: Disease Management*; Maheshwari, D. K., Ed.; Springer: Heidelberg, New York; Dordrecht: London, 2013; pp 111–152.

Tariq, M.; Yasmin, S.; Hafeez, F. Y. Biological Control of Potato Black Scurf by Rhizosphere Associated Bacteria. *Braz. J. Microbiol.* **2010**, *41*(2), 439–451.

Tazawa, J.; Watanabe, K.; Yoshida, H.; Sato, M.; Homma, Y. Simple Method of Detection of the Strains of Fluorescent *Pseudomonas* spp. Producing Antibiotics, Pyrrolnitrin and Phloroglucinol. *Soil Microorg.* **2000**, *54*, 61–67.

Teixido, N.; Usall, J.; Palou, L.; Asensio, A.; Nunes, C.; Vinas, I. Improving Control of Green and Blue Molds of Oranges by Combining Pantoea Agglomerans (CPA-2) and Sodium Bicarbonate. *Eur. J. Plant Pathol.* **2001**, *107*(7), 685–694.

Thilagavathi, R.; Saravanakumar, D.; Ragupathi, N.; Samiyappan, R. A Combination of Biocontrol Agents Improves the Management of Dry Root Rot (*Macrophomina phaseolina*) in Green Gram. *Phytopathol. Mediterr.* **2007**, *46*, 157–167.

Thomashow, L S.; Weller, D. M.; Bonsall, R. F.; Pierson, L. S. Production of the Antibiotic Phenazine-1-carboxylic Acid by Fluorescent *Pseudomonas* Species in the Rhizosphere of Wheat. *Appl. Environ. Microbiol.* **1990**, *56*, 908–912.

Thrane, C.; Nielsen, T. H.; Nielsen, M. N.; Olsson, S.; Sorensen, J. Viscosinamide Producing *Pseudomonas fluorescens* DR54 Exerts Biocontrol Effect on *Pythium ultimum* in Sugar Beet Rhizosphere. *FEMS Microbiol. Ecol.* **2000**, *33*, 139–146.

Tian, S. P.; Fan, Q.; Xu, Y.; Qin, G. Z.; Liu, H. B. Effect of Biocontrol Antagonists Applied in Combination with Calcium on the Control of Postharvest Diseases in Different Fruit. *Bull. OILB/SROP*, **2002**, *25*(10), 193–196.

Tian, S. P.; Qin, G. Z.; Xu, Y. Survival of Antagonistic Yeasts under Field Conditions and Their Biocontrol Ability Against Postharvest Diseases of Sweet Cherry. *Postharvest Biol. Technol.* **2004**, *33*(3), 327–331.

Tian, S. P.; Qin, G. Z.; Xu, Y. Synergistic Effects of Combining Biocontrol Agents with Silicon Against Postharvest Diseases of Jujube Fruit. *J. Food Prot.* **2005**, *68*(3), 544–550.

Tokala, R. K.; Strap, J. L.; Jung, C. M.; Crawford, D. L.; Salove, M. H.; Deobald, L. A.; Bailey, J. F.; Morra, M. J. Novel Plant-microbe Rhizosphere Interaction Involving *Streptomyces lydicus* WYEC108 and the Pea Plant (*Pisum sativum*). *Appl. Environ. Microbiol.* **2002**, *68*, 2161–2171.

Tomczyk, A. Increasing Cucumber Resistance to Spider Mites by Biotic Plant Resistance Inducers. *Biol. Lett. (Warsaw)*, **2006**, *43*, 381–387.

Tong-Kwee L.; Rohrbach, K. G. Role of *Penicillium funiculosum* Strains in the Development of Pineapple Fruit Diseases. *Ecol. Epidemiol.* **1980**, *70*(7), 663–665.

Torres, R.; Nunes, C.; Garcia, J. M.; Abadias, M.; Vinas, I.; Manso, T.; Olmo, M.; Usall, J. Application of *Pantoea agglomerans* CPA-2 in Combination with Heated Sodium Bicarbonate Solutions to Control the Major Postharvest Diseases Affecting Citrus Fruit at Several Mediterranean Locations. *Eur. J. Plant Pathol.* **2007**, *118*(1), 73–83.

Torres, R.; Teixido, N.; Vinas, I.; Mari, M.; Casalini, L.; Giraud, M.; Usall, J. Efficacy of Candida Sake CPA-1 Formulation for Controlling *Penicillium expansum* Decay on Pome Fruit from Different Mediterranean Regions. *J. Food Prot.* **2006**, *69*(11), 2703–2711.

Tran, H. T. T.; Ficke, A.; Asiimwe, T.; Hofte, M.; Raaijmakers, J. M. Role of the Cyclic Lipopeptide Surfactant Massetolide A in Biological Control of *Phytophthora infestans* and Colonization of Tomato Plants by *Pseudomonas fluorescens*. *New Phytol.* **2007**, *175*, 731–742.

Trewevas, A. Malthus Foiled Again and Again. *Nature* **2001**, *418*, 668–670.

Tronsmo, A. Biological and Integrated Controls of *Botrtis cinerea* on Apple with *Trichoderma harzianum*. *Biol. Control* **1991**, *1*, 59–62.

Tronsmo, A.; Denis, C. The Use of *Trichoderma* species to Control Strawberry Fruit Rots. *Neth. J. Plant Pathol.* **1977**, *83*, 449–455.

Ukey, R. C.; Meshram, M. K. Biological Control in Plant Disease Management. In *Role of Biocontrol Agents for Disease Management in Sustainable Agriculture*; Pub. Res.: India, 2009; pp 388–391.

Umesha, S.; Dharmesh, S. M.; Shettyt, S. A.; Krishnappa, M.; Shetty, H. S. Biocontrol of Downy Mildew Disease of Pearl Millet Using *Pseudomonas fluorescens*. *Crop Prot.* **1998**, *17*, 387–392.

Umezawa, H.; Okami, T.; Hashimoto, T.; Suhara, Y.; Hamada, M.; Takeuchi, T. A New Antibiotic, Kasugamycin. *J. Antibiot. Ser. A.* **1965**, *18*, 101–103.

Usall, J.; Teixido, N.; Torres, R.; Ochoa de Eribe, X.; Vinas, I. Pilot Tests of *Candida sake* (CPA-1) Applications to Control Postharvest Blue Mold on Apple Fruit. *Postharvest Biol. Technol.* **2001**, *21*(2), 147–156.

Utkhede, R. S.; Sholberg, P. L. In vitro Inhibition of Plant Pathogens: *Bacillus subtilis* and *Enterobacter aerogenes* In Vivo Control of Two Postharvest Cherry Diseases. *Can. J. Microbiol.* **1986**, *32*, 963–967.

Utkhede, R. S.; Smith, E. M. Development of Biological Control of Apple Replant Disease. *Acta Hortic.* **1994**, *363*, 129–134.

Valasubramanian, R. Biological Control of Rice Blast with *Pseudomonas fluorescens* Migula: Role of Antifungal Antibiotic in Disease Suppression. Ph.D. Dissertation, University of Madras, Chennai, 1994.

Valois, D.; Fayad, K.; Barasubiye, T.; Garon, T.; Dery, C.; Brzezinski, R.; Beaulieu, C. Glucanolytic Actinomycetes Antagonistic to *Phytophthora fragariae* var. Rubi, the Causal Agent of Raspberry Root Rot. *Appl. Environ. Microbiol.* **1996**, *62*, 1630–1635.

van Dijk, K.; Nelson, E. B. Fatty Acid Competition as a Mechanism by Which Enterobacter Cloacae Suppresses *Pythium ultimum* Sporangium Germination and Damping-off. *Appl. Environ. Microbiol.* **2000**, *66*, 5340–5347.

Van Esse, H. P.; Van't Klooster, J. W.; Bolton, M. D.; Yadeta, K. A.; van Baarlen, P.; Boeren, S.; Vervoort, J.; de Wit, P. J. G. M.; Thomma, B. P. The *Cladosporium fulvum* Virulence Protein Avr2 Inhibits Host Proteases Required for Basal Defense. *Plant Cell* **2008**, *20*, 1948–1963.

Van Loon, J. C. Induced Resistance. In *Mechanisms of Resistance to Plant Diseases*; Slusarenko, A. J., Fraser, R. S. S., Van Loon, J. C., Eds.; Kluwer Academic Publishers: Dordrecht, NL, 2000; pp 521–574.

Van Peer, R.; Niemann, G. J.; Schippers, B. Induced Resistance and Phytoalexin Accumulation in Biological Control of Fusarium wilt of Carnation by *Pseudomonas* sp. Strain WCS417r. *Phytopathology* **1991**, *81*, 728–734.

Van Wees, S. C. M.; Pieterse, C. M. J.; Trijssenaar, A.; Vant Westende, Y.; Hartog, F. Differential Induction of Systemic Resistance in Arabidopsis by Biocontrol Bacteria. *Mol. Plant-Microbe Interact.* **1997,** *10,* 716–724.

Vasudevan, P. Isolation and Characterization of *Bacillus* spp. from the Rice Rhizosphere and Their Role in Biological Control of Bacterial Blight of Rice Caused by *Xanthomonas oryzae* pv. *oryzae.* Ph.D. Dissertation, University of Madras, Chennai, 2002.

Vasudevan, P.; Kavitha, S.; Priyadarisini, V. B.; Babujee, L.; Gnanamanickam, S. S. Biological Control of Rice Diseases. In *Biological Control of Crop Diseases*; Gnanamanickam, S. S., Ed.; Marcel Dekker Inc.: New York, 2002; pp 11–32.

Veerabhadrasamy, A. L.; Garampalli, R. H. Effect of Arbuscular Mycorrhizal Fungi in the Management of Black Bundle Disease of Maize Caused by *Cephalosporium acremonium. Sci. Res. Rep.* **2011,** *1*(2), 96–100.

Velusamy, P.; Gnanamanickam, S. S. Plant Associated Bacteria, 2,4-diacetylphloroglucinol (DAPG) Production and Suppression of Rice Bacterial Blight in India. *Curr. Sci.* **2003,** *85*, 1270–1273.

Velusamy, P.; Immanuel, J. E.; Gnanamanickam, S. S.; Thomashow, L. Biological Control of Rice Bacterial Blight by Plant-Associated Bacteria Producing 2,4-diacetylphloroglucinol. *Can. J. Microbiol.* **2006,** *52*, 56–65.

Verhagen, B. W. M.; Trotel-Aziz, P.; Couderchet, M.; Hofte, M.; Aziz, A. *Pseudomonas* spp.-induced Systemic Resistance to *Botrytis cinerea* is Associated with Induction and Priming of Defence Responses in Grapevine. *J. Exp. Bot.* **2010,** *61,* 249–260.

Verma, D. K.; Mohan, M.; Asthir, B.; Chandra, S. Organic Agriculture: An Approach for Sustainable Food Production and Environmental Security. In *Organic Farming and Management of Biotic Stresses*; Biswas, S. K., Pal, S., Eds.; Pub. Biotech Books: New Delhi, 2014; pp 34–43.

Verma, D. K.; Srivastava, S.; Kumar, V.; Srivastav, P. P. Biological Agents in Fusarium wilt (FW) Diagnostic for Sustainable Pigeon Pea Production, Opportunities and Challenges. In *Current Trends in Plant Disease Diagnostics and Management Practices*; Kumar, P., Gupta, V. K., Tiwari, A. K., Kamle, M., Eds.; Part of Book Series on "Fungal Biology," Springer Science+Business Media: New York, USA, 2016.

Vidhyasekaran, P.; Kamala, N.; Ramanathan, A.; Rajappan, K.; Paranidharan, V.; Velazhahan, R. Induction of Systemic Resistance by *Pseudomonas fluorescens* Pf1 against *Xanthomonas oryzae* pv. *oryzae* in Rice Leaves. *Phytoparasitica* **2001,** *29,* 155–166.

Vinale, F.; Sivasithamparamb, K.; Ghisalbertic, E. L.; Marraa, R.; Wooa, S. L.; Moritoa L. *Trichoderma*–plant–pathogen Interactions. *Soil Biol. Biochem.* **2008,** *40,* 1–10.

Viswanathan, R.; Samiyappan, R. Role of Chitinases in *Pseudomonas* spp. Induced Systemic Resistance against Colletotrichum Falcatum in Sugarcane. *Indian Phytopathol.* **2001,** *54*, 418–423.

Vivekananthan, R.; Ravi, M.; Ramanathan, A.; Samiyappan, R. Lytic Enzymes Induced by *Pseudomonas fluorescens* and Other Biocontrol Organisms Mediate Defense against Anthracnose Pathogen in Mango. *World J. Microbiol. Biotechnol.* **2004,** *20*, 235–244.

Wakelin, S. A.; Walter, M.; Jaspers, M.; Stewart, A. Biological Control of *Aphanomyces euteiches* Root-rot of Pea with Spore-forming Bacteria. *Aust. Plant Pathol.* **2002,** *31,* 401–407.

Wang, H.; Hwang, S. F.; Chang, K. F.; Turnbull, G. D.; Howard, R. J. Suppression of Important Pea Diseases by Bacterial Antagonists. *Biol. Control.* **2003,** *48,* 447–460.

Weller, D. M. Biological Control of Soil Borne Plant Pathogens in the Rhizosphere with Bacteria. *Annu. Rev. Phytopathol.* **1988,** *26,* 379–469.

Weller, D. M.; Cook, R. J. Suppression of Take-all of Wheat by Seed Treatments with Fluorescent *Pseudomonads. Phytopathology.* **1983,** *73,* 463–469.

Weller, D. M.; Raaijmakers, J. M.; Gardener, B. B.; Thomashow, L. S. Microbial Populations Responsible for Specific Soil Suppressiveness to Plant Pathogens. *Annu. Rev. Phytopathol.* **2002,** *40,* 309–348.

Whipps, J. M.; McQuilken, M. P. Biological Control Agents in Plant Disease Control. In *Disease Control in Crops-Biological and Environmentally Friendly Approaches*; Walters, D., Ed.; Wiley Blackwell, New Jersey, USA, 2009; pp 27–61.

Wie, G.; Kloepper, J. W.; Tuzun, S. Induction of Systemic Resistance of Cucumber to *Colletotrichum orbiculare* by Select Strains of Plant Growth-promoting *Rhizobacteria. Phytopathology* **1991,** *81,* 1508–1512.

Wilson, C. L.; Chalutz, E. Postharvest Biocontrol of Penicillium Rots of Citrus with Antagonistic Yeasts and Bacteria. *Sci. Hortic.* **1989,** *40,* 105–112.

Wilson, C. L.; Franklin, J. D.; Pusey, P. L. Biological Control of Rhizopus Rot of Peach with Enterobacter Cloacae. *Phytopathology* **1987,** *77,* 303–305.

Windstam, S.; Nelson, E. B. Differential Interference with *Pythium ultimum* Sporangium Activation and Germination by *Enterobacter cloacae* in the Corn and Cucumber Spermospheres. *Appl. Environ. Microbiol.* **2008a,** *74,* 4285–4291.

Windstam, S.; Nelson, E. B. Temporal Release of Fatty Acids and Sugars in the Spermosphere: Impacts on *Enterobacter cloacae*-induced Biological Control. *Appl. Environ. Microbiol.* **2008b,** *74,* 4292–4299.

Xi, L.; Tian, S. P. Control of Postharvest Diseases of Tomato Fruit by Combining Antagonistic Yeast with Sodium Bicarbonate. Sci. *Agric. Sin.* **2005,** *38*(5), 950–955.

Xiao, K.; Samac, D. A.; Kinkel, L. L. Biological Control of *Phytophthora* Root Rots on Alfalfa and Soybean with *Streptomyces. Biol. Control.* **2002,** *23,* 285–295.

Xue, A. G.; Chen, Y.; Voldeng, H. D.; Fedak, G.; Savard, M. E.; Längle, T.; Zhang, J.; Harman, G. E. Concentration and Cultivar Effects on Efficacy of CLO-1 Biofungicide in Controlling *Fusarium* Head Blight of Wheat. *Biol. Control.* **2014,** *73,* 2–7.

Yan, Z.; Reddy, M. S.; Ryu, C.-M.; McInroy, J. A.; Wilson, M.; Kloepper, J. W. Induced Systemic Protection against Tomato Late Blight Elicited by Plant Growth-Promoting *Rhizobacteria. Phytopathology* **2002,** *92,* 1329–1333.

Yang, C.-H.; Menge, J. A.; Cooksey, D. A. Mutations Affecting Hyphal Colonization and Pyoverdine Production in *Pseudomonads antagonistic* Toward *Phytophthora parasitica. Appl. Environ. Microbiol.* **1994,** *60,* 473–481.

Yang, D. M.; Bi, Y.; Chen, X. R.; Ge, Y. H.; Zhao, J. Biological Control of Postharvest Diseases with *Bacillus subtilis* (B1 Strain) on Muskmelons (*Cucumis melo* L. cv. Yindi). *Acta Hortic.* **2006,** *712*(2), 735–739.

Yang, L.; Xie, J.; Jiang, D.; Fu, Y.; Li, G.; Lin, F. Antifungal Substances Produced by *Penicillium oxalicum* Strain PY-1 Potential Antibiotics against Plant Pathogenic Fungi. *World J. Microbiol. Biotechnol.* **2008,** *24*(7), 909–915.

Yao, H. J.; Tian, S. P. Effects of a Biocontrol Agent and Methyl Jasmonate on Posthar-vest Diseases of Peach Fruit and the Possible Mechanisms Involved. *J. Appl. Microbiol.* **2005,** *98*(4), 941–950.

You, M. P.; Sivasithamparam, K.; Kurboke, D. J. Actinomycetes in Organic Mulch Used in Avocado Plantations and Their Ability to Suppress *Phytophthora cinnamomi. Biol. Fertil. Soils* **1996,** *22*, 237–242.

Yun-feng, Y. E.; Qi-qin, L. I.; Gang, F. U.; Gao-qing, Y. U. A. N.; Jian-hua, M. I. A. O.; Wei, L. I. N. Identification of Antifungal Substance (IturinA2) Produced by *Bacillus subtilis* B47 and Its Effect on Southern Corn Leaf Blight. *J. Integr. Agric.* **2012,** *11*(1), 90–99.

Zarandi, M. E.; Bonjar, G. H. S.; Dehkaei, F. P.; Moosavi, S. A. A.; Farokhi, P. R.; Aghighi, S. Biological Control of Rice Blast (*Magnaporthe oryzae*) by Use of *Streptomyces sinde-neusis* Isolate 263 in Greenhouse. *Am. J. Appl. Sci.* **2009,** *6*(1), 194–199.

Zhang, B.; Xie, C.; Yang, X. A Novel Small Antifungal Peptide from *Bacillus strain* B-TL2 Isolated from Tobacco Stems. *Peptides* **2008a,** *29*, 350–355.

Zhang, H.; Wang, L.; Dong, Y.; Jiang, S.; Cao, J.; Meng, R. Postharvest Biological Control of Gray Mold Decay of Strawberry with *Rhodotorula glutinis. Biol. Control* **2007a,** *40*(2), 287–292.

Zhang, H.; Wang, L.; Dong, Y.; Jiang, S.; Zhang, H.; Zheng, X. Control of Postharvest Pear Diseases Using *Rhodotorula glutinis* and its Effects on Postharvest Quality Parameters. *Int. J. Food Microbiol.* **2008b,** *126*(1–2), 167–171.

Zhang, H.; Wang, L.; Ma, L.; Dong, Y.; Jiang, S.; Xu, B.; Zheng, X. Biocontrol of Major Postharvest Pathogens on Apple Using *Rhodotorula glutinis* and Its Effects on Posthar-vest Quality Parameters. *Biol. Control* **2009,** *48*(1), 79–83.

Zhang, H.; Zheng, X. D.; Yu, T. Biological Control of Postharvest Diseases of Peach with *Cryptococcus laurentii. Food Control* **2007c,** *18*(4), 287–291.

Zhang, H.; Zheng, X.; Fu, C.; Xi, Y. Biological Control of Blue Mold Rot of Pear by *Cryp-tococcus laurentii. J. Hortic. Sci. Biotechnol.* **2003,** *78*, 888–893.

Zhang, H.; Zheng, X.; Fu, C.; Xi, Y. Postharvest Biological Control of Gray Mold Rot of Pear with *Cryptococcus laurentii. Postharvest Biol. Technol.* **2005,** *35*(1), 79–86.

Zhang, H.; Zheng, X.; Wang, L.; Li, S.; Liu, R. Effect of Antagonist in Combination with Hot Water Dips on Postharvest Rhizopus Rot of Strawberries. *J. Food Eng.* **2007b,** *78*, 281–287.

Zhang, S.; White, L. W.; Martinez, M. C.; McInroy, J. A.; Kloepper, J. W.; Klessen, W. Evaluation of Plant Growth-promoting *Rhizobacteria* for Control of *Phytophthora* Blight on Squash Under Greenhouse Conditions. *Biol. Control.* **2010,** *53*, 129–135.

Zhang, Y.; Fernando, W. G. D. Zwittermicin A Detection in *Bacillus* spp. Controlling *Scler-otinia sclerotiorum* on Canola. *Phytopathology* **2004,** *94*, S116.

Zhang, Z.; Yuen, G. Y.; Sarath, G.; Penheiter, A. Chitinases from the Plant Disease Biocon-trol Agent, *Stenotrophomonas maltophilia* C3. *Phytopathology* **2001,** *91*, 204–211.

Zhao, Y.; Shao, X. F.; Tu, K.; Chen, J. K. Inhibitory Effect of *Bacillus subtilis* B10 on the Diseases of Postharvest Strawberry. *J. Fruit Sci.* **2007,** *24*(3), 339–343.

Zhao, Y.; Tu, K.; Shao, X.; Jing, W.; Su, Z. Effects of the Yeast Pichia Guilliermondii against *Rhizopus nigricans* on Tomato Fruit. *Postharvest Biol. Technol.* **2008,** *49*(1), 113–120.

Zhou, T.; Chu, C. L.; Liu, W. T.; Schneider, K. E. Postharvest Control of Blue Mold and Gray Mold on Apples Using Isolates of *Pseudomonas syringae*. *Can. J. Plant Pathol.* **2001,** *23*, 246–252.

Zhou, T.; Northover, J.; Schneider, K. E. Biological Control of Postharvest Diseases of Peach with Phyllosphere Isolates of *Pseudomonas syringae*. *Can. J. Plant Pathol.* **1999,** *21*, 375–381.

Zhou, T.; Northover, J.; Schneider, K. E.; Lu, X. W. Interactions between *Pseudomonas syringae* MA-4 and Cyprodinil in the Control of Blue Mold and Gray Mold of Apples. *Can. J. Plant Pathol.* **2002,** *24*(2), 154–161.

Zinada, D. S.; Shaabana, K. A.; Abdalla, M. A.; Islam, M. T.; Schuffler, A.; Laatsch, H. Bioactive Isocoumarins from a Terrestrial *Streptomyces sp.* ANK302. *Nat. Prod. Commun.* **2011,** *6*, 45–48.

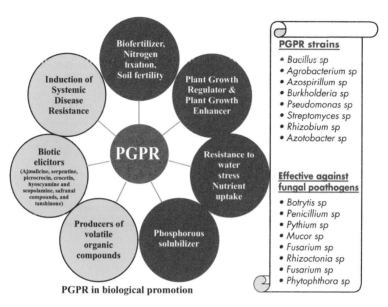

FIGURE 1.4 Role of PGPR in biological growth promotion.

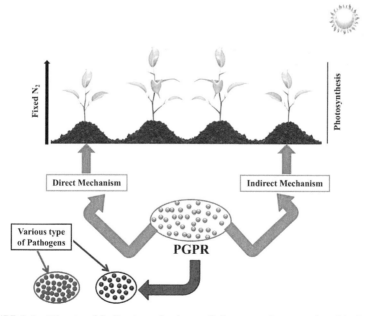

FIGURE 2.1 Direct and indirect mechanisms of plant growth–promoting rhizobacteria (PGPR) on plant. (*Source*: Picture adapted from García-Fraile et al., 2015)

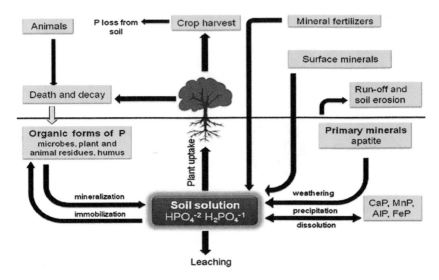

FIGURE 2.2   Movement of phosphorus in soil. (Reprinted from Ahemad and Kibret, 2014. © 2004 with permission from Elsevier.)

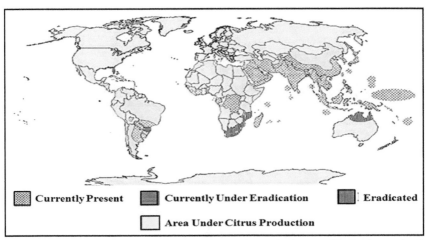

FIGURE 4.1   Worldwide distribution of citrus canker and where eradication is practiced. (Reprinted from Schubert et al., 2001; http://apsjournals.apsnet.org/doi/abs/10.1094/ PDIS.2001.85.4.340. Open access.)

FIGURE 4.2   Leaves are demonstrating citrus canker lesions and chlorotic halos of (A) Kagzi lime; (B) Kinnow (mandarin); (C) Ruby nuclear (sweet orange); and (D) Trovita (sweet orange).

FIGURE 4.3   Citrus canker infection in fruits and plant. (A) and (B) Citrus canker lesions on fruit of grapefruit (cv. Red blush) and (C) grapefruit (cv. Red blush) tree showing fruit drop.

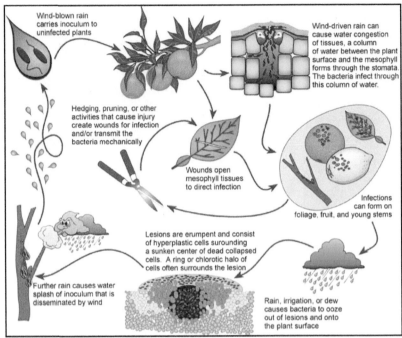

FIGURE 4.5 Diagrammatic representation on disease cycle of citrus canker. (Reprinted from Schubert et al., 2001; http://apsjournals.apsnet.org/doi/abs/10.1094/PDIS.2001.85.4.340. Open access.)

FIGURE 4.6 Leaf miner larvae gallery, which facilitate infection of *X. citri* bacteria.

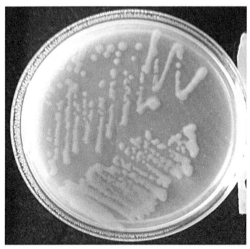

FIGURE 4.7   Yellow pigmented culture plate of *Xanthomonas citri*.

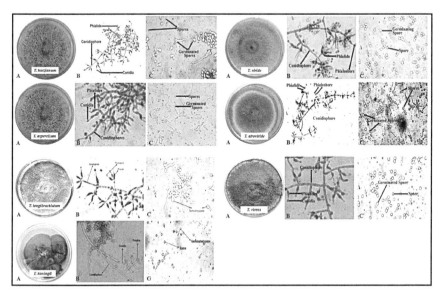

FIGURE 5.1   Morphological and cultural characteristic and molecular identification of *Trichoderma* spp., (A) growth on PDA medium, (B) microscopic observation (40×), and (C) spore germination (40×).

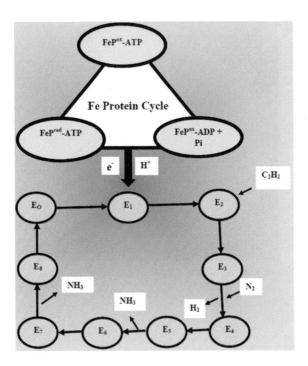

FIGURE 6.2 Fe and MoFe protein cycle (three state cycle Fe protein at top) and (an eight state cycle MoFe protein) at bottom. Fe protein (abbreviated FeP) can exist in the 1^+ reduced state (Red) or the 2^+ oxidized state (Ox) and two MgATP molecules bound as ATP or two MgADP with two Pi as ADP+Pi. When FeP gets associated to MoFe protein, then the exchange of an electron occurs. In the MoFe protein cycle, the MoFe protein is successively reduced by one electron, with reduced states represented by En, where n is the total number of electrons donated by the Fe protein. Acetylene (C_2H_2) binds to E_2 and N_2 binds to E_3 and E_4. The N_2 binding results in displacement of H_2 and two ammonia molecules being liberated from later E states. (*Source*: Adapted from Seefeldt et al. 2009).

FIGURE 8.1 Distribution of pesticides on target and nontarget areas (*Source*: Adapted from Gavrilescu, 2005).

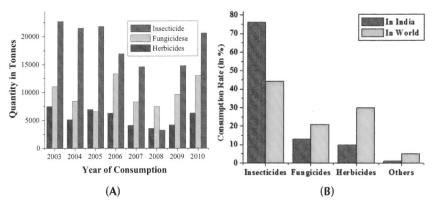

FIGURE 8.3 General and comparative presentation of chemical consumption. (A) Pesticides consumption in India from 2003 to 2010 (adapted from FAOSTAT, 2016) and (B) chemical consumption pattern for agriculture India vs World (adapted from Aktar et al., 2009).

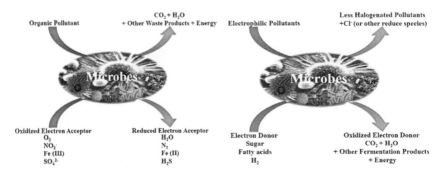

FIGURE 8.4   General process of organic contaminant degradation (*Source*: Adapted from Rockne and Reddy, 2003). (A) Oxidative biodegradation and (B) reductive biodegradation.

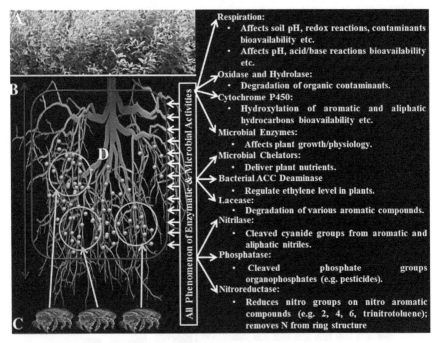

FIGURE 8.5   Enzymatic and microbial activities responsible for the enhanced remediation in rhizospheric zone (*Source*: Adapted from Abhilash et al., 2009).

(A) Standing plants in field indicated above the yellow line; (B) rhizospheric zone inside the ground surface indicated by blue arrow below the yellow line; (C) commonly found microorganisms in the rhizosphere, namely, *Pseudomonas* spp., *Flavobacterium* spp., and *Alcaligenes* spp. are involved in microbial activity of remediation (*Sources*: Adapted from Barber and Martin, 1976; Barber, 1984); and (D) involved microorganisms in root of the plants inside the rhizosphere.

CHAPTER 4

Xanthomonas citri: THE PATHOGEN OF CITRUS CANKER DISEASE AND ITS MANAGEMENT PRACTICES

ADESH KUMAR[1*], ROOMI RAWAL[2], AJAY KUMAR[3], NIRDESH KUMAR[3], DEEPAK KUMAR VERMA[4], and DIGANGGANA TALUKDAR[5]

[1]*Department of Plant Pathology, Punjab Agriculture University, Ludhiana 141004, Punjab, India*

[2]*Department of Entomology, Chaudhary Charan Singh Haryana Agricultural University, Hisar 125004, Haryana, India*

[3]*Department of Plant Protection, Chaudhary Charan Singh University, Meerut 200005, Uttar Pradesh, India*

[4]*Agricultural and Food Engineering Department, Indian Institute of Technology, Kharagpur 721302, West Bengal, India*

[5]*Plant Pathology and Microbiology, College of Horticulture Under College of Agricultural Engineering and Post-harvest Technology, Central Agricultural University, Ranipool 737135, East Sikkim, India*

*Corresponding author. E-mail: fruitpatho@gmail.com/ adeshpp@gmail.com

4.1 INTRODUCTION

Citrus canker is one of the most important diseases in citrus caused by a bacterium *Xanthomonas citri* (syn. *Xanthomonas campestris* pv. *citri*). The importance of this disease can be ascertained with its dominance in almost all the citrus growing countries worldwide specially in countries like Asia, South America, Oceania, Africa as well as in the United States (OEPP/ EPPO, 1990; CABI, 2006; EPPO, 2006). The disease causes tremendous yield losses in citrus cultivars as well as citrus relatives (Stall and Civerola, 1991; Schubert and Miller, 2000). The pathogen produces necrotic lesions almost in all aerial parts of plant and initiates severe infections on premature fruit that results in dropping of the fruits. Naturally, bacterial infection is caused by wounds or by mechanical damages and is disseminated by wind and rainfall splashes which assist infection much faster. Citrus leaf miner is also one of the vital aspects causing infestation of citrus plants and indirectly raises the disease to many folds (Bergamin-Filho et al., 2000). The disease first occurred in an endemic form in India, Japan, and other Southeast Asian countries and then spread to other citrus growing continents except Europe. The pathogen survive nonsystemically and have numerous pathotypes specifically pathotypes A, B, and C. Pathotype A causes Asiatic citrus canker with broad host range, and pathotype B is restricted to only lemon, Mexican lime, sour orange, and pummelo, while pathotype C causes disease only in Mexican lime. The *X. citri* is a rod shaped, Gram-negative bacterium having single polar flagellum. The pathogen is obligated aerobic and growing well at temperature range between 28°C and 30°C (Mehrotra, 1980; Whiteside et al., 1988).

The aim of the chapter entitled "*Xanthomonas citri:* The Pathogen of Citrus Canker Disease and Their Management Practices" is to focus and review on *Xanthomonas citri* causing economic losses in India and many other countries by producing poor quality fruits and reduction in yield; management practices for controlling canker diseases; and thus improving quality and production of citrus species.

4.2   HISTORICAL BACKGROUND

Citrus canker disease has debatable thoughts regarding their history. Lee (1918) reported that the disease may have first arisen in southern China. However, Fawcett and Jenkins (1933) gave evidence about canker disease to be originated in India and Java. Many authors also supported this proclamation (Loucks, 1934; Dopson, 1964; Civerolo, 1981, 1984; Schoulties et al., 1987; Schubert et al., 1996; CABI and EPPO, 1997; Gottwald et al., 1997; Schubert et al., 2001; Gottwald et al., 2002a,b; Yang et al., 2002; Das, 2003). In 1910, canker disease was first time reported and described in the United States, in the Gulf Coast States (Dopson, 1964). The disease also reported to be emerged in the same century in Australia (Garnsey et al., 1979), South Africa (Doidge, 1916) and South America (Rossetti, 1977). In the Gulf Coast States, from Florida to Texas, approximately, 20 million trees in nurseries and groves had been devastated and destroyed in 1934 due to citrus canker. Consequently, epidemics had occurred in Argentina, Australia, Brazil, Oman, Reunion Island, Saudi Arabia, Uruguay, and in USA.

In India, Luthra and Sattar (1942) first reported citrus canker from Punjab, while Govinda Rao (1954) and Chowdhury (1951) reported it from Andhra Pradesh and Assam, respectively. This incidence was further reported from Tamil Nadu (Ramakrishnan, 1954), Karnataka (Venkatakrishnaiah, 1957; Aiyappa 1958), Madhya Pradesh (Parsai, 1959), Rajasthan (Prasad, 1959), and Uttar Pradesh (Nirvan, 1960). Recently canker disease had been detected in Kinnow mandarin nursery in the state of Punjab (Thind and Singh, 2015).

Dopson in the year 1964 reported that Citrus canker was established in Florida around 1910 and depicted that it was the beginning of this disease in North Florida and other Gulf States on different trifoliate rootstock that was imported from Japan. But the fact was that, up to the year 1915, no one was conscious about that causal organism of the disease. A quarantine regulation on all citrus plants was enacted in 1915 and eradication program began by Florida Division of Plant Industry and the U.S. Department of Agriculture's Animal and Plant Health Inspection Service. The last canker infected tree in Florida was removed in 1927 and officially eradicated in 1933. The disease eradication was confirmed from the United States in 1947. In June 1986, citrus canker was detected in residential citrus in Hillsborough, Pinellas, Sarasota, and Manatee counties. All the diseased plants were removed and second eradication plan completed with the last detection in January 1992 (Schubert, 1991). During September 1995, canker was again discovered for a third time in Florida near the Miami International Airport and immediate eradication attempt begun instantaneously by the authorities. Genetic characterization of the isolate of *Xac* causing canker in the Miami area revealed that the pathogen was visibly dissimilar from the archived isolates from the 1986–1992 outbreaks in west-central Florida (Schubert et al., 1996). As of March 2001, disease was managed before spread by means of far-reaching eradication endeavor.

4.3 *Xanthomonas Citri:* THE PATHOGEN OF CITRUS CANKER

Different pathovars of *Xanthomonas citri* (syns. *X. campestris, X. axonopodis*) have been reported to cause citrus canker disease. Both the pathogens have approximately similar symptoms; nevertheless, separation can be made based on host series, cultural and biochemical characteristics, bacteriophage activity (Civerolo, 1984a,b), DNA–DNA similarity (Egel et al., 1991), plasmid fingerprints (Pruvost et al., 1992), serology (Alvarez et al., 1991), and by various restriction fragment length polymorphism and polymerase chain reaction (PCR) analysis (Hartung, 1992; Hartung and Civerolo, 1989; Hartung et al., 1993; Hartung et al., 1996). The studies on molecular and pathogenicity of this disease suggested that the strains of *X. campestris* and *X. axonopodis* are distinctive with respect to genetical and pathological characteristics (Table 4.1).

Table 4.1   Scientific classification and Nomenclature of *Xanthomonas citri*.

Domain	Bacteria
Phylum	Proteobacteria
Class	Gammaproteobacteria
Order	Xanthomonadales
Family	Xanthomonadaceae
Genus	*Xanthomonas*
Species	*Citri*

Sources: Hasse (1915) and Gabriel et al. (1989).

The Asiatic type of canker (Canker A) strain originally found in Asia and widely distributed in the world as compared to other four strains, namely, B, C, D, and E (Civerolo, 1985). Gabriel et al. (1989) pooled B, C, and D strains in *Xanthomonas campestris* pv. *aurantifolii* but unsuccessful to execute the standards necessary for officially designation of pathovars. They also designated group E strains as *Xanthomonas campestris* pv. *citrumelo*, but botched to differentiate this pathogen from earlier named *Xanthomonas campestris* pv. *alfalfae* (Young et al., 1991). Vauterin et al. (1995) allocated the pathovar citri group A strains to *Xanthomonas axonopodis* pv. *citri* and used the faulty names pv. *Aurantifolii* and pv. *citrumelo* without modification. Two strains namely A* and A^W are genetically similar with group A but have limited host range (Sun et al., 2000; Verniere et al., 1998). Except group A, the other groups (B, C, D, and E) have week pathogenicity characters which are widely acknowledged (Stall and Civerolo, 1991). At present, we don't have any satisfactory nomenclature for these dissimilar pathogens; therefore, the discussion here refers to the group A. The current preferred name for this pathogen is *Xanthomonas citri*.

4.3.1   GLOBAL DISTRIBUTION OF DISEASE AND PATHOGEN

Citrus canker has been reported to cause devastation in all the major citrus growing countries of the world causing extensive damage to citrus plantation (Civerolo, 1984; Gottwald et al., 2002a,b) (Fig. 4.1). The geographical distribution of the pathogen present in different outline based on their location. Canker A (Asiatic canker) is established in Asia, South America, Oceania, and USA (Carrera, 1933); canker B in South America (Carrera 1933); canker C in Brazil (Namekata and Oliveira, 1972); and canker D present in Mexico (Rodriguez et al., 1985).

Currently, canker disease occurs in almost 30 countries of Asia, Indian Ocean islands, South America, United States, Arab countries (Ibrahim and Bayaa, 1989), and also in some countries where the disease was eradicated few years back with quarantine measures such as New Zealand and Australia (IMI, 1996), Thursday Island (Jones, 1991), and South Africa (IMI, 1996). An outbreak in Queensland, Australia in 2004 was later on declared eradicated (IPPC, 2009). The disease was considered to be of minor importance before, but with the extent of its serious spreading proportions in all the citrus growing countries by the 20th century, it is now considered as the most important disease causing significant loss in yield and marketable quality. The most recent outbreak of the disease was reported in Florida in 1995 and after a 10-year eradication effort, the USDA avowed that eradication was no longer possible. However, widespread occurrence of the disease in many areas is a permanent nuisance to all citriculture especially in canker-free areas. The disease continues to raise its geographic array in spite of the heightened regulations imposed by many countries to prevent its introduction. The disease reduces the income sources of many citrus-growing countries like Brazil (Leite and Mohan, 1990), Argentina (Canteros, 2000), and USA (Gottwald and Irey, 2007).

4.3.2   SYMPTOMS AND INFECTION

All the aerial parts of the plant are susceptible to the bacterium during the initial tender growth stage (Gottwald and Graham, 1992; Stall et al., 1980) (Fig. 4.2). The infected plants are identified by the emergence of raised necrotic lesions on plant parts, that is, on leaves, twigs, thorns, and fruits (Gottwald and Graham, 1992). Canker lesions initiate as light yellow,

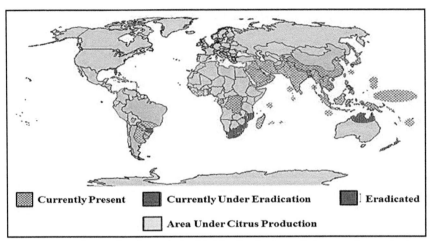

FIGURE 4.1 (See color insert.) Worldwide distribution of citrus canker and where eradication is practiced. (Reprinted from Schubert et al., 2001; http://apsjournals.apsnet. org/doi/abs/10.1094/PDIS.2001.85.4.340. Open access.)

raised, malleable eruptions on the surface of leaves, twigs, and fruits. The margins of the lesions are bordered by a chlorotic yellow halo or rings which aggregate to form bigger patches and with age, spots become white or grayish and finally rupture in the center giving a rough, corky, and crater-like appearance (Das, 2003). Under severe disease conditions, plants show deformity of fruit with premature fruit drop (Rossetti, 1977; Civerolo, 1981; Chand and Pal, 1982; Stall and Seymour, 1983) (Fig. 4.3). Superficial lesions on the fruits can appear when they are immature and do not penetrate deep into the rind.

Bacterial ooze serves as inoculum for further spread of infection from the existing lesions (Timmer et al., 1991). The bacterium enters through stomata (Gottwald and Graham, 1992; Graham et al., 1992; Koizumi and Grierson, 1979) and wounds (Koizumi and Grierson, 1979; Takahishi and Doke, 1984) (Fig. 4.4). The earliest symptoms on leaves appeared around 7 days after inoculation with favorable conditions (20°C and 30°C) (Koizumi, 1985). The infection might be restricted when plant parts become mature for which it becomes more resistant to infection (Stall et al., 1982). The galleries of the citrus leaf miner (*Phyllocnistis citrella* Stainton) provided support the disease amplification (Sinha et al., 1972; Sohi and Sandhu, 1968). Stomatal infection occurs by windblown rain

FIGURE 4.2   (See color insert.) Leaves are demonstrating citrus canker lesions and chlorotic halos of (A) Kagzi lime; (B) Kinnow (mandarin); (C) Ruby nuclear (sweet orange); and (D) Trovita (sweet orange).

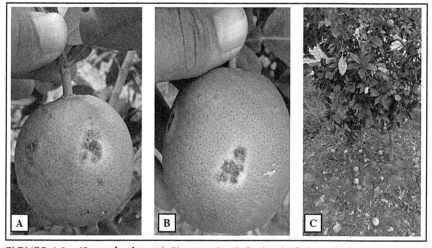

FIGURE 4.3   (See color insert.) Citrus canker infection in fruits and plant. (A) and (B) Citrus canker lesions on fruit of grapefruit (cv. Red blush) and (C) grapefruit (cv. Red blush) tree showing fruit drop.

splashes with wind speeds of 17 to 18 mph (Serizawa and Inoue, 1974). Unsterilized equipment and persons can also facilitate dissemination of the bacteria (Fig. 4.5).

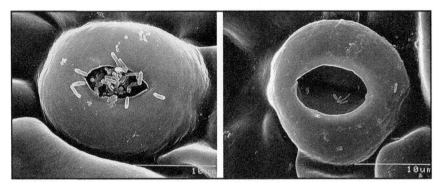

FIGURE 4.4 Electron microscopy photograph of *Xac* bacteria entering stomatal chamber on the leaf of grapefruit. (Reprinted from Gottwald et al. 2002b; http://www. plantmanagementnetwork.org/pub/php/review/citruscanker/ Photos courtesy J. Cubero.)

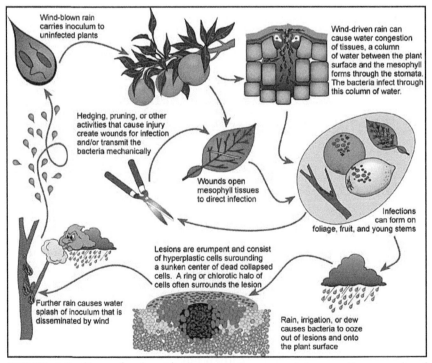

FIGURE 4.5 (See color insert.) Diagrammatic representation on disease cycle of citrus canker. (Reprinted from Schubert et al., 2001; http://apsjournals.apsnet.org/doi/ abs/10.1094/PDIS.2001.85.4.340. Open access.)

4.3.3   FAVORABLE ENVIRONMENTAL CONDITIONS

The weather factors play a pivotal role in the spread of the disease. Warm and humid condition has been most favorable for the pathogen to cause infection and spread the disease. Aiyappa (1958) studied that if environmental conditions are favorable for pathogen than all the cultivated/ wild species of citrus showed susceptible reaction against this bacterium in Karnataka. Prasad (1959) also reported similar observations from Rajasthan. Peltier and Frederich (1926) determined that the optimum temperature for citrus canker is ranged between 28°C and 30°C with high humidity. Similarly, Ramakrishnan (1954) also reported that temperature between 20°C and 30°C with evenly distributed rains were most suitable for the disease development. Many authors also widely acknowledged the relation of citrus canker disease with environment (Kuhara, 1978; Reddy, 1984; Southiosorubini et al., 1986; Kalita et al., 1995; Arora et al., 2013). Gottwald et al. (1989) studied that with the increase in rainfall, there is an increase in spread of the disease. Arora et al. (2013) studied that canker is more severe during July to August and September to October months due to weather parameters like high rainfall, relative humidity, and optimum temperature.

4.3.4   SURVIVAL OF X. axonopodis pv. citri

The existing lesions on plant parts act as the primary source of inoculum for initiation of new infection because of the fact that the bacterium survives in those diseased tissues lesions with variable durations. Vasudeva (1958) found that the bacterium survives in unsterilized soils for 9 days only, but on the contrary, the bacterium can survive up to as many as 52 days on properly sterilized soils. He also observed that the infected leaf carry bacterium for more than 6 months. The bacterium has short life in soil or in fallen leaves because of antagonism organism and other soil microflora present in the soil which have competition for food and shelter (Graham et al., 1989). Just like infected leaves, Chakravarti et al. (1966) found that the bacterium can survive on diseased twigs for more than 6 years. It was also observed that the pathogen can also survive in roots of some grass species belonging to the genera Trichachne, Zoysia, and Panicum. The pathogen has the capability to survive for a number of weeks on nonhost plant like some grasses root zone under eradicated diseased trees in Japan (Pereira et al., 1978) and Brazil (Pereira et al., 1976). Kalita et al. (1997) discovered goat weed as a new host of the pathogen. The pathogen also survives on various inanimate surfaces such as metal, plastics, cloth, and processed wood without sun light condition but inoculum dies within 24–72 h under sun light conditions (Graham et al., 2000). The extracellular polysaccharide (EPS) slime layer of the bacterium acts as a pathogenic factor that prop up bacterial cells for survival (Goto and Hyodo, 1985) in various environmental conditions.

4.3.5 PATHOGEN DISSEMINATION

The bacterium is dispersed with the help of rainwater which run over the surfaces of lesions and then splashing on the disease free plant parts. The pathogen is easily suspended in water and disseminate with the help of mucilaginous coat. The cfu/mL value of bacteria is dependent on lesions age with a maximum of a million cells/drop (Stall et al., 1980). The bacterium is spread by rain and wind with short distance within tree or neighboring trees, but rainstorms can transport bacteria up to 100 m (Goto, 1992). Disease can spread up to 7 mi with help of tropical storms, cyclone, and tornadoes (Gottwald et al., 2001). Propagating material infected with disease have significant role in long-distance pathogen dissemination. But till date, there is no any single evidence regarding seed transmission of the canker bacterium. The bacterium also transmitted by farm workers clothes, equipment, and contaminated budwood. Citrus leaf miner (P. citrella) is very important for the dissemination of the disease and they carry bacteria on their body (Nirvan, 1961; Sohi and Sandhu, 1968; Sinha et al., 1972; Cook, 1988).

4.3.5.1 LEAF MINER INTERACTION

Asian leaf miner, P. citrella Stainton, plays an important role for transmission of canker disease. The insect infests younger leaves, stems, and young fruit, especially grapefruit. Leaf miner larvae are making feeding galleries in the epidermal tissue of new plant growth and tearing the cuticle (Achor et al., 1996). It is facilitating X. citri infection by tearing the cuticle, opening the mesophyll cells for direct infection, and in addition to this, contaminated leaf miner larvae extend bacterium by means of feeding galleries (Graham et al., 1996; Bergamin-Filho et al., 2000) (Fig. 4.6). The exudates containing bacterial ooze from leaf miner-induced lesions helps spreading and accelerating the dispersal process (Gottwald et al., 1997; Bergamin-Filho et al., 2000). No evidence has been presented about the adult insect serving as vector. Leaf miner lesions are allowing bacterium infection for 7–14 days, while it takes 24 h in case of wounds created by wind (Filho and Hughes, 2000).

4.3.6 HOST RANGE OF PATHOGEN

Xanthomonas citri have broad host range, encompassing many citrus species and citrus relative trifoliate orange *Ponciru strifoliata* (Leite and Mohan, 1984; Gottwald et al., 1993). The bacterium is more severe on grapefruit, some sweet oranges (such as hamlin, navel, and pineapple) Mexican lime and hybrids of trifoliate rootstocks. Therefore, these cultivars are not easy to grow in moist subtropical and tropical climates region

FIGURE 4.6 (See color insert.) Leaf miner larvae gallery, which facilitate infection of *X. citri* bacteria.

(Leite and Mohan, 1984; Graham, 2001). Civerolo (1984) studied on hosts of the bacterium and observed the Rutaceae plants other than Citrus and *Poncirus* can serve as host of *X. citri*. However, these plants do not have any crucial role in epidemiology. In Indian condition, the disease is more sever on acid lime and less on mandarin and sweet orange (Ramakrishnan, 1954). Nirvan (1961) studied that Kaghzi Lime was more susceptible as compared to grape fruit and sweet oranges in Karnataka. However, Aiyappa (1958) observed that all the cultivars of citrus and some wild species in Karnataka are susceptible to canker with respective environment condition. Prasad (1959) also made similar observations in Rajasthan. Many other authors also studied susceptible and resistance reaction of citrus cultivars against this pathogen (Naik, 1949; Jain, 1959; Goto, 1992; Gottwald et al., 1993; Graham, 2001). Mundkur (1961) reported that Jambheri, sour orange, and Kaghzi lime were very susceptible as compared to sweet orange and pummel. Moreover, Das (2003) studied that mandarins and lemons were resistant and Kumquats showed immune character under Uttar Pradesh conditions. Kalita et al. (1997) reported a non-Rutaceous natural host goat weed (*Ageratum conyzoides* L.) in India that serves as a host of *X. citri* which commonly present in citrus orchards in India. However, pathogenicity tests of *X. citri* on *Ageratum conyzoides* in Brazil were found to be negative.

4.4   VIRULENCE FACTORS OF *Xanthomonas citri*

Virulence factors are the most fundamental characteristics of the bacterial pathogens, which establish disease severity. The major virulence factors include secretion system, polysaccharides, extracellular enzymes, toxins, and plant hormones. Two bacterium, *Pseudomonas* and *Xanthomonas*, species usually do not use the plant hormones as virulence factors.

4.4.1.   *pthA GENE CANKER PATHOGENICITY DETERMINANT*

The gene is the basic unit of heredity which governs many characters. *pthA* gene in *X. citri* plays important role in pathogenicity which belongs to avrBs3/pthA gene family (Swarup et al., 1991). Initially, majority of the genes belonging to this family were isolated as *avr* genes without knowing properly their function. The genes members *pthA, pthB, pthC,* and *pthW* are helping canker disease development and two involved in cotton blight (*pthN* and *pthN2*) (Chakrabarty et al., 1997). Some other members of the avrBs3/pthA gene family are also capable to help the bacterium for amplifying and release of the disease on the leaf surface. For example, *avrb6* helps X. *campestris* pv. *malvacearum* for dispersal and release of 240-fold bacteria to the leaf surface (Yang et al., 1994).

4.4.2. EXTRACELLULAR POLYSACCHARIDES

EPSs secreted by microorganisms into their environment are the natural polymers having high molecular weight. It plays a significant role in pathogenicity for plant pathogenic bacteria by interfering with host cells and also providing resistance to oxidative stress. *X. citri* generates abundant EPS in the host tissue which helps the bacterium transmission easily by rain splash. The EPS molecules play an important defensive role against the "dilution effect" in water and dehydration in air (Goto and Hyodo, 1985). In vascular plant pathogens, xanthan (exopolysaccharides) is the most important virulent factor that causes wilting in the host plants by blocking the xylem vessels (Denny, 1995; Chan and Goodwin, 1999). Many authors also widely studied about the production of EPS and its role in pathogenicity of *Xanthomonas* spp. (Jansson et al., 1975; Becker et al., 1998; Denny, 1999). Gum genes of numerous *Xanthomonas* spp. including *X. campestris* pv. *campestris, X. axonopodis* pv. *citri, Xanthomonas oryzae* pv. *oryzae,* and *X. axonopodis pv. Manihotis* play a vital role in epiphytic survival and symptomatological formations (Chou et al., 1997; Katzen et al., 1998; Dharmapuri and Sonti, 1999; Kemp et al., 2004; Dunger et al., 2007; Rigano et al., 2007). However, xanthan may or may not help in canker disease development depending on the host and environmental conditions (Dunger et al., 2007; Rigano et al., 2007).

4.4.3 PILI AND ATTACHMENT

The bacteria gets exposed to several proteinaceous structures on their surfaces, which have several functions (Fronzes et al., 2008; Erhardt et al., 2010; Epstein and Chapman, 2008; Ayers et al., 2010; Craig and Li, 2008). In plant pathogenic bacteria, pilus is one of the important virulent factors that works as mediator between bacteria and target cell and involved in avoidance of the host immune systems or biofilm formation. Pili are made from many classes of proteins, associated with adhesion in enterobacteria. It is assuming that *X. citri* has four gene clusters and two independently located genes which are involving in type IV pilus formation and regulation. For adhesion, tips of the pili bind to specific receptors on the surface of host cells (Sauer et al., 2000).

4.4.4 QUORUM SENSING (QS)

Quorum sensing (QS) is a route that allows bacterial cell-to-cell communication (cross-talk) in a wide range of bacteria. The bacterial pathogens multiply and buildup their population with the help of cross talks and then attack on their host cell (Bassler, 1999). Diffusible signal factor (DSF) producing by *Xanthomonas campestris* bacterium for QS production is created by two Rpf proteins, that is, RpfB and RpfF (Barber et al., 1997). When the DSF concentrations are reaching at a threshold level, then after that other Rpf proteins respond to transcription of pathogenicity determinant (Slater et al., 2000). *X. citri* has seven Rpf encoding genes including rpfB and rpfF which shows QS, plays a key role in canker disease pathogenicity (Da Silva et al., 2002) and also xanthan gum production (Tang et al., 1991), thereby supporting in progress of the pathogen from the canker lesion.

4.4.5 SECRETION SYSTEM

Plant pathogenic bacteria uses various secretion systems to deliver effector proteins and nucleic acids into the host cells. *Xanthomonas* spp. also has six types of protein secretion systems which have different compositions and functions (Gerlach and Hensel, 2007). Types I and II secretion systems secrete proteins to the host intercellular spaces, and types III and IV systems can transport proteins or nucleic acids directly into the host plant cell (Ponciano et al., 2003).

Type 1 secretion system (T1SS) has the simplest structure and found in almost all phytopathogenic bacteria. Type I secretion contains adenosine triphosphate (ATP)-binding cassette proteins and carry out the export and import of numerous compounds via energy formed by the hydrolysis of

ATP (Hennecke and Verma, 1991). Till date, no any information about virulence function of T1SS has been published in *Xanthomonas* spp. However, T1SS is always required in *X. oryzae* pv. *oryzae*, which is the causal agent of bacterial blight of rice for the elicitation of a resistance response in rice plants carrying the disease resistance gene *Xa21* (Shen et al., 2002; Da Silva et al., 2004).

Type 2 secretion system (T2SS) is common in all Gram-negative bacteria especially important plant pathogenic bacteria such as *Xanthomonas* spp., *Ralstonia solanacearum*, and *Erwinia* spp. (Finlay and Falkow, 1997; Cianciotto, 2005; Jha et al., 2005). It is involved in the delivery of various proteins, toxins, enzymes, and other virulence factors (Finlay and Falkow, 1997; Lee and Schneewind, 2001). T2SS apparatus consists of 12–15 components, most of which are associated with the inner membrane of bacterium (Sandkvist, 2001). *Xanthomonas* and *Ralstonia* use T2SS for delivery of virulence factors such as pectinolytic and cellulolytic enzymes outside the bacterium.

Type 3 secretion system (T3SS) plays an important role in pathogenicity determination in many Gram-negative bacteria genera, namely, *Xanthomonas*, *Pseudomonas*, *Ralstonia*, *Erwinia*, and *Pantoea* mainly due to their potentiality to produce a T3SS (Desvaux et al., 2004). The main function of the T3SS is to deliver T3SS effectors proteins into plant cells. It is a needle-like structure with a channel in which bacteria transfer their effectors into host cell (Galan and Collmer, 1999; Blocker et al., 2003). The T3SS is conserved between pathogens of plant and animal (Cornelis and Van Gijsegem, 2000) which encoded by hypersensitive response and pathogenicity (*hrp*) gene cluster which is compulsory for both hypersensitive response and pathogenicity. In *Xanthomonas*, the expression of these *hrp* genes is efficiently controlled by the regulators HrpG and HrpX, which are encoded by genes situated outside of *hrp* gene cluster (Schulte and Bonas, 1992). The HrpX protein, an AraC-type transcriptional activator, regulates the expression of genes in *hrp* cluster and other effectors genes (Wengelnik and Bonas, 1996). For more information to know type III effectors in *Xanthomonas* spp., we referred to outstanding reviews (Mudgett, 2005; Schornack et al., 2006; Kay and Bonas, 2009).

In the cases of Type 4 secretion system (T4SS) and Type 5 secretion system (T5SS), the bacterium transfers their macromolecules in to the host cell with the help of T4SS, which is deeply studied in a soil bacterium *Agrobacterium tumefaciens*. *A. tumefaciens* uses *vir* gene operon to deliver transfer DNA (T-DNA) to plant cell, while in case of animal pathogens such as *Legionella pneumophilia* and *Bordetella pertussis*, it uses their system to deliver a combination of deoxyribonucleic acid (DNA) and effector proteins into animal hosts (Burns, 2003; Christie et al., 2005). Three other plant-pathogenic bacteria, *X. axonopodis* pv. *citr*, *Xylella fastidiosa*, and Eca1043, carry gene clusters homologous to *virB* (Bell et al., 2004). In T4SS, an autotransporter is found in *Xylella* and *Xanthomonas* and consists of genes that encode surface associated adhesions (Hahn, 1996).

4.5 CHARACTERIZATION AND GENOME ANALYSIS OF PATHOGEN

4.5.1 PHENOTYPIC AND PHYSIOLOGICAL CHARACTERIZATION

Xanthomonas citri is a Gram-negative, rod-shaped bacterium with size 1.5–2.0 × 0.5–0.75 μm. It is motile having a single polar flagellum. It shares numerous physiological and biochemical feature with other members of the genus *Xanthomonas*. *X. citri* is chemoorganotrophic and obligatorily aerobic with an oxidative metabolism of glucose. The yellow pigment is xanthomonadin (Fig. 4.7). Some of the biochemical characteristics that identify *X. citri* are listed in Table 4.2 (Goto, 1992). Strains of groups B, C, and D are similar to group A but the differences being detected by the exploitation of some carbohydrates (Goto et al., 1980).

4.5.2 MOLECULAR CHARACTERIZATION

Disease eradication and control is possible only if we have rapid, accurate, and sensitive detection method. The PCR is reliable method for detection of many plant pathogens (Henson and French, 1993). Plasmid (*pthA* gene family) and chromosomal DNA of *X. campestris* pv. *citri* (Xcc) have been used to design specific PCR primers (Hartung et al., 1993; Kingsley and Fritz, 2000; Mavrodieva et al., 2004) for detection of *X. axonopodis* pv. *citri*. Lertsuchatavanich et al. (2007) identified 23 strains of *X. axonopodis* pv. *citri* and 34 strains of other xanthomonads including *X. alfalfae* subsp. *citrumelonis*, *X. campestris* pv. *campestris*, *X. campestris* pv. *glycines*, *X. citri* subsp. *malvacearum*, *X. fuscans* subsp. *aurantifolii*, and *X. fuscans*

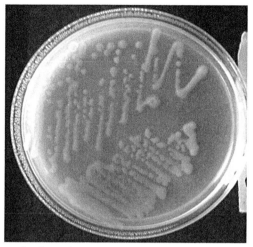

FIGURE 4.7 **(See color insert.)** Yellow pigmented culture plate of *Xanthomonas citri.*

TABLE 4.2 Key Biochemical Characteristics of *X. citri.*

Test	Results
Catalase	+
Oxidase	**– or weak**
Nitrate reduction	–
Hydrolysis	
Starch	+
Casein	+
Tween 80	+
Aesculin	+
Gelatin liquefaction	+
Pectate gel liquefaction	+
Utilization of asparagines	–
Growth require	
Methionine	+
Cysteine	+
0.02% Triphenyl tetrazolium chloride (TTC) (w/v)	–

subsp. *fuscans* with the help of classical PCR. The results showed that 354 F/R primers specifically amplified all of Xcc strains but no other xanthomonad strains (Fig. 4.8). The similarity of nucleotide sequences were 99.7% to gene XAC2443 of Xac strain 306 (Accession AE011881). The many other primer pairs also evaluated against Xanthomonads characterization, namely, KF (5′-TCCACTGCATCCCACAT CTG3′) and KR (5′-CAGGTGTACTGCGCTC TTCTTG-3′); VM3 (5′GCATTT-GATGACGCCATGAC-3′) and VM4 (5′-TCCCTGATGCCTGGAG GATA-3′); 2 (5′-CACGGGTGCAAAA AATCT-3′) and 3 (5′-TGGT-GTCGTCGCTT GTA T-3) (Kingsley and Fritz, 2000; Mavrodieva et al., 2004; Hartung et al., 1993). Cubero and Graham (2002) developed 16S and 23S rDNAs specific primers (J-Rxg: 5′-GCGTTGAGGCTGAGA-CATG-3′ J-RXc2: 5′-CAAGTTGCCTCGGAGCTATC-3′) for character-ization of *X. citri* subsp. *citri* strains (A* and Aw). Similarly, Coletta-Filho et al. (2006) developed primers Xac01: 5′-CGCCATCCCCACCACCAC-CACGAC-3′ Xac02: 5′ AACCGCTCAATGCCATCCACTTCA-3′ based on the *rpf* gene cluster for *X. citri* characterization. Same year, Park et al. (2006) developed primers based on the *hrpW* gene sequences for pathogen characterization.

To find the phylogenetic relationships of species within a genus and family, multilocus sequence analysis (MLSA) is presently and extensively used method. In the case of MLSA studies, partial sequences of genes coding for specific proteins with conserved functions ("housekeeping genes") are used to create phylogenetic trees ultimately assume phylog-enies. MLSA approach has been used for the precise identification of *X. citri* subsp. *citri* by many authors (Almeida et al., 2010; Bui Thi Ngoc et al., 2010; Young et al., 2008). Repetitive element sequence-based PCR

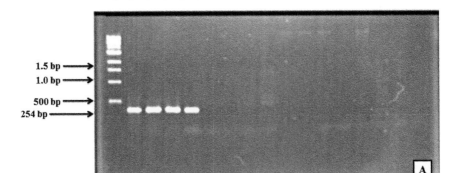

FIGURE 4.8   1% agarose gel electrophoresis amplification of PCR product with 354 F/R primers.

(RES-PCR) is a typing method that facilitates the generation of DNA fingerprinting that helps in distinguishing bacterial strains. It can be used for strain recognition and characterization under specific PCR conditions (Cubero and Graham, 2002).

4.5.3   ANALYSIS OF WHOLE GENOME

The information about nucleotide sequencing of pathogens genome has become an imperative step to understand the mechanisms of pathogenesis and the process that limits the host range of the strain. The genomes sequences of several phytopathogenic bacteria, such as *Agrobacterium tumifaciens*, *Pseudomonas syringae*, *R. solanacearum*, *X. fastidiosa*, and two *X. oryzae*, and many species of soft rot *Erwinia* were recently determined (Simpson et al., 2000; Buell et al., 2003; Lee et al., 2005; Salanoubat et al., 2002; Wood et al., 2001; Pritchard et al., 2012).

Jalan et al. (2013) determined complete genome sequences of *X. citri* subsp. *citri* strain AW12879 with help of using 454 pyrosequencing and Illumina technology, resulting in average coverage of 24× and 410×, respectively. The final sequence of AW12879 strain contains a chromosome (5.32 Mb and 64.71% GC) and plasmids, pXcaw19 (18,869 bp and 63.07% GC) and pXcaw58 (58,317 bp and 61.85% GC content). Cunnac et al. (2013) reported genome sequences of two strains, namely, X18 and X20 of *X. citri* pv. *malvacearum* with help of Illumina Hi-Seq2000 platform (GATC Biotech, Germany). Midha et al. (2012) reported about the 5.1-Mb genome sequence of *X. citri* pv. *mangiferaeindicae* strain LMG941, pathogen of bacterial black spot of mango. It was observed that *X. citri* pv. *mangiferaeindicae* have >99% identical to *X. axonopodis* pv. *citri* and *X. axonopodis* pv. *punicae*, based on 16S rRNA and *rpoB* gene sequences (Sharma et al., 2012).

4.6   MANAGEMENT PRACTICES TO CONTROL DISEASE

The citrus canker has been managed by employing integrated disease control methods. The regions where citrus canker is in endemic form, integrated control actions have great scope to control the disease by applying resistant or tolerant cultivars. The disease has been reported to be more severe on acid lime and less on mandarin and sweet orange (Ramak-rishnan, 1954). Nirvan (1961) found that Kaghzi Lime was more suscep-tible to the disease as compared to that of grape fruit and sweet oranges in Karnataka. Many other authors also studied about susceptible and resis-tance reaction of citrus cultivars against this pathogen (Naik, 1949; Jain, 1959; Goto, 1992; Gottwald et al., 1993; Graham, 2001). Mandarins and lemons are resistant, while Kumquats cultivar showed immune response under Uttar Pradesh conditions (Das, 2003). Cultural practices including windbreaks, and pruning of diseased shoots which are considered as stan-dard control measures of the disease throughout the world (Kuhara, 1978; Leite and Mohan, 1990). Many authors are also widely acknowledged about infected twigs pruning which should be done before monsoon and spraying of 1% Bordeaux mixture (Fawcett, 1936; Naik, 1949; Cheema et al., 1954; Ramakrishnan, 1954; Govinda Rao, 1954; Prasad, 1959; Paracer, 1961). Patel and Desai (1970) revealed that the disease can be reduced via pruning during November–December with 3–4 sprays of Bordeaux mixture (1%) in a year. In the case of other chemicals, effec-tiveness against this pathogen was also studied by many authors, namely, perenox (Chowdhury, 1951), Ultrasulfur (Nirvan, 1961), sodium arsenate + copper sulfate (Patel and Padhya, 1964), Blitox, and nickel chloride (Ram et al., 1972). Rangaswami et al. (1959) studied that 500–1000 ppm streptomycin sulfate was effective when sprayed with 1% glycerine on acid lime. Six sprays of streptomycin sulfate @1000 ppm and two pruning reduced the canker disease in acid lime (Balaraman and Purushotman, 1981). Many others also widely studied the role of copper (alone) and with combination of antibiotics to reduce the citrus cankers diseases (Sawant et al., 1985; Mathur et al., 1973; Krishna and Nema, 1983; Kale et al., 1988; Dakshinamurthi and Rao, 1959; Reddy and Rao, 1960; Rani, 1998; Das and Singh, 2001; Ravikumar et al., 2002; Gopal et al., 2004; Thind and Aulakh, 2007; Thind and Singh, 2015; Scapin et al., 2015). However, the effectiveness of the chemicals is dependent upon many factors, namely, germplasm susceptibility, environmental conditions, and agreement of other control measures (Kuhara, 1978; Leite, 1990; Leite and Mohan, 1990; Leite et al., 1987). Copper reduces infection of the bacterium while rain water with wind save bacterium and introduce directly into stomata (Gottwald and Graham, 1992; Gottwald and Timmer, 1995). The another issues regarding copper is that the bacterium have become resistant to copper due to prolific use by the farmers (Rinaldi and Leite, 2000) and the buildup of copper metal in soils with potential phytotoxic and environmental effects (Alva et al., 1995). Antibiotics are not popular like that of copper-based bactericides (Leite and Mohan, 1990; Timmer, 1988) as the main drawback is the development of the resistance within xanthomonads population against the antibiotics (Ritchie and Dittspongpitch, 1991).

The other splendid method of disease control is using of induced systemic resistance (ISR) agents to innate the immune system of the plants (Kessmann et al., 1994). Numerous mechanisms for ISR may function simultaneously to control the disease and avoiding pathogen resistance (Tally et al., 1999). Several compounds, such as benzothiadiazoles, harpin protein, and salicylic acid, are well-known inducers of plant resistance to Xanthomonads diseases (Romero et al., 2001; Wei et al., 1992).

The canker disease is reduced by spraying of insecticides to control the leaf minor which assist bacterial infections. Shivanker (2000) recommended the application of Monocrotophos (0.05%), Fenvalrate (0.005%), Bifenthrin (0.005%), Trizophos (0.05%), Quinalphos (0.05%), and Neem (50 g/100 mL) to effectively check the extend of disease by leaf miner. Studies on biological control of citrus canker are still in a commencement stage of initiation. Several strains of bacteria, namely, *Bacillus subtilis*, *Erwinia herbicola, Pseudomonas fluorescence*, and *P. syringae* isolated from citrus phylloplane were reported to be antagonistic in vitro to the canker pathogen (Goto et al., 1979; Ota, 1983; Unnimalai and Gnanamanickam, 1984; Kalita et al., 1996). Recently, Das et al. (2014) used an inhibitory strain of *B. subtilis* (S-12) for spraying (2.7×10^9 cells/mL) during peak period of disease (in the month of July) in the field condition and found satisfactory decline of the citrus canker disease. Exclusion of existing disease inoculums by elimination and demolition of infected trees is the most accepted practice to stop further spread (Stall et al., 1987). Eradication programs have taken place in many locations where disease exist previously, namely, Florida, Argentina, Uruguay, Brazil, Paraguay, Fiji Island, and at least five times in Australia. Disease free citrus producing areas, strict quarantine regulations are adopted which aimed at excluding the pathogen. When the canker pathogen is introduced into such an areas (as it was in Florida, USA in 1910, 1984, and 1995), eradication operation was conducted by uprooting and burning all infected trees (Gottwald et al., 2002a,b).

4.7   SUMMARY AND CONCLUSION

Citrus canker is one of the most important diseases, which is present world-wide in entire citrus growing areas in the world. The disease is occurring in severe form during rainy season in India. The *X. citri* is having varieties of virulence factors such as *PthA* gene, QS, EPS, secretion system, and pilus. The bacterium uses these factors for disease development and due to these factors the bacterium is having multi host range. The bacterium survives in the infected lesions and rain water helps the bacterium for further expansion to the healthy plants. The pathogen can be characterized by biochemical, pathological, and molecular tactics. At present, virulence gene specific primers, rDNA specific primers, and MLSA are frequently used for the characterization of *X. citri*. Pathogen can be controlled in the field with the help of some chemicals but still there is need to work more on development of new resistant germplasm as well as improvement of already existed germplasm and also formulation of new chemical molecules which would effectively control this disease.

KEYWORDS

- **citrus canker**
- *Xanthomonas campestris*
- *Phyllocnistis citrella* Stainton
- **quorum sensing**
- *Xanthomonas citri*

REFERENCES

Achor, D. S.; Browning, H. W.; Albrigo, L. G. *Anatomical and Histological Modification in Citrus Leaves Caused by Larval Feeding of Citrus Leaf-Miner (Phyllocnistis citrella Staint)*, Proceeding of International Conference on Citrus Leafminer, Orlando, Florida, 23–25 April 1996. University of Florida, Gainesville, 1996, p 69.

Aiyappa, K. M. Citrus Canker - *Xanthomonas citri* (Hasse) Dowson. *Mysore J. Agric. Sci.* **1958**, *13*, 164–167.

Almeida, N. F.; Yan, S; Cai, R.; Clarke, C. R.; Morris, C. E.; Schaad, N. W.; Schuenzel, E. L.; Lacy, G. H.; Sun, X.; Jones, J. B.; Castillo, J. A.; Bull, C. T.; Leman, S.; Guttman, D. S.; Setubal, J. C.; Vinatzer, B. A. PAMDB, A Multilocus Sequence Typing and Analysis Database and Website for Plant-Associated Microbes. *Phytopathology* **2010**, *100*(3), 208–215.

Alva, A. K.; Graham, J. H.; Anderson, C. A. Soil pH and Copper Effects on Young 'Hamlin' Orange Trees. *Soil Sci. Soc. Am. J.* **1995**, *59*, 481–487.

Alvarez, A. M.; Benedict, A. A.; Mizumoto, C. Y.; Pollard, L. W.; Civerolo, E. L. Analysis of *Xanthomonas campestris* pv. *citri* and *X. c. citrumelo* with Monoclonal Antibodies. *Phytopathology* **1991**, *81*, 857–865.

Arora, A.; Kaur, S.; Rattanpal, H. S.; Singh, J. Development of Citrus Canker in Relation to Environmental Conditions. *Plant Dis. Res.* **2013**, *28*, 100–101.

Ayers, M.; Howell, P. L.; Burrows, L. L. Architecture of the Type II Secretion and Type IV Pilus Machineries. *Future Microbiol.* **2010**, *5*, 1203–1218.

Balaraman, K.; Purushotman, R. Control of Citrus Canker on Acid Lime. *South Indian Hortic.* **1981**, *29*, 175–177.

Barber, C. E.; Tang, J. L.; Feng, J. X.; Pan, M. Q.; Wilson, T. J.; Slater, H.; Dow, J. M.; Williams, P.; Daniels, M. J. A Novel Regulatory System Required for Pathogenicity of *Xanthomonas campestris* is Mediated by a Small Diffusible Signal Molecule. *Mol. Microbiol.* **1997**, *24*, 555–566.

Bassler, B. L. How Bacteria Talk to Each Other: Regulation of Gene Expression by Quorum Sensing. *Curr. Opin. Microbiol.* **1999**, *2*, 582–587.

Becker, A.; Katzen, F.; Puhler, A.; Ielpi, L. Xanthan Gum Biosynthesis and Application: A Biochemical/Genetic Perspective. *Appl. Microbiol. Biotechnol.* **1998**, *50*, 145–152.

Bell, K. S.; Sebaihia, M.; Pritchard, L.; Holden, M. T. G.; Hyman, L. J. Genome Sequence of the Enterobacterial Phytopathogen *Erwinia carotovora* subsp. *atroseptica* and Characterization of Virulence Factors. *Proc. Natl. Acad. Sci. U.S.A.* **2004**, *101*, 11105–11110.

Bergamin-Filho, A.; Amorim, L.; Laranjeira, F.; Gottwald, T. R. *Epidemiology of Citrus Canker in Brazil With and Without The Asian Citrus Leafminer*, Proceedings of the International Citrus Canker Research Workshop. June 20–22, 2000. Division of Plant Industry, Florida Department of Agriculture and Consumer Services, 2000, p 70.

Blocker, A.; Komoriya, K.; Aizawa, S. I. Type III Secretion Systems and Bacterial Flagella: Insights into Their Function from Structural Similarities. *Proc. Natl. Acad. Sci. U.S.A.* **2003**, *100*, 3027–3030.

Buell, C. R.; Joardar, V.; Lindenberg, M.; Selengut, J.; Paulsen, I. T.; Gwnn, M. L.; Dodson, R. J.; Deboy, R. T. The Complete Genome Sequence of the Arabidopsis and Tomato Pathogen *Pseudomonas syringae* pv. *tomato* DC3000. *Proc. Natl. Acad. Sci. U.S.A.* **2003**, *100*, 10181–10186.

Bui Thi Ngoc, L.; Vernière, C.; Jouen, E.; Ah-You, N.; Lefeuvre, P.; Chiroleu, F.; Gagnevin, L.; Pruvost, O. Amplified Fragment Length Polymorphism and Multilocus Sequence Analysis Based Genotypic Relatedness Among Pathogenic Variants of *Xanthomonas citri* pv. *citri* and *Xanthomonas campestris* pv. *bilvae. Int. J. Syst. Evol. Microbiol.* **2010**, *60*, 515–525.

Burns, D. L. Type IV Transporters of Pathogenic Bacteria. *Curr. Opin. Microbiol.* **2003**, *6*, 29–34.

CABI. Crop Protection Compendium. CABI: Wallingford, UK, 2006.

CABI and EPPO *Xanthomonas axonopodis* pv. citri. In: Quarantine Pests for Europe. CABI Publishing: Wallingford, Oxford, UK, 1997, pp. 1101–1108.

Canteros, B. I. Citrus Canker In *Argentina—Control, Eradication, and Current Management.* International Citrus Canker Research Workshop, June 20–22, 2000. Ft. Pierce, Florida, 2000, pp 10–11.

Carrera, C. Informe preliminar sobre una enfermedad nueva comprobada en los citrus de Bella Vista (Corrientes). *Bol. Mens. Min. Agric. Nac. Buenos Aires* **1933,** *34,* 275–280.

Chakrabarty, P. K.; Duan, Y. P.; Gabriel, D. W. Cloning and Characterization of a Member of the *Xanthomonas* avr/pth Gene Family That Evades all Commercially Utilized Cotton R. Genes in the U.S. *Phytopathology* **1997,** *87,* 1160–1167.

Chakravarti, B. P.; Porwal, S.; Rangarajan, M. Studies on Citrus Canker in Rajasthan. In: Disease Incidence and Survival of the Pathogen. *Labdev J. Sci. Technol.* **1966,** *4,* 262–265.

Chan, J. W. Y. F.; Goodwin, P. H. The Molecular Genetics of Virulence of *Xanthomonas campestris. Biotechnol. Adv.* **1999,** *17,* 489–508.

Chand, J. N.; Pal, V. Citrus Canker in India and Its Management. In *Problems of Citrus Diseases in India*; Raychaudhuri, S. P., Ahlawat, Y. S., Eds.; Surabhi Printers and Publishers: New Delhi, 1982, pp 21–26.

Cheema, G. S.; Bhat, S. S.; Naik, K. C. *Commercial Fruits of India*; Macmillan and Co.: Bombay, 1954, p 422.

Chou, F. L.; Chou, H. C.; Lin, Y. S.; Yang, B. Y.; Lin, N. T.; Weng, S. F.; Tseng, Y. H. The *Xanthomonas campestris* Gum D Gene Required for Synthesis of Xanthan Gum Is Involved in Normal Pigmentation and Virulence in Causing Black Rot. *Biochem. Biophys. Res. Commun.* **1997,** *233,* 265–269.

Chowdhury, S. Citrus Canker in Assam. *Plant Protect. Bull.* **1951,** *3,* 78–79.

Christie, P. J.; Atmakuri, K.; Krishnamoorthy, V.; Jakubowski, S.; Cascales, E. Biogenesis, Architecture, and Function of Bacterial Type IV Secretion Systems. *Annu. Rev. Microbiol.* **2005,** *59,* 451–485.

Cianciotto, N. P. Type II Secretion: A Protein Secretion System for All Seasons. *Trends Microbiol.* **2005,** *13,* 581–588.

Civerolo, E. L. Citrus Bacterial Canker Disease: An Overview. *Proc. Int. Soc. Citricult.* **1981,** *1,* 390–394.

Civerolo, E. L. Bacterial Canker Disease of Citrus. *J. Rio Grande Valley Horticult. Assoc.* **1984,** *37,* 127–146.

Civerolo, E. L. Indigenous Plasmids in *X. campestris* pv. *citri. Phytopathology* **1985,** *75,* 524–528.

Coletta-Filho, H. D.; Takita, M. A.; Souza, A. A.; Neto, J. R.; Destefano, S. A. L.; Hartung, J. S.; Machado, M. A. Primers based on the rpf Gene Region Provide Improved Detection of *Xanthomonas axonopodis* pv. citri in Naturally and Artificially Infected Citrus Plants. *J. Appl. Microbiol.* **2006,** *100,* 279–285.

Cook, A. A. Association of Citrus Canker Pustules with Leaf Miner Tunnels in North Yemen. *Plant Dis.* **1988,** *66,* 231–236.

Cornelis, G. R.; Van Gijsegem, F. Assembly and Function of Type III Secretory Systems. *Annu. Rev. Microbiol.* **2000,** *54,* 735–774.

Craig, L.; Li, J. Type IV Pili: Paradoxes in Form and Function. *Curr. Opin. Struct. Biol.* **2008,** *18,* 267–277.

Cubero, J.; Graham, J. H. Genetic Relationship Among Worldwide Strains of *Xanthomonas* Causing Canker in Citrus Species and Design of New Primers for Their Identification by PCR. *Appl. Environ. Microbiol.* **2002,** *68,* 1257–1264.

Cunnac, S.; Bolot, S.; Forero, S. N.; Ortiz, E.; Szurek, B.; Noël, L. D.; Arlat, M.; Jacques, M. A.; Gagnevin, L.; Carrere, S.; Nicole, M.; Koebnik, R. High-Quality Draft Genome

Sequences of Two *Xanthomonas citri* pv. *malvacearum* Strains. *Genome Announc.* **2013,** *1*(4), e00674-13.doi:10.1128/genomeA.00674-13.

Da Silva, A. C.; Ferro, J. A.; Reinach, F. C. Comparison of the Genomes of Two *Xanthomonas* Pathogens with Differing Host Specificities. *Nature* **2002,** *417*, 459–463.

Da Silva, F. G.; Shen, Y.; Dardick, C.; Burdman, S.; Yadav, R. C.; De Leon, A. L.; Ronald, P. C. Bacterial Genes Involved in Type I Secretion and Sulfations Are Required to Elicit the Rice Xa21-Mediated Innate Immune Response. *Mol. Plant–Microbe Interac.* **2004,** *17*, 593–601.

Dakshinamurthi, V.; Rao, D. K. Preliminary Studies on the Control of Citrus Canker on Acid Lime. *Andhra Agric. J.* **1959,** *6*, 145–148.

Das, A. K.; Singh, S. Managing Citrus Bacterial Diseases in the State of Maharashtra. *Indian Hortic.* **2001,** *46*, 11–13.

Das, A. K.. Citrus Canker—A Review. *J. Appl. Hortic.* **2003,** *5*, 52–60.

Das, R.; Mondal, B.; Mondal, P.; Khatua, D. C.; Mukherjee, N. Biological Management of Citrus Canker on Acid Lime through *Bacillus subtilis* (S-12) in West Bengal, India. *J. Biopest.* **2014,** *7*, 38–41.

Denny, T. P. Involvement of Bacterial Polysaccharide in Plant Pathogenesis. *Annu. Rev. Phytopathol.* **1995,** *32*, 173–197.

Denny, T. P. Autoregulator-Dependent Control of Extracellular Polysaccharide Production in Phytopathogenic Bacteria. *Eur. J. Plant Pathol.* **1999,** *105*, 417–430.

Desvaux, M.; Parham, N. J.; Scott-Tucker, A.; Henderson, I. R. The General Secretory Pathway: A General Misnomer? *Trends Microbiol.* **2004,** *12*, 306–309.

Dharmapuri, S.; Sonti, R. V. A Transposon Insertion in the Gum G Homologue of *Xanthomonas oryzae* pv. *oryzae* Causes Loss of Extracellular Polysaccharide Production and Virulence. *FEMS Microbiol. Lett.* **1999,** *179*, 53–59.

Doidge, E. M. Citrus Canker in South Africa. *South Afr. Fruit Grower* **1916,** *3*, 265–268.

Dopson, R. N. The Eradication of Citrus Canker. *Plant Disease Rep.* **1964,** *48*, 30–31.

Dunger, G.; Relling, V. M.; Tondo, M. L.; Barreras, M.; Ielpi, L.; Orellano, E. G.; Ottado, J. Xanthan Is Not Essential for Pathogenicity in Citrus Canker but Contributes to *Xanthomonas* Epiphytic Survival. *Arch. Microbiol.* **2007,** *188*, 127–135.

Egel, D. S.; Graham, J. H.; Stall, R. E. Genomic Relatedness of *Xanthomonas campestris* Strains Causing Diseases of Citrus. *Appl. Environ. Microbiol.* **1991,** *57*, 2724–2730.

EPPO (European and Mediterranean Plant Protection Organization). *PQR Database (version 4.5)*. Paris, EPPO, 2006.

Epstein, E. A.; Chapman, M. R. Polymerizing the Fibre Between Bacteria and Host Cells: The Biogenesis of Functional Amyloid Fibres. *Cell Microbiol.* **2008,** *10*, 1413–1420.

Erhardt, M.; Namba, K.; Hughes, K. T. Bacterial Nanomachines: The Flagellum and Type III Injectisome. *Cold Spring Harbor Persp. Biol.* **2010,** *2*, a000299.

Fawcett, H. S.; Jenkins, A. E. Records of Citrus Canker from Herbarium Specimens of the Genus Citrus in England and the United States. *Phytopathology* **1933,** *23*, 820–824.

Fawcett, H. S. *Citrus Diseases and Their Control*; McGraw-Hill Book Co. Inc., New York, 1936, p 656.

Filho, A. B.; Hughes, G. In *Citrus Canker Epidemiology Methodologies and Approaches*; Proceeding of International Citrus Canker Research Workshop; 20–22 June 2000. Florida, 2000, pp 24–25.

Finlay, B. B.; Falkow, S. Common Themes in Microbial Pathogenicity Revisited. *Microbiol. Mol. Biol. Rev.* **1997**, *61*, 139–169.

Fronzes, R.; Remaut, H.; Waksman, G. Architectures and Biogenesis of Non-Flagellar Protein Appendages in Gram Negative Bacteria. *EMBO J.* **2008**, *27*, 2271–2280.

Gabriel, D. W.; Kingsley, M. T.; Hunter, J. E.; Gottwald, T. Reinstatement of *Xanthomonas citri* (ex Hasse) and *X. phaseoli* (ex Smith) to Species and Reclassification of all *X. campestris* pv. *citri* strains. *Int. J. Syst. Bacteriol.* **1989**, *39*, 14–22.

Galan, J. E.; Collmer, A. Type III Secretion Machines: Bacterial Devices for Protein Delivery into Host Cells. *Science* **1999**, *284*, 1322–1328.

Garnsey, S. M.; Ducharme, E. P.; Lightfied, J. W.; Seymour, C. P.; Griffiths, J. T. Citrus Canker. *Citrus Ind.* **1979**, *60*, 5–6.

Gerlach, R. G.; Hensel, M. Protein Secretion Systems and Adhesins: The Molecular Armory of Gram-Negative Pathogens. *Int. J. Med. Microbiol.* **2007**, *297*, 401–415.

Gopal, K.; Reddy, M. R. S.; Babu, P. G. Management of Bacterial Canker in Acid Lime (*Citrus aurantifolia* Swingle) in Andhra Pradesh. *Indian J. Plant Prot.* **2004**, *32*, 111–113.

Goto, M.; Hyodo, H. Role of Extracellular Polysaccharides of *Xanthomonas campestris* pv. *citri* in The Early Stage of Infection. *Ann. Phytopathol. Soc. Jpn.* **1985**, *51*, 22–31.

Goto, M.; Tadanchi, Y.; Okabe, N. Interaction between *Xanthomonas citri* and *Erwinia herbicola* In Vitro and In Vivo. *Ann. Phytopathol. Soc. Jpn.* **1979**, *45*, 618–624.

Goto, M.; Toyoshima, A.; Messima, M. A. A Comparative Study of The Strains of *Xanthomonas campestris* pv. *citri* Isolates from Citrus Canker in Japan and Cancrosis B in Argentina. *Ann. Phytopathol. Soc. Jpn.* **1980**, *46*, 329–338.

Goto, M. Citrus Canker. In: *Plant Diseases of International Importance: Diseases of Fruit Crops*; Kumar, J. et al. Eds.; Prentice Hall: Englewood Cliffs, NJ, 1992, Vol. III, pp 170–208.

Gottwald, T. R.; Graham, J. H. A Device for Precise and Nondisruptive Stomatal Inoculation of Leaf Tissue with Bacterial Pathogens. *Phytopathology* **1992**, *82*, 930–935.

Gottwald, T. R.; Timmer, L. W. The Efficiency of Windbreaks in Reducing The Spread of Citrus Canker Caused by *Xanthomonas campestris* pv. *citri. Trop. Agric.* **1995**, *72*, 194–201.

Gottwald, T. R.; Irey, M. Post-Hurricane Analysis of Citrus Canker II: Predictive Model Estimation of Disease Spread and Area Potentially Impacted by Various Eradication Protocols Following Catastrophic Weather Events. *Plant Health Progress* **2007**. DOI:10.1094/PHP-2007-0405-01-RS.

Gottwald, T. R.; Graham, J. H.; Egel, D. S. Analysis of Foci of Asiatic Citrus Canker in a Florida Citrus Orchard. *Plant Dis.* **1992**, *76*, 389–396.

Gottwald, T. R.; Graham, J. H.; Schubert, T. S. An Epidemiological Analysis of The Spread of Citrus Canker in Urban Miami, Florida, and Synergistic Interaction with the Asian Citrus Leafminer. *Fruits* **1997**, *52*, 383–390.

Gottwald, T. R.; Graham, J. H.; Civerolo, E. L.; Barret, H. C.; Hearn, C. J. Differential Host Range Reaction of Citrus and Citrus Relatives to Citrus Canker and Citrus Bacterial Spot Determined by Leaf Mesophyll Susceptibility. *Plant Dis.* **1993**, *77*, 1004–1009.

Gottwald, T. R.; Hughes, G.; Graham, J. H.; Sun, X.; Riley, T. The Citrus Canker Epidemic in Florida: the Scientific Basis of Regulatory Eradication Policy for an Invasive Species. *Phytopathology* **2001**, *91*, 30–34.

Gottwald, T. R.; Sun, X.; Riley, T.; Graham, J. H.; Ferrandino, F.; Taylor, E. L. Georeferenced Spatiotemporal Analysis of The Urban Citrus Canker Epidemic in Florida. *Phytopathology* **2002a**, *92*, 361–377.

Gottwald, T. R.; Timmer, L. W.; McGuire, R. G. Analysis of Disease Progress of Citrus Canker in Nurseries in Argentina. *Phytopathology* **1989,** *79,* 1276–1283.

Gottwald, T. R.; Graham, J. H.; Schubert, T. S. An Epidemiological Analysis of the Spread of Citrus Canker in Urban Miami, Florida, and Synergistic Interaction with the Asian Citrus Leafminer. *Fruits* **1997,** *52,* 371–378.

Gottwald, T. R.; Graham, J. H.; Schubert, T. S. Citrus Canker: The Pathogen and Its Impact. Online. *Plant Health Progress* 2002b. Doi:10.1094/PHP-2002-0812-01-RV.

Govinda Rao, P. Citrus Diseases and Their Control in Andhra State. *Andhra Agric. J.* **1954,** *1,* 187–192.

Graham, J. H.; Gottwal, T. R.; Browning, H. W.; Achor, D. S. In *Citrus Leaf Miner Exacerbated the Outbreak of Asiatic Citrus Canker in South Florida*; International Conference on Citrus Leafminer; 1996. Proc Orlando, University of Florida, 1996, p 83.

Graham, J. H.; Gottwald, T. R.; Civerolo, E. L.; McGuire, R. G. Population Dynamics and Survival of *Xanthomonas campestris* in Soil in Citrus Nurseries in Maryland and Argentina. *Plant Dis.* **1989,** *73,* 423–427.

Graham, J. H.; Gottwald, T. R.; Riley, T. D.; Achor, D. Penetration Through Leaf Stomata and Growth of Strains of *Xanthomonas campestris* in Citrus Cultivars Varying in Susceptibility to Bacterial Diseases. *Phytopathology* **1992,** *82,* 1319–1325.

Graham, J. H.; Gottwald, T. R.; Riley, T. D.; Cubero, J.; Drouillard, D. L. In *Survival of Xanthomonas campestri pv. citri (Xcc) on Various Surfaces and Chemical Control of Asiatic Citrus Canker (ACC). (Abstr.)*; Proc. Int. Citrus Canker Res. Workshop, June 20–22, 2000. Florida Department of Agriculture and Consumer Services. Published online, 2000.

Graham, J. H. Varietal Susceptibility to Citrus Canker: Observations from Southern Brazil. *Citrus Ind.* **2001,** *82,* 15–17.

Hahn, M. G. Microbial Elicitors and Their Receptors in Plants. *Annu. Rev. Phytopathol.* **1996,** *34,* 387–411.

Hartung, J. S.; Daniel, J. F.; Pruvost, O. P. Detection of *Xanthomonas campestris* pv. *citri* by the Polymerase Chain Reaction. *Appl. Environ. Microbiol.* **1993,** *59,* 1143–1148.

Hartung, J. S.; Daniel, J. F.; Pruvost, O. P.; Civerolo, E. L. Detection of *Xanthomonas campestris* pv. *citri* by the Polymerase Chain Reaction Method. *Appl. Environ. Microbiol.* **1993,** *59*(4), 1143–1148.

Hartung, J. S.; Civerolo, E. L. In *Restriction Fragment Length Polymorphims Distinguish Xanthomonas campestris Strains Isolated from Florida Citrus Nurseries from Xanthomonas campestris pv. citri*; 7th International Conference of Plant Pathology; 1989 Akademiai Kiado, Budapest, Hungary, 1989, pp 503–508.

Hartung, J. S.; Pruvost, O. P.; Villemot, I.; Alvarez, A. Rapid and Sensitive Colorimetric Detection of *Xanthomonas axonopodis* pv. *citri* by Immunocapture and Nested-Polymerase Chain Reaction. *Phytopathology* **1996,** *86,* 95–101.

Hartung, J. S. Plasmid-Based Hybridisation Probes for Detection and Identification of *Xanthomonas campestris* pv. *citri. Plant Dis.* **1992,** *76,* 889–893.

Hasse, C. H. *Pseudomonas citri*, The Cause of Citrus Canker. *J. Agric. Res.* **1915,** *4,* 97–100.

Hennecke, H.; Verma, D. P. S. *Advances in Molecular Genetics of Plant–Microbe Interactions*; Kluwer Academic Publishers: Dordrecht, Netherlands, 1991, pp 3–9.

Henson, J. M.; French, R. The Polymerase Chain Reaction and Plant Disease Diagnosis. *Annu. Rev. Phytopathol.* **1993,** *31,* 81–109.

Ibrahim, G.; Bayaa, B. Fungal, Bacterial and Nematological Problems of Citrus, Grape and Stone Fruits in Arab Countries. *Arab J. Plant Protect.* **1989,** *7*(2), 190–197.

IMI. *Distribution Maps of Plant Diseases, Map No. 11*; CAB International: Wallingford, UK, 1996.

IPPC. Eradication of Citrus Canker from Australia. *IPPC Official Pest Report, No. AU-18/1*; FAO: Rome, Italy, 2009. https://www.ippc.int/IPP/En/default.

Jain, S. S. In *Citrus Canker*; Proc Seminar on Diseases of Horticultural Plants, Simla, 1959, pp 104–177.

Jalan, N.; Kumar, D.; Yu, F.; Jones, J. B.; Graham, J. H.; Wang, N. Complete Genome Sequence of *Xanthomonas citri* subsp. *citri* Strain AW12879, A Restricted-Host-Range Citrus Canker-Causing Bacterium. *Genome Announcement* **2013,** *1*(3), e00235-13. DOI:10.1128/genomeA.00235-13.

Jansson, P. E.; Kenne, L.; Lindberg, B. Structure of Extracellular Polysaccharide from *Xanthomonas campestris. Carbohydr. Res.* **1975,** *45*, 274–282.

Jha, G.; Rajeshwari, R.; Sonti, R. V. Bacterial Type Two Secretion System Secreted Proteins: Double-Edged Swords for Plant Pathogens. *Mol. Plant-Microbe Interact.* **2005,** *18*, 891–898.

Jones, D. R. Successful Eradication of Citrus Canker from Thursday Island. *Austr. Plant Pathol.* **1991,** *20*(3), 89–91.

Kale, K. B.; Raut, J. G.; Ohekar, G. B. Efficacy of Fungicides and Antibiotics Against Acid Lime Canker. *Pesticides* **1988,** *22*(1), 26–27.

Kalita, P.; Bora, L. C.; Bhagabati, K. N. Influence of Environmental Parameters on Citrus Canker Incidence in Assam. *J. Agric. Sci. Soc. North-East India* **1995,** *8*, 33–35.

Kalita, P.; Bora, L. C.; Bhagabati, K. N. Phylloplane Microflora of Citrus and Their Role in Management of Citrus Canker. *Indian Phytopathol.* **1996,** *49*, 234–237.

Kalita, P.; Bora, L. C.; Bhagabati, K. N. Goat Weed-Host of Citrus Canker (*X. campestris* pv. *citri*). *J. Mycology Plant Pathol.* **1997,** *27*, 96–97.

Katzen, F.; Ferreiro, D. U.; Oddo, C. G.; Ielmini, M. V.; Becker, A.; Puhler, A.; Ielpi, L. *Xanthomonas campestris* pv. *campestris* Gum Mutants: Effects on Xanthan Biosynthesis and Plant Virulence. *J. Bacteriol.* **1998,** *180*, 1607–1617.

Kay, S.; Bonas, U. How *Xanthomonas* Type III Effectors Manipulate the Host Plant. *Curr. Opin. Microbiol.* **2009,** *12*, 1–7.

Kemp, B. P.; Horne, J.; Bryant, A.; Cooper, R. M. *Xanthomonas axonopodis* pv. *manihotis* gumD Gene is Essential for EPS Production and Pathogenicity and Enhances Epiphytic Survival on Cassava (*Maniho esculente*). *Physiol. Mol. Plant Pathol.* **2004,** *64*, 209–218.

Kessmann, H.; Stauv, T.; Hoffmann, C.; Maetzke, T.; Herzog, J. Induction of Systemic Acquired Disease Resistance in Plants by Chemicals. *Annu. Rev. Phytopathol.* **1994,** *32*, 439–459.

Kingsley, M. T.; Fritz, L. K. Identification of Citrus Canker Pathogen *Xanthomonas axonopodis* pv. *citri* A by Fluorescent PCR Assays (Abstr.). *Phytopathology* **2000,** *90*(Suppl.), S42.

Koizumi, M.; Grierson, W. Relation of Temperature to the Development of Citrus Canker Lesions in the Spring. *Proc. Int. Soc. Citricult.* **1979,** *3*, 924–928.

Koizumi, M. Citrus Canker: The World Situation. In *Citrus Canker: An International Perspective*; Timer, L.W., Ed. Proceedings Symptoms Institute of Food Agriculture Sciences University of Florida, 1985, pp 2–7.

Krishna, A.; Nema, A. G. Evaluation of Chemicals for the Control of Citrus Canker. *Indian Phytopathol.* **1983**, *36*, 348–350.

Kuhara, S. Present Epidemic Status and Control of The Citrus Canker Disease (*Xanthomonas citri* (Hasse) Dowson in Japan. *Rev. Plant Protect. Res.* **1978**, *11*, 132–142.

Lee, B. M. et al. The Genome Sequence of *Xanthomonas oryzae* pv. *oryzae* KACC10331, The Bacterial Blight Pathogen of Rice. *Nucleic Acids Res.* **2005**, *33*, 577–586.

Lee, H. A. Further Data on the Susceptibility of Rutaceous Plants to Citrus Canker. *J. Agric. Res.* **1918**, *15*, 661–665.

Lee, V. T.; Schneewind, O. Protein Secretion and the Pathogenesis of Bacterial Infections. *Genes Dev.* **2001**, *15*, 1725–1752.

Leite Jr, R. P.; Mohan, S. K. Evaluation of Citrus Cultivars for Resistance to Canker Caused by *Xanthomonas campestris* pv. *citri* (Hasse) Dye in the State of Paraná, Brazil. *Proc. Int. Soc. Citricult.* **1984**, *1*, 385–389.

Leite Jr, R. P.; Mohan, S. K.; Pereira, A. L. G.; Campacci, C. A. Integrated Control of Citrus Canker: Effect of Genetic Resistance and Application of Bactericides. *Fitopatologia-Brasileira* **1987**, *12*, 257–263.

Leite Jr., R. P.; Mohan, S. K. Integrated Management of The Citrus Bacterial Canker Disease Caused by *Xanthomonas campestris* pv. *citri* in the State of Paraná, Brazil. *Crop Protect.* **1990**, *9*, 3–7.

Leite, Jr., R. P.. Citrus Canker: Prevention and Control in the State of Parana. Fundacao IAPAR, Circular Instituto Agronomico do Paraná. No. 61, 1990.

Lertsuchatavanich, U.; Paradornuwat, A.; Chunwongse, J.; Schaad, N. W.; Thaveechai, N. Novel PCR Primers for Specific Detection of *Xanthomonas citri* subsp. *citri* the Causal Agent of Bacterial Citrus Canker. *Kasetsart J. Nat. Sci.* **2007**, *41*, 262–273.

Loucks, K. W. Citrus Canker and its Eradication in Florida. Unpublished Manuscript in the Archives of the Florida Department of Agriculture, Division of Plant Industry, Gainesville, 1934.

Luthra, J. C.; Sattar, A. Citrus Canker and Its Control in Punjab. *Punjab Fruit J.* **1942**, *6*(1), 179–182.

Mathur, A. S.; Irulappan, I.; Godhar, R. B. Efficacy of Different Fungicides and Antibiotics in The Control of Citrus Canker Caused by *Xanthomonas citri* (Hasse) Dowson. *Mysore Agric J.* **1973**, *60*, 626.

Mavrodieva, V.; Levy, L.; Gabriel, D. W. Improved Sampling Methods for Real-Time Polymerase chain Reaction Diagnosis of Citrus Canker from Field Samples. *Phytopathology* **2004**, *94*, 61–68.

Mehrotra, R. S. Bacteria and Bacterial Diseases. *Plant Pathology*; Tata McGraw–Hill pub. Co. Ltd.: New Delhi, 1980, pp 636–638.

Midha, S.; Ranjan, M.; Sharma, V.; Pinnaka, A. K.; Patil, P. B. Genome Sequence of *Xanthomonas citri* pv. *mangiferae indicae* Strain LMG 941. *J. Bacteriol.* **2012**, *194*(11), 30–31.

Mudgett, M. B. New Insights to the Function of Phytopathogenic Bacterial Type III Effectors in Plants. *Annu. Rev. Plant Physiol.* **2005**, *56*, 509–531.

Mundkur, B. B. *Fungi and Plant Disease*; Macmillan and Co. Ltd: New York, 1961, p. 246.

Naik, K. C. *South Indian Fruits and Their Culture*; Varadachary and Co.: Madras, 1949, p 335.

Namekata, T.; Oliveira, A. R. Comparative Serological Studies Between *Xanthomonas citri* and a Bacterium Causing Canker on Mexican Lime. Proceeding of International Conference on Plant Pathogens; Netherlands, 1972, p 365.

Nirvan, R. S. Effect of Antibiotic Sprays on Citrus Canker. *Hortic. Adv.* **1960**, *4*, 155–160.

Nirvan, R. S. Citrus Canker and Its Control. *Hortic. Adv.* **1961**, *5*, 171–175.

OEPP/EPPO. EPPO Standards PM 3/27 Quarantine Procedure for *Xanthomonas campestris* pv. *citri*. Bulletin OEPP/EPPO Bulletin **1990**, *20*, 263–272.

Ota, T. Interaction In Vitro and In Vivo Between *Xanthomonas campestris* pv. *citri* and Antagonistic *Pseudomonas* sp. *Ann. Phytopathol. Soc. Jpn.* **1983**, *49*, 308.

Paracer, C. S. Some Important Diseases of Fruit Trees. *Punjab Hortic. J.* **1961**, *1*(1), 45–47.

Park, D.; Hyun, J.; Park, Y.; Kim, J.; Kang, H.; Hahn, J.; Go, S. Sensitive and Specific Detection of *Xanthomonas axonopodis* pv. *citri* by PCR Using Pathovar Specific Primers Based on *hrpW* Gene Sequences. *Microbiol. Res.* **2006**, *161*(2), 145–149.

Parsai, P. S. In *Citrus Canker*. Proceedings, Seminar on Diseases of Horticultural Plants, Simla, 1959, pp 91–95.

Patel, M. K.; Padhya, A. C. Sodium Arsenite, Copper Sulphate Spray for the Control of Citrus Canker. *Curr. Sci.* **1964**, *33*, 87–88.

Patel, R. S.; Desai, M. V. Control of Citrus Canker. *Indian J. Hortic.* **1970**, *27*, 93–98.

Peltier, G. L.; Frederich, W. J. Effects of Weather on the World Distribution and Prevalence of Citrus Canker and Citrus Scab. *J. Agric. Res.* **1926**, *32*, 147–164.

Pereira, A. L.; Watanabe, K.; Zagato, A. G.; Cianciulli, P. L. Survival of *Xanthomonas citri* (Hasse) Dowson [the Causal Agent of Citrus Canker] on Sourgrass (*Trichachne insularis* (L.) Nees) from Eradicated Orchards in the State of Sao Paulo, Brazil. *Biologico* **1976**, *42*, 217–221.

Pereira, A. L.; Watanabe, K.; Zagatto, A. G.; Cianciulli, P. L. Survival of *Xanthomonas citri* (Hasse) Dowson, the Causal Agent of "Citrus Canker" in the Rhizosphere of Guineagrass (*Panicum maximum* Jacq.). *Biologico* **1978**, *44*, 135–138.

Ponciano, G.; Ishihara, H.; Tsuyumu, S.; Leach, J. E. Bacterial Effectors in Plant Disease and Defense: Keys to Durable Resistance? *Plant Dis.* **2003**, *87*(11), 1272–1282.

Prasad, N. In *Citrus Canker*; Proceedings, Seminar on Disease of Horticultural Plants, Simla, 1959, pp 87–88.

Pritchard, L.; Humphris, S.; Saddler, G.; Parkinson, N. M.; Bertrand, V.; Elphinstone, J. G.; Toth, I. K. Detection of Phytopathogens of the Genus *Dickeya* Using a PCR Primer Prediction Pipeline for Draft Bacterial Genome Sequences. *Plant Pathol.* **2013**, *62*, 587–596.

Pruvost, O.; Hartung, J. S.; Civerolo, E. L.; Dubois, C.; Perrier, X. Plasmid DNA Fingerprints Distinguish Pathotypes of *Xanthomonas campestris* pv. *citri*, the Causal Agent of Citrus Bacterial Canker Disease. *Phytopathology* **1992**, *82*, 485–490.

Ram, G.; Nirvan, R. S.; Saxena, M. L. Control of Citrus Canker. *Prog. Hortic.* **1972**, *12*, 240–243.

Ramakrishnan, T. S. Common Diseases of Citrus in Madras State. Govt. of Madras Publication: Chennai, India, 1954.

Rangaswami, G.; Rao, R. R.; Lakshaman, A. R. Studies on Control of Citrus Canker with Streptomycin. *Phytopathology* **1959**, *49*, 224–226.

Rani, U. Investigations on Black Spot of Pomegranate (*Punica granatum* L.) and Its Management. M.Sc. Thesis, Punjab Agricultural University Ludhiana, Punjab, India, 1998.

Ravikumar, M. R.; Jahagirdar, S.; Khan, A. N. A. In *Management of Bacterial Leaf Spot of Grape through Chemicals and Antibiotics in Northern Karnataka.* Paper Presented, Ann Meet Symposium Plant Disease Scenario in Southern India, Bangalore, India, Dec 19–21, 2002, p 50.

Reddy, B. C. Incidence of Bacterial Canker of Citrus in Relation to Weather. *Geobios New Rep.* **1984,** *3,* 918–924

Reddy, G. S.; Rao, A. P. Control of Canker in Citrus Nurseries. *Andhra Agric. J.* **1960,** *7*(3), 11–13.

Rigano, L. A.; Siciliano, F.; Enrique, R. Biofilm Formation, Epiphytic fitness, and Canker Development in *Xanthomonas axonopodis* pv. *citri. Mol. Plant–Microbe Interact.* **2007,** *20,* 1222–1230.

Rinaldi, D. A. M. F.; Leite, R. P. Jr. Adaptation of *Xanthomonas axonopodis* pv. *citri* Population to the Presence of Copper Compounds in Nature. *Proc. Int. Soc. Citricult.* **2000,** *2,* 1064.

Ritchie, D. F.; Dittspongpitch, V. Copper and Streptomycin Resistant Strains and Host Differentiated Races of *Xanthomonas campestris* pv. *vesicatoria* in North Carolina. *Plant Dis.* **1991,** *75,* 733–736.

Romero, A. M.; Kousik, C. S.; Ritchie, D. F. Resistance to Bacterial Spot in Bell Pepper induced by Acibenzolar-*S*-Methyl. *Plant Dis.* **2001,** *85,* 189–194.

Rossetti, V. Citrus Canker in Latin America. *Proc. Int. Soc. Citricult.* **1977,** *3,* 918–924.

Salanoubat, M. et al. Genome Sequence of the Plant Pathogen *Ralstonia solanacearum. Nature* **2002,** *415,* 497–502.

Sandkvist, M. Biology of Type II Secretion. *Mol. Microbiol.* **2001,** *40,* 271–283.

Sauer, F. G.; Mulvey, M. A.; Schilling, J. D.; Martinez, J. J.; Hultgren, S. J. Bacterial Pili: Molecular Mechanisms of Pathogenesis. *Curr. Opin. Microbiol.* **2000,** *3,* 65–72.

Sawant, D. M.; Ghawte, A. G.; Jadhav, J. V.; Chaudhari, K. G. Control of Citrus Canker in Acid Lime. *Maharashtra J. Hortic.* **1985,** *2,* 55–58.

Scapin, M. S.; Behlau, F.; Scandelai, L. H. M.; Fernandes, R. S.; Silva Junior, G. J.; Ramos, H. H. Tree-Row-Volume-Based Sprays of Copper Bactericide for Control of Citrus Canker. *Crop Protect.* **2015,** *77,* 119–126.

Schornack, S.; Meyer, A.; Romer, P.; Jordan, T.; Lahaye, T. Gene-for-Gene-Mediated Recognition of Nuclear-Targeted AvrBs3-Like Bacterial Effector Proteins. *J. Plant Physiol.* **2006,** *163,* 256–272.

Schoulties, C. L.; Civerolo, E. L.; Miller, J. W.; Stall, R. E.; Krass, C. J.; Poe, S. R.; Ducharme, E. P. Citrus Canker in Florida. *Plant Dis.* **1987,** 71, 388–395.

Schubert, T. S.; Miller, J. W. Bacterial Citrus Canker. Fla. Dep. Agric. Conservation Serv. – Div. Plant Ind. *Plant Pathology Circ.* **2000,** *377,* revised.

Schubert, T. S.; Miller, J. W.; Gabriel, D. W. Another Outbreak of Bacterial Canker on Citrus in Florida. *Plant Dis.* **1996,** *80,* 1208.

Schubert, T. S.; Rizvi, S. A.; Sun, X.; Gottwald, T. R.; Graham, J. H.; Dixon, W. N. Meeting the Challenge of Eradicating Citrus Canker in Florida-Again. *Plant Dis.* **2001,** *85,* 340–356.

Schubert, T. S. Recent History of Citrus Canker Eradication Programs in Florida. *Newslett. Florida Phytopathol. Soc.* **1991,** *2,* 1–6.

Schubert, T. S.; Miller, J. W.; Gabriel, D. W. Another Outbreak of Bacterial Canker on Citrus in Florida. *Plant Dis.* **1996,** *80*, 1208.

Schulte, R.; Bonas, U. Expression of the *Xanthomonas campestris* pv. *vesicatoria* hrp Gene Cluster, which Determines Pathogenicity and Hypersensitivity on Pepper and Tomato, Is Plant Inducible. *J. Bacteriol.* **1992,** *174*, 815–823.

Serizawa, S.; Inoue, K. Studies on Citrus Canker, *Xanthomonas citri*. III. The Influence of Wind on the Infection of Citrus Canker. Bull. Shizuoka Prefect. Citrus Exp. Stn. Komagoe Shimizu City, Japan **1974,** *11*, 54–67.

Sharma, V.; Midha, S.; Ranjan, M.; Pinnaka, A.; Patil, P. B. Genome Sequence of *Xanthomonas axonopodis* pv. *punicae*, Strain LMG859. *J. Bacteriol.* **2012,** *194*, 2395.

Shen, Y.; Sharma, P.; da Silva, F. G.; Ronald, P. The *Xanthomonas oryzae* pv. *oryzae* raxP and raxQ Gene Encode an ATP Sulphurylase and Adenosine-5′-Phosphosulphate Kinase That Are Required for AvrXa21 Avirulence Activity. *Mol. Microbiol.* **2002,** *44*, 37–48.

Shivanker, V. J. In *Recent Trends in Insect-Pest Management in Citrus,* Proceedings National Symposium on Citriculture. NRC for Citrus, Nagpur, Maharashtra, 2000, pp 762–784.

Simpson, A. J.; Reinach, F. C.; Arruda, P.; Abreu, F. A.; Acencio, A. R.; Alves, L. M.; Araya, J. E.; Baia, G. S.; Baptista, C. S.; Barros, M. H. The Genome Sequence of the Plant Pathogen *Xylella fastisiosa. Nature* **2000,** *406*, 151–157.

Sinha, M. K.; Batra, R. C.; Uppal, D. K. Role of Citrus Leaf-Miner (*Phyllocnistis citrella* Staintan (sic) on the Prevalence and Severity of Citrus Canker [*Xanthomonas citri* (Hasse) Dowson]. *Madras Agric. J.* **1972,** *59*, 240–245.

Slater, H.; Alvarez-Morales, A.; Barber, C. E.; Daniels, M. J.; Dow, J. M. A Two-Component System Involving an HD-GYP Domain Protein Links Cell–Cell Signalling to Pathogenicity Gene Expression in *Xanthomonas campestris. Mol. Microbiol.* **2000,** *38*, 986–1003.

Sohi, G. S.; Sandhu, M. S. Relationship Between Citrus Leafminer (*Phyllocnistis citrella* Stainton) Injury and Citrus Canker [*Xanthomonas citri* (Hasse) Dowson] Incidence on Citrus Leaves. *J. Res.* **1968,** *5*, 66–69.

Sothiosorubini, N.; Sundersan, R. S. V.; Sivapalan, A. Studies on *Xanthomonas* pv. *citri,* Causing Canker Disease of Citrus. *Vingnaman J. Sci. Sri Lanka* **1986,** *1*, 19–25.

Stall, R. E.; Civerolo, E. L. Research Relating to the Recent Outbreak of Citrus Canker in Florida. *Annu. Rev. Phytopathol.* **1991,** *29*, 399–420.

Stall, R. E.; Seymour, C. P. Canker, A Threat to Citrus in the Gulf-Coast States. *Plant Dis.* **1983,** *67*, 581–585.

Stall, R. E.; Civerolo, E. L.; Ducharme, E. P.; Krass, C. J.; Poe, S. R.; Miller, J. W.; Schoulties, C. L. Management of Citrus Canker by Eradication of *Xanthomonas campestris* pv. *citri*. In *Plant Pathogenic Bacteria, Current Plant Science and Biotechnology in Agriculture*; Civerolo, E. L., Collmer, A., Davis, R. E., Gillaspie, A. G., Eds.; Martinus Nijhoff Publishers: Dordrecht, The Netherlands, 1987, pp 900–905.

Stall, R. E.; Marcó, G. M.; Canteros de Echenique, B. I. Importance of Mesophyll in Mature-Leaf Resistance to Cancrosis of Citrus. *Phytopathology* **1982,** *72*, 1097–1100.

Schubert, T. S.; Gottwald, T. R.; Rizvi, S. A.; Graham, J. H.; Sun, X.; Dixon, W. N. Meeting The Challenge of Eradicating Citrus Canker in Florida Again. *Plant Dis.* **2001,** *85*(4), 340–356.

Stall, R. E.; Miller, J. W.; Marco, G. M.; Canteros de Echenique, B I. Population Dynamics of *Xanthomonas citri* Causing Cancrosis of Citrus in Argentina. *Proc. Florida* State *Hortic. Soc.* **1980,** *93,* 10–14.

Sun, X.; Stall, R. E.; Cubero, J.; Gottwald, T. R.; Graham, J. H.; Dixon, W. D.; Schubert, T. S.; Peacock, M. E.; Dickestein, E. R.; Chaloux, P. H. Detection of a Unique Isolate of Citrus Canker Bacterium from Key Lime in Wellington and Lake Worth, Florida. Proceedings of the International Citrus Canker Research Workshop. Fort Pierce (US), 2000. http://doacs.state.fl.us/canker

Swarup, S.; De Feyter, R.; Brlansky, R. H.; Gabriel, D. W. A Pathogenicity Locus from *Xanthomonas citri* Enables Strains from Several Pathovars of *X. campestris* to Elicit Canker Like Lesions on Citrus. *Phytopathology* **1991,** *81,* 802–809.

Takahishi, T.; Doke, N. A Role of Extracellular Polysaccharides of *Xanthomonas campestris* pv. *citri* in Bacterial Adhesion to Citrus Leaf Tissues in Preinfectious Stage. *Ann. Phytopathol. Soc. Jpn.* **1984,** *50,* 565–573.

Tally, A.; Oostendorp, M.; Lawton, K.; Staub, T.; Bassi, B. Commercial Development of Elicitors of Induced Resistance to Pathogens. In: *Induced Plant Defenses Against Pathogens and Herbivores. Biochemistry, Ecology and Agriculture*; Agrawal, A. A., Tuzun, S., Bent, E., Eds., St. Paul, MN: American Phytopathological Society Press, 1999, pp 357–370.

Tang, J. L.; Liu, Y. N.; Barber, C. E.; Dow, J. M.; Wootton, J. C.; Daniels, M. J. Genetic and Molecular Analysis of a Cluster of rpf Genes Involved in Positive Regulation of Synthesis of Extracellular Enzymes and Polysaccharide in *Xanthomonas campestris* Pathovar *campestris*. *Mol. Genet. Genomics* **1991,** *226,* 409–417.

Thind, S. K.; Aulakh, P. S. Efficacy of Agrochemicals for the Control of Citrus Canker in Kinnow Mandarin. *J. Res.* **2007,** *44,* 312–313.

Thind, S. K.; Singh, D. Comparative Efficacy of Inoculation Methods and Agrochemicals in Managing Citrus Canker (*Xanthomonas axonopodis* pv. *citri*). *Plant Dis. Res.* **2015,** *30,* 142–146.

Timmer, L. W.; Gottwald, T. R.; Zitko, S. E. Bacterial Exudation from Lesions of Asiatic Citrus Canker and Citrus Bacterial Spot. *Plant Dis.* **1991,** *75,* 192–195.

Timmer, L. W. Evaluation of Bactericides for Control of Citrus Canker in Argentina. *Proc. Florida State Hortic. Soc.* **1988,** *101,* 6–9.

Unnamalai, N.; Gnanamanikam, S. S. *Pseudomonas flurenscence* Is an Antogonist to *Xanthomonas citri*, the Incitant of Citrus Cankcr. *Curr. Sci.* **1984,** *53,* 703–704.

Vasudeva, R. S. Scientific Report Indian Agricultural Research Institute: New Delhi, India, 1958; p 93.

Vauterin, L.; Hoste, B.; Kersters, K.; Swings, J. Reclassification of *Xanthomonas*. *Int. J. Syst. Evol. Microbiol.* **1995,** *45,* 472–489.

Venkatakrishnaiah, N. S. Canker Disease of Sour Lime and Its Control. *J. Mysore Hortic. Sci.* **1957,** *2,* 40–44.

Verniere, C.; Hartung, J. S.; Pruvost, O. P.; CIverolo, E. L.; Alvarez, A. M.; Maestri, P. Luisetti, J. Characterization of Phenotypically Distinct Strains of *Xanthomonas axonopodis* pv. *citri* from Southwest Asia. *Eur. J. Plant Pathol.* **1998,** *104,* 477–487.

Wei, Z. M.; Laby, R. J.; Zumoff, C. H.; Bauer, D. W.; He, S. Y.; Collmer, A.; Beer, S. V. Harpin, Elictor of the Hypersensitive Response Produced by the Plant Pathogen *Erwinia amylovora. Science* **1992,** *257,* 85–88.

Wengelnik, K.; Bonas, U. HrpXv, an AraC-Type Regulator, Activates Expression of Five of the Six Loci in the hrp Cluster of *Xanthomonas campestris* pv. *vesicatoria*. *J. Bacteriol.* **1996,** *178,* 3462–3469.

Whiteside, J. O.; Garney, S. M.; Timmer, L W. *Compendium of Citrus Diseases*; The American Phytopathological Society, 1988, pp 80.

Wood, D. W.; Settubal, J. C.; Kaul, R.; Monks, D. E.; Kitajima, J. P.; Okura, V. K.; Zhou, Y.; Chen, L.; Wood, G. E. The Genome of the Natural Genetic Engineer *Agrobacterium tumefaciens* C58. *Science* **2001,** *294,* 2317–2323.

Yang, Y.; De Feyter, R.; Gabriel, D. W. Host-Specific Symptoms and Increased Release of *Xanthomonas citri* and *X. campestris* pv. *malvacearum* from Leaves are Determined by the 102-bp Tandem Repeats of pthA and avrb6, Respectively. *Mol. Plant-Microbe Interact.* **1994,** *7,* 345–355.

Yang, X. J.; Chen, F. R.; Xie, S. Y. Advances in Research on the Occurrence and Control of Citrus Canker. *China Fruits* **2002,** *5,* 46–50.

Young, J. M.; Bradbury, J. F.; Gardan, L.; Gvozdyak, R. I.; Stead, D. E. Comment on the Reinstatement of *Xanthomollas citri* (ex Hasse) Gabriel et al 1989 and *X. phaseoli* (ex Smith) Gabriel et al 1989: Indication of the Need from Minimal Standards for the Genus *Xanthomonas*. *Int. J. Syst. Evol. Microbiol.* **1991,** *41,* 172–177.

Young, J. M.; Park, D. C.; Shearman, H. M.; Fargier, E. A Multilocus Sequence Analysis of the Genus *Xanthomonas*. *System. Appl. Microbiol.* **2008,** *31*(5), 366–377.

CHAPTER 5

Trichoderma spp.: IDENTIFICATION AND CHARACTERIZATION FOR PATHOGENIC CONTROL AND ITS POTENTIAL APPLICATION

VIPUL KUMAR[1*], DEEPAK KUMAR VERMA[2*], ABHAY K. PANDEY[3*], and SHIKHA SRIVASTAVA[4]

[1]School of Agriculture, Lovely Professional University, Phagwara 144411, Punjab, India

[2]Agricultural and Food Engineering Department, Indian Institute of Technology, Kharagpur 721302, West Bengal, India

[3]PHM-Division, National Institute of Plant Health Management, Ministry of Agriculture and Farmers Welfare, Government of India, Rajendranagar 500030 Hyderabad, Telangana, India

[4]Department of Botany, Deen Dayal Upadhyay Gorakhpur University, Gorakhpur 273009, Uttar Pradesh, India

*Corresponding authors. E-mail: vipulpathology@gmail.com; deepak.verma@ agfe.iitkgp.ernet.in/rajadkv@rediffmail.com; abhaykumarpandey.ku@gmail.com

5.1 INTRODUCTION

In agriculture, green revolution has brought in the necessary impetus to making India self-sufficient in food grains and great improvement in sustainable crop production. However, due to requirement of high input that we reobserved at how sustainable technologies can be setup which will improve the crop productivity (Zadoks and Waibel, 1999). Diseases are major problems in crops that can be caused by a number of plant pathogens such as fungi, bacteria, virus, nematode, and mycoplasma. Among these plant pathogenic organisms, fungi are major pathogens causing damage of agricultural crops worldwide resulted in the reduction of yield. Total crop loss estimated by fungal pathogens amounts to 15%

(Oerke and Dehne, 2004) during cropping and harvesting. Fungal infection in food commodities also leads to mycotoxin production. Continuous use of fungicides against the plant disease may cause several side effects to the environment, human health, crops, and beneficial microorganisms (Zadoks and Waibel, 1999). Plant diseases do have many negative impacts on our daily lives that we sometimes may not be aware of. If our crops aren't successful, we have to struggle for more food and sometimes forced to die. That's why people must diverse our food and develop agricultural technologies to provide better and healthier crop.

In such a way, *Trichoderma* have long been recognized as biological agents which is a substitute to the commercial fungicides against a broad range of fungal pathogens (Papavizas, 1985; Chet, 1987; Latunde-Dada, 1991; Shamim et al., 1997; Harman and Björkman, 1998; Adekunle et al., 2001; Howell, 2003; Harman, 2000; Ranasingh et al., 2006; Harman and Shoresh, 2007; Saba et al., 2012) and reported as effective against several root rot, soilborne, and foliar pathogens (Weller, 1988; Whipps et al., 1993; Elad et al., 1998; Elad, 2000; Chaube et al., 2002; Harman, 2006) for disease management and also for their ability to enhance the root and shoot development, crop productivity, resistance to abiotic stresses, and nutrients uptake (Monte and Llobell, 2003; Harman, 2006; Ranasingh et al., 2006; Saba et al., 2012). They enhance to crop productivity without causing ecological imbalance and social harm and also beneficial for sustainable food security. *Trichoderma* spp. are extensively applied for the various plant disease management effectively (Howell, 2003; Harman, 2006), since the first recognized application in the early of 1930, and have also been controlled and added many diseases to the list in subsequent years (Howell, 2003), some of them are wilt disease, dry root rot, damping off, collar rot caused by *Fusaium* spp., *Rhizoctonia* spp., *Pythium* spp., *Phytophthora* spp., and *Sclerotium rolfsii*, respectively. This chapter is addressing on identification and characterization of *Trichoderma* spp., its role in disease management for control of pathogens and its potential application in agriculture.

5.2 *Trichoderma* spp. AT A GLANCE

Trichoderma is a genus of free living, anamorphic, filamentous, and asexually reproducing fungi commonly available in soil and root ecosystems

(Kubicek et al., 2002; Monte and Llobell, 2003; Harman, 2006). The characteristic features of fungus include rapid growth; conidia those are mostly bright green and conidiophore branched (Saba et al., 2012). Most of the interaction occurs in soil and root ecosystems followed by foliar environments as plant organic debris forms (Ranasingh et al., 2006). *Trichoderma* spp. is avirulent plant symbionts, very resourceful as well as being parasites of other fungi with direct confrontation reported in the review of Harman et al. (2004). In the last few decades, *Trichoderma* spp. is gaining immense importance due to its effective features and biological ability to control against several plant pathogens all over the world (Harman and Kubicek, 1998; Howell, 2003; Ha, 2010); these have been demonstrated in literatures by many researchers (Haran et al., 1996; Kubicek et al., 2002; Brimner and Boland, 2003; Wang et al., 2003).

Over 200 years ago in the end of 17th century, *Trichoderma* was first proposed and described as a genus in Germany by Persoon (1794), and in India, Thakur and Norris first time isolated it during the year 1928 from Madras (Pandya et al., 2011). In the genus of *Trichoderma*, there were 89 species (Samuels, 2006), but today, about 13 (from India) and 104 (internationally) species have been recorded and isolated from various substrates and locations with the high levels of diversity (www.isth.info. in; Pandya et al., 2011). Growing *Trichoderma* on *Trichoderma* selective media is helpful in the identification of the genus. Colonies have key characteristics such as growth pattern, growth rate, odor, and color that can be used to identify them as *Trichoderma* (Gams and Bissett, 1998).

5.2.1   MORPHOLOGY

Trichoderma spp. shows rapid growth which matures in 3–5 days. Initially, the colony grows as fluffy white tufts which later change into greenish color due to production of conidia and characterized by presence of concentric rings on the agar plate. The reverse of the colony is whitish yellow or light tan to yellow or pale orange. Morphological characteristic of the genus includes conidiophore branched, apex coiled, straight, or undulate, presence of phialides (Bisset, 1991; Rifai, 1969). Identification of the bioagent is made by us by comparing the cultural and morphological characters of the fungus with that of described by Rifai (1969) for *Trichoderma viride* the growth rate, colony character, color of the surface, and reverse of the colony, texture of the colony and conidiation, conidio-phore branching, shape, and size; and color was studied on potato dextrose agar (PDA) for its identification. The pure culture of *Trichoderma* was then obtained by adopting single spore technique. Maintaining slides of the cultures were observed through microscopic observation and were confirmed as *Trichoderma asperellum* by The Indian Type Culture Collection (ITCC), Division of Plant Pathology, Indian Agricultural Research Institute, New Delhi, India.

5.2.2   ECOLOGY

Trichoderma species is universal in occurrence and is commonly soil inhabitant *Trichoderma* spp. degrade cellulose. We have studied the influence of temperatures on the linear hyphal growth of *T. viride* in in vitro condition. In the study, fungal growth was excellent at temperature range of 25–30°C, thereafter the growth decreases. Maximum average dry weight was observed at 30°C. pH is also an important factor that effect on the growth of fungi. The maximum dry weight was recorded at pH 7, followed by at pH 6.5. The minimum dry weight was recorded at pH 8.5. The highest radial hyphal growth of the purified *Trichoderma* was recorded on potato dextrose agar media followed by malt agar media, rose Bengal agar media, and Sabouraud's agar media were good but growth on Czapek's dox medium was found poor. Pandey and Upadhyay (1997b) tested several synthetic and semisynthetic growth media against various isolates of *Trichoderma* spp. and *Gliocladium virens* and reported that potato dextrose agar medium prepared from fresh potatoes was best for radial growth and sporulation of *T. viride*. The excellent sporulation was observed in potato dextrose broth (PDB) and Malt broth. Good and fair sporulation was observed on Rose Bengal and Sabouraud's, respectively, but it was fairly poor on Czapek's dox broth.

5.2.3   HABITATS AND ENVIRONMENTAL CONDITION

Water, decaying wood, and soil (rhizosphere of the plants) and even nonnatural materials (rubber foam or kerosene tanks) are the source of isolation of *Trichoderma* spp. (Klein and Eveleigh, 1998). *Trichoderma* spp. is a rapid growing fungus extremely common in agricultural, prairie, forest, salt marsh, and desert soils in all climatic zones (Danielson and Davey, 1973a; Domsch et al., 1980; Roiger et al., 1991; Wardle et al., 1993). Generally, they are present in the litter of humid, mixed hardwood forests, but more dominant in the H and F horizons. *Trichoderma* is generally saprophytes, with the exception that they can attack other fungi. They give coconut odor when present in the soil which is due to the volatile 6-pentyl-*a*-pyrone (Collins and Halim, 1972; Kikuchi et al., 1974; Moss et al., 1975). *T. viride* occurs in cool temperature regions, while *Trichoderma harzianum* occurs in warm climates. This correlates with optimal temperature requirements for each species (Danielson and Davey, 1973b). In general, *Trichoderma* spp. appears to be more prevalent in acidic soils.

5.3 IDENTIFICATION OF *Trichoderma spp.*

Morphological and cultural characteristic, and molecular identification of different species of *Trichoderma*, namely, *T. harzianum*, *T. asperellum*, *T. viride*, *T. atroviride*, *T. longibrachiatum*, *T. koningii*, and *T. virens* are described (Table 5.1 and Fig. 5.1) in the following section.

5.3.1 T. harzianum

5.3.1.1 MORPHOLOGICAL AND CULTURAL CHARACTERISTIC

- **Mycelium:** Changes from watery white to light green in color. Reverse side of petri-plate shows uncolored ring-like zones.
- **Colonies:** Grow rapidly from 7 to 8 cm in diameter in 5 days, smooth surface, mycelial mat develop with white aerial hyphae.
- **Conidiophores:** Highly branched and forming loose tufts.
- **Phialides:** Short-skittle shaped, bulged at the middle, and narrower at the base, arise singly. Phialide size lies within a range of 7.2–11.2 × 2.5–3.1 μm.
- **Phialospores:** Subglobose or short ovoid often with truncate base, perfectly smooth walled, size ranging from 2.8–3.2 × 2.5–2.9 μm.
- **Spore germination time:** 12 h.

TABLE 5.1 *Trichoderma* spp. with Their Isolate Name, Location of Isolate, Field of Crop, and ITCC No.

Trichoderma spp.	Isolate name	Location of isolate	Field of crop	ITCC no.
T. harzianum	*Th* Azad	Kanpur Nagar, UP, India	Chickpea	6796/11
T. asperellum	8CP	Kanpur Nagar, UP, India	Chickpea	8305/11
T. viride	T_{asp}/CSAU	Kanpur Nagar, UP, India	Pigeon pea	8940/11
T. atroviride	71L	Hardoi, UP, India	Lentil	7445/09
T. longibrachiatum	21PP	Kaushambi, UP, India	Pigeonpea	7437/09
T. koningii	T_k/CSAU	Kanpur Nagar, UP, India	Pigeon pea	5201/11
T. virens	T_{vi}/CSAU	Kanpur Nagar, UP, India	Pigeon pea	4177/11

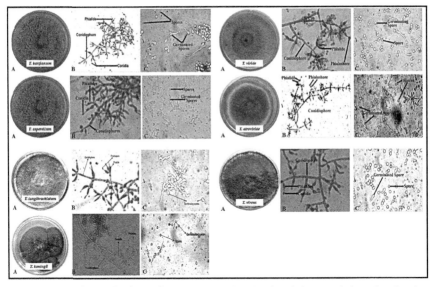

FIGURE 5.1 **(See color insert.)** Morphological and cultural characteristic and molecular identification of *Trichoderma* spp., (A) growth on PDA medium, (B) microscopic observation (40×), and (C) spore germination (40×).

5.3.1.2 MOLECULAR IDENTIFICATION

- **Locus:** KC800922.
- **Definition:** *T. harzianum* isolate *Th* Azad/CSAU 6796 18S ribosomal RNA gene, partial sequence; internal transcribed spacer (ITS) 1, 5.8 S ribosomal RNA gene, and ITS4, partial sequence.

- **Primers used:** ITS1-AGAGTTTGATCCTGGCTCAG and ITS4-GGTTACCTGTTACGACTT.
- **Sequence:** 546 bp.

5.3.2 T. asperellum

5.3.2.1 MORPHOLOGICAL AND CULTURAL CHARACTERISTIC

- **Mycelium:** Forms a smooth hairy yellowish green cotton pattern usually in the form of 1–2 ringed concentrics.
- **Colonies:** Changes cottony white to yellowish green and after 2 days from to deep chrysolite green, emitting coconut odor.
- **Conidiophores:** Arise highly branched in compact form and the phialides arise singly or in opposite pairs along the branches.
- **Phialides:** Appear in ninepin shape attenuated into long neck, usually $6.8–7.2 \times 3.0–3.4$ µm.
- **Phialospores:** Globose or short obovoid in shape, green colored with an approximate size of $3.6–4.0 \times 3.4–4.0$ µm.
- **Spore germination time:** 12–13 h.

5.3.2.2 MOLECULAR IDENTIFICATION

- **Locus:** KC800921.
- **Definition:** *T. asperellum* isolate T_{asp} (CSAU)-8940 18S ribosomal RNA gene, partial sequence; ITS1, 5.8 S ribosomal RNA gene, and ITS4, partial sequence.
- **Primers used:** ITS1-TCCGTAGGTGAACCTGCGG and ITS2-TCCTCCGCTTATTGATATGC.
- **Sequence:** 1200 bp.

5.3.3 T. Viride

5.3.3.1 MORPHOLOGICAL AND CULTURAL CHARACTERISTIC

- **Mycelium:** Changes green to dark yellowish green after 2–3 days, no odor.

- **Colonies:** Smooth surface, cottony white mycelial mat with aerial hyphae.
- **Conidiophores:** Conidia production is less in center than toward the margins with green conidia distributed throughout.
- **Phialides:** Long, swollen in middle, like slender, and horn shaped with a size range of 6.2–10.5 × 3.1–3.9 µm.
- **Phialospores:** Usually globose or obovoid often, perfectly smooth walled, with a size ranging from 2.6–3.0 × 2.0–2.4 µm.
- **Spore germination time:** 13 h.

5.3.3.2 MOLECULAR IDENTIFICATION

- **Locus:** KC800921.
- **Definition:** *T. asperellum* isolate T_{asp} (CSAU)-8940 18S ribosomal RNA gene, partial sequence; ITS1, 5.8 S ribosomal RNA gene, and ITS4, partial sequence.
- **Primers used:** ITS1-TCCGTAGGTGAACCTGCGG and ITS4-TCCTCCGCTTATTGATATGC.
- **Sequence:** 641 bp.

5.3.4 T. Atroviride

5.3.4.1 MORPHOLOGICAL AND CULTURAL CHARACTERISTIC

- **Mycelium:** The mycelium mat appears watery white, submerged composed of translucent smooth and appear floccose on PDA.
- **Colonies:** The colony color changes after 2 days from yellowish green to artemisia green and reverse remaining dull yellowish and odorless.
- **Conidiophores:** Highly branched arise in compact form and the phialides solitary paired on the terminals, swollen at the middle and narrow at the tips.
- **Phialides:** Phialides appear in ampulliform and oblong shaped, curved, and constricted at the base, with a varying size of 5.2–10.5 × 2.4–2.8 µm.
- **Phialospores:** Globose in shape, green colored with an approximate size of 2.4–3.6 µm.
- **Spore germination time:** 12–13 h.

5.3.4.2 MOLECULAR IDENTIFICATION

- **Locus:** KC008065.
- **Definition:** *Trichoderma atroviride* strain TAU818S ribosomal RNA, partial sequence; ITS1, 5.8S ribosomal RNA, and ITS2 complete sequences; and 28S ribosomal RNA partial sequence.
- **Primers used:** ITS1-tcctccgcttattgatatgc and ITS2-ggaagtaaaagtcgtaacaagg.
- **Sequence:** 627 bp.

5.3.5 *T. longibrachiatum*

5.3.5.1 MORPHOLOGICAL AND CULTURAL CHARACTERISTIC

- **Mycelium:** Mostly submerged translucent or watery white.
- **Colonies:** Changes after 2 days from yellowish green to lily green, no smell. Smooth conidiophores arise from substratum and form irregular tufts or arise from aerial hyphae.
- **Conidiophores:** Smooth conidiophores arise from substratum and form irregular tufts or arise from aerial hyphae.
- **Phialides:** Arises singly or in verticils of 2–3, usually lageniform, slightly constricted at the base, usually 3.4–5.2 × 2.3–3.0 µm.
- **Phialospores:** Smooth walled, obovoid to ellipsoidal, dilute green, apex broadly rounded, with an approximate size of 2.4–3.6 µm.
- **Spore germination time:** 12–13 h.

5.3.5.2 MOLECULAR IDENTIFICATION

- **Locus:** JX978542.
- **Definition:** *Trichoderma longibrachiatum* strain 21 PP 18S ribosomal RNA gene, partial sequence; ITS1, 5.8S ribosomal RNA gene, and ITS2, complete sequence; and 28S ribosomal RNA gene, partial sequence.
- **Primers used:** ITS1-TCCTCCGCTTATTGATATGC and ITS2-GGAAGTAAAAGTCGTAACAAGG.
- **Sequence:** 664 bp.

5.3.6 *T. koningii*

5.3.6.1 *MORPHOLOGICAL AND CULTURAL CHARACTERISTIC*

- **Mycelium:** Creamy white charges from white to terreverte in color.
- **Colonies:** Crysty, compact, and glaucous like.
- **Chlamydospores or conidiophores:** Formed intercalary or terminally, much branched.
- **Phialides:** Narrow at the base, alternate to conical apices (*Note:* Most phialides of *T. koningii* arise singly and laterally and appears Nine-pin bowling shaped singly rather than pyramidal, size range of 3.8–7.6 × 2.5–3.2 µm).
- **Phialospores:** Ellipsoidal or oblonged, with a rounded apex and acute base, measuring 2.5–4.2 × 1.8–2.6 µm.
- **Spore germination time:** 14 h.

5.3.6.2 *MOLECULAR IDENTIFICATION*

- **Locus:** KC800924.
- **Definition:** *Trichoderma koningii* T_k (CSAU) 5201 18S ribosomal RNA gene, partial sequence; ITS1, 5.8S ribosomal RNA gene, and ITS4, partial sequence.
- **Primers used:** ITS1-TCTGTAGGTGAACCTGCGG and ITS4-GGAAGTAAAAGTCGTAACAAGG.
- **Sequence:** 206 bp.

5.3.7 *T. Virens*

5.3.7.1 *MORPHOLOGICAL AND CULTURAL CHARACTERISTIC*

- **Mycelium:** Changes watery white to nice green color with dull blackish green shades granules, no characteristic odor.
- **Colonies:** On potato dextrose agar medium, colonies grow rapidly from 7 to 8 cm in diameter in 5 days at 25°C temperature.
- **Conidiophores:** Branched irregularly near the apex with each branch terminated by a cluster of 3–6 closely bunched phialides.

- **Phialides:** Lageniform to ampuliform, swollen at the middle, attenuated to apex, broadly attached to the conidia in ranges between 4.4–12.8 × 2.6–4.2 μm of size.
- **Phialospores:** Broadly ellipsoidal to obovoid, both ends rounded dark green in color and with a size range of 3.2–5.6 × 2.5–3.9 μm.
- **Spore germination time:** 13 h.

5.3.7.2 MOLECULAR IDENTIFICATION

- **Locus:** KC800923.
- **Definition:** *Trichoderma virens* isolate T_{vi} (CSAU)-417718S ribosomal RNA gene, partial sequence; ITS1, 5.8S ribosomal RNA gene, and ITS4, partial sequence.
- **Primers used:** ITS1-tcctccgcttattgatatgc and ITS4-ggaagtaaaagtcgtaacaagg.
- **Sequence:** 635 bp.

5.4 CHARACTERIZATION OF *Trichoderma spp.*

5.4.1 PHYSIOLOGICAL CHARACTERIZATION

5.4.1.1 SOLID MEDIA STUDY

Growth, sporulation, and conidial discharge of *Trichoderma* are found to be significantly affected with the use of different culture media and temperature range. Different agar media, namely, potato dextrose agar, Czapek's dox, Sabouraud, rose Bengal, and corn meal agar media were prepared separately and used to evaluate the mycelial growth and sporulation of the *Trichoderma*.

5.4.1.1.1 Significant Achievement

The highest radial hyphal growth of the purified *Trichoderma* was recorded on potato dextrose agar media followed by malt agar media, rose Bengal agar media, and Sabouraud's agar media were good but growth on Czapek's dox medium was found poor. Pandey and Upadhyay

(1997b) tested several synthetic and semisynthetic growth media against various isolates of *Trichoderma* spp. and *G. virens* and reported that potato dextrose agar medium prepared from fresh potatoes was best for radial growth and sporulation of *T. viride*.

5.4.1.2   LIQUID MEDIA STUDY

To determine the growth of mycelial produced of *T. viride* in five different liquid media, namely, potato dextrose, rose Bengal, corn meal, Sabouraud, and Czapek's dox broth was tested. The isolated *Trichoderma* spp. was inoculated separately in the respective media and incubated at 28°C temperature for 9 days.

5.4.1.2.1   Significant Achievement

Highest mycelial weight was recorded in potato dextrose broth as followed by malt broth, rose Bengal broth, Sabouraud's broth, and Czapek's dox broth. The excellent sporulation was observed in PDB and Malt broth. Good and fair sporulation was observed on rose Bengal and Sabouraud's, respectively, but it was fairly poor on Czapek's dox broth.

5.4.1.3   EFFECT OF TEMPERATURE ON GROWTH AND SPORULATION

Influence of temperatures on the linear hyphal growth of *T. viride* was studied in vitro on PDA (15 mL) in 9 cm petri plates adjusted for each temperature. The fungal growth was observed by placing a bit (9 mm) of pure culture at the center of the Petri plates with the help of sterile cork borer and three replications were taken for each treatment. The cultures were incubated at 15, 20, 25, 30, and 35 in bio-oxygen demand (BOD) incubator and daily observation on mycelial growth of *T. viride* was recorded at every 24 h up to 7 days.

5.4.1.3.1   Significant Achievement

It can be concluded that the growth of the fungi was excellent at temperature range of 25–30°C, thereafter the growth decreases. Maximum average dry weight was observed at 30°C.

5.4.1.4 EFFECT OF pH ON GROWTH AND SPORULATION

Potato dextrose broths were set at different pH levels 4.5, 5.5, 6.5, 7.5. The pH was adjusted according to the requirement. All the treatments were carried out in triplicates. The mycelial mat after 7 days was then harvested from the flask by collecting the culture filtrate through sterilized filter paper (Whatman No. 4). The harvested mycelium was kept in hot-air oven at 35°C for 48 h and final weight was measured.

5.4.1.4.1 Significant Achievement

It was also found that pH has marked effect on the growth and sporulation of fungi. The maximum weight was recorded at pH 7, followed by at pH 6.5. The degree of sporulation of fungi was determined according to standard methods (Sharma and Pandey, 2011).

5.4.2 BIOCHEMICAL CHARACTERIZATION

5.4.2.1 NITROGEN ESTIMATION

In *Trichoderma* (*T. viride*), total nitrogen content is analyzed by KEL PLUS Nitrogen Analyzer. By multiplying a factor with the observed nitrogen values, total protein content can be determined by using following formula:

$$\text{Nitrogen \% (in 100 g)} = \frac{1.4 \times N \times V \times 100}{W}$$

$$\text{Protein \%} = \text{nitrogen \%} \times 6.25$$

where N = normality of HCl

V = titer value of sample − titer value of blank

W = weight of sample

5.4.2.2 PROTEIN QUANTIFICATION

5.4.2.2.1 Lowry Method

Lowry method is a biochemical assay for determining the total content of protein in a solution.

Methodology

1. 0.5 mg/mL of bovine serum albumin (BSA) standard was used.
2. Different dilutions of the standard and test samples were made.
3. To each tube 2 mL of complex forming reagent was added and keep for 10 min at room temperature (RT).
4. After 10 min of incubation period, 0.2 mL of Folin–Ciocalteu re-agent solution was added to each tube and incubates for 20–30 min.
5. After incubation period sample absorbance was taken at 660 nm.
6. Calibration curve was constructed by plotting absorbance reading on Y axis against standard protein concentration (mg/mL).
7. Sample concentration was calculated using standard graph as a reference.

5.4.2.2.2   Bradford Method

Procedure

1. 0.5 mg/mL of the BSA is used as standard.
2. Different dilutions of the standard and samples were made.
3. Into tube 1 aliquot 100 µL of distilled water, this was served as blank.
4. To each tube add 1 mL of the Bradford reagent was added and mixed. All tubes were incubated at RT for 2 min.
5. After incubation period optical density was taken at 595 nm.
6. Calibration curve was constructed by plotting absorbance reading on Y axis against standard protein concentration (mg/mL).
7. Sample conc. was calculated using standard graph as a reference.

5.4.3   MOLECULAR CHARACTERIZATION

5.4.3.1   DNA ISOLATION OF TRICHODERMA

Pure culture of the target fungi was grown in liquid potato dextrose broth medium for the isolation of genomic DNA using a method described by Doyle and Doyle (1990).

5.4.3.2   PURIFICATION OF PCR PRODUCT

The total genomic DNA was extracted from isolate of *Trichoderma* based on cetrimide tetradecyl trimethyl ammonium bromide extraction method of Vipul et al. (2011) with little modification. Ten microliter of the reaction mixture was then analyzed by submarine gel electrophoresis using 1.0% agarose with ethidium bromide, and the reaction product was visualized under Gel' doc/UV transilluminator. The polymerase chain reaction (PCR) product was purified by Qiagen gel extraction kit using the protocol of Williams et al. (1990) and Dodd et al. (2004).

5.4.3.3   INTERNAL TRANSCRIBED SPACER REGION

The ITS regions of the recombinant DNA (rDNA) repeat from the 3'-end of the 18S and the 5'-end of the 28S gene were amplified using the two primers, ITS1 and ITS4 which were synthesized on the basis of conserved regions of the eukaryotic rRNA gene (White et al., 1990).

5.4.3.4   SEQUENCE ANALYSIS

A comparison of the 18S rRNA gene sequence with the test strain against nucleotide collection based on database was done using Basic Local Alignment Search Tool (BLAST) (Zhang et al., 2000). A number of sequences of *Trichoderma* were selected on the basis 95% similarity score of the determined sequence with a reference sequence. Multiple sequence alignment of these selected homologous sequences and 18S rRNA gene sequence of test strain were performed using clusters (Thompson et al., 1994).

5.5   BENEFITS OF TRICHODERMA

Globally, many novel microorganism species are found. Among them, *Trichoderma* spp. is one which is evaluated because of its potential alternative of the commercial fungicides like thiram, bavistin, benomyl, etc., against many soilborne and foliar fungal pathogen. In recent years, several species of *Trichoderma* are used as biocontrol agents/biopesticide against fungal diseases of plants (Upadhyay and Rai, 1981; Bhatnagar, 1996; Somasekhara et al., 1996, 1998; Gundappagal and Bidari, 1997; Biswas and Das, 1999; Prasad et al., 2002; Khan and Khan, 2002; Anjaiah et al., 2003; Sawant et al., 2003; Roy and Sitansu, 2005; Dhar et al., 2006; Maisuria et al., 2008; Ram and Pandey, 2011). *Trichoderma* usually grows in its natural habitat on the root surface, and so affects root disease in particular, but is also effective against foliar diseases of the crops. *Trichoderma* mechanism of action includes antibiosis, parasitism, inducing host-plant resistance, and competition. *T. harzianum, T. viride,* and *T. hamatum* are the most common species of *Trichoderma* being used in agriculture.

5.5.1   PLANT GROWTH PROMOTER

This is the indirect mechanisms carried out by *Trichoderma* spp. resulted in the improvement of plant health. The ability to produce siderophores, phosphate-solubilizing enzymes, and phytohormones of *Trichoderma* spp. keep them in the category of plant growth promoting fungi (Doni et al., 2013). *Trichoderma* treatment is found to be increased root and shoot growth, thereby reduces the bioactivity of harmful microbes in the plants' rhizosphere and improves the nutrient status of the plant. The mechanisms of *Trichoderma* spp. help in plant growth promotion include mycoparasitism, degradation of toxins, antibiosis, pathogenic enzymatic pathways inactivation, nutrients uptake enhancement, resistance to pathogens, solubilization, inorganic nutrients sequestration, and root hair development enhancement (Harman, 2000; Harman and Shoresh, 2007). *Trichoderma* has been reported to promote growth responses in pepper, radish, tomato, and cucumber (Chang et al., 1986). Mishra and Salokhe (2011) observed significant contribution of *Trichoderma* spp. to enhance the root and shoot growth and also considered as a key factor to the rice plant root inoculated by strain SL2 of *Trichoderma* spp. and recorded large mass compared to control prolonged photosynthetic activity and delayed senescence in rice plants. Furthermore, Jiang et al. (2011) also found that *Trichoderma* spp. having the ability to degrade cellulose. A large amount of nitrogen may release in rhizospheric zone of rice plant after degradation of cellulose. It has been reported that when N concentration is high, the uptake has positive correlation with photosynthetic rate. In another study by Doni et al. (2014) in rice for plant growth promotion by *Trichoderma* spp., it is observed that *Trichoderma* strain SL2 enhances leaf number, growth components, plant height, root length, and root fresh weight as well as tiller number. Similar, observation was found in strawberry, tomato, soybean, apple, cotton, and gray mangroves and reported to increase plant growth (Morsy et al., 2009). Moreover, several plant nutrients are solubilized by *T. harzianum* (Altomare et al., 1999), and cucumber roots colonization by *T. asperellum* has been resulted in enhancement of availability of P and Fe to plants, with significant increases in shoot length, leaf area, and dry weight (Yedidia et al., 2001). On the contrary, the interaction of two tomato lines with two bioagents *T. atroviride* and *T. harzianum* did not exert any plant growth promotion effect and was even seen to be detrimental (Tucci et al., 2011).

Apart from this, Chowdappa et al. (2013) and Martínez-Medina et al. (2011) also found that *Trichoderma* releases the hormones such as auxins and gibberellins which enhances rice seedling during seed germination Th3 strain of *T. harzianum* (Sharma et al., 2012) that has promoted the plant growth in wheat. The growth elevation effect of this strain on tillers, rootlets, grains weight, and grain yield was evaluated by using it at three stages of wheat, namely, seed, flowering, and preharvesting @4 g/kg and @4 mL/L in soil treatment along with mixture of farm yard manure and formulation was @50:1 before sowing. Compared to the first year (2008–2009) where the farmers were unaware of *Trichoderma*, a significant increase in wheat yield from 36.25 to 46.73 Q/ha (29% in Jaipur) and from 36.88 to 50.12 Q/ha (36% in Kota) is reported after regular application for 3 years (2008–2011) for growth stimulating capacity of strain Th3 in the popular wheat variety Raj 3765. Ratnakumari et al. (2014) recently proved the effect of strain NFCCI 2689 of *T. ovalisporum* and found more significant enhancement of the plant growth in *Mentha arvensis*. More than 100% increase in the oil yield and menthol content were recorded treated with *Trichoderma* strain NFCCI 2241. Baker (1988) and Lynch et al. (1991a,b) have reported the effect of *Trichoderma* in some vegetable and bedding plant crops such as lettuce and found that this species helps in growth stimulation with its ability to manage damping off diseases caused by *Pythium ultimum* and *Rhizoctonia solani*.

In soybean, mustard, maize, beans, chili, and many agricultural crops, research is conducted to improve the seed germination by treating the seeds with *Trichoderma* spp. (Asaduzzaman et al., 2010; Okoth et al., 2011; Lalitha et al., 2012; Tančić et al., 2013). Zheng and Shetty (2000) investigated that during seed germination, *Trichoderma* spp. induced some phenolic compounds in the plant which led to the enhancement of seed vigor. Some secondary metabolites such as harzianolide secreted by *Trichoderma* spp. can influence the early stage of plant development through root length enhancement (Cai et al., 2013).

5.5.2 *PRODUCED BIOCHEMICAL ELICITORS FOR DISEASE RESISTANCE*

Trichoderma spp. is having very important role in induce systemic resistance in plants (Table 5.2). *Trichoderma* produces three classes of enzymes which help the induce resistance in plants which induce the production of ethylene, hypersensitive responses, and other defense related reactions in crops. Introduction of endochitinase, glucanase, xylanase, chitinase genes from *Trichoderma* into plants (tobacco and potato) has increased their resistance to fungal growth (Woo et al., 2006). *Trichoderma*'s plant resistance mechanism ability to systemically activate against plant pathogens like *Alternaria* spp., *Botrytis cinerea*, *Phytophthora* spp., *Colletotrichum* spp., *Magnoporthe grisea*, and *R. solani* is reported in several dicots and monocots families including Gramineae, Solanaceae, and Cucurbitaceae (Woo et al., 2006). In red pepper plants, induced systemic resistance against *Phytophthora capsici* has been reported where when plants were treated with different *T. harzianum* isolates, increase in glucanase activity and phenolic content was found. *Trichoderma* spp. released elicitors as peptides or proteins (Harman et al., 2004) with 6–42 kDa molecular mass and perhaps include serine protease, xylanase, and induce the phytoalexins, terpenoid, and peroxidase in plants. Treatment of seedling root of red pepper by root dipping with elicitors isolated from *T. harzianum* reduced the infection in plants at extent of 23% (Sriram et al., 2009). Some of the enzymes like chitinases and glucanase involved in mycoparasitism are also act as elicitors (Woo et al., 2006). There are certain proteins like elicitors like Avr like proteins similar to those present in avirulent pathogens and those that are present in *Trichoderma* spp. (Woo et al., 2006; Harman et al., 2004). Hanania and Avni (1997) reported that challenging tomato or tobacco varieties with ethylene inducing xylanase (EIX) from *T. viride* caused quick stimulation of plant-defense responses, resulted in programed cell death. *T. harzianum* also showed significant reduction of stem necrosis due to *P. capsici* in pepper plants because the capsidiol induction in the stems of treated plants was high (Ahmed et al., 2000). Cucumber roots inoculated with *T. harzianum* T-203 have also been induced peroxidase and chitinase activity (Yedidia et al., 1999). Higher activities of cellulose, β-1,3-glucanase, chitinase, and peroxidase are observed in cucumber roots treated with *T. harzianum* (Yedidia et al., 2000). Elad (2000) also showed the phenomenon of induced systemic resistance in foliar pathogens control like *B. cinerea* in cucumber using *Trichoderma* isolates. Thus, identification and integration of induced systemic resistance (ISR) eliciting *Trichoderma* strains in the management of disease are important and will help in a sustainable way.

5.5.3   SECONDARY METABOLITES PRODUCER

Secondary metabolites are low-molecular-weight organic molecules that are not involved in the normal growth and development of an organism they often synthesized after active growth has ceased. T22 and T39 are the major metabolites produced by *Trichoderma* spp. commercially. These have pronounced antifungal and antagonistic activity. Secondary metabolites are broadly divided into many characteristic groups such as polyketides, terpenes, phenols, alkaloids—that reflect their origin and biosynthesis. The mechanism of biological control of *Trichoderma* spp. is a complex route mediated by extracellular enzymes secretion such as β-glucanases, chitinases, and proteinases as well as secondary metabolites.

Trichoderma spp. is known to be extracellular volatile compound producer which is found to be toxic to wilt pathogen (Pandey and Upadhaya, 1997a,b). Somasekhara et al. (1998) found that nonvolatile metabolites from *T. viride* are highly toxic toward the plant pathogens followed by *T. harzianum*, and *T. koningii*. Harzianic acid is a secondary metabolite extracted from *T. harzianum* is able to promote plant growth and strongly bind iron (Vinale et al., 2014). It has been reported that in cucurbits, ISR induction is correlated to the upregulation of diverse pathogenesis related and defense-related proteins (glucanase, peroxidases, chitinases, and specific phytoalexins) and activities of enzymes especially phenylalanine ammonia lyase and synthesis of other phenols and related proteins (Khan et al., 2004). *T. atroviride* isolated from root of *Salvia miltiorrhiza* is found to be produced tanshinone I and tanshinone IIA. These both secondary metabolites play a key role in the survival of host plant from the diseases (Ming et al., 2012). Secondary metabolites, namely, harzianolide

TABLE 5.2   *Trichoderma* spp. Induced Resistance in Plants for Pathogen Control.

Trichoderma strains	Plants	Pathogen
T. virens G-6, G-6-5 and G-11	Cotton	*Rhizoctonia solani*
T. harzianum T-39	Bean	*Colletotrichum lindemuthianum, Botrytis cinerea*
T. harzianum T-39	Tomato, pepper, tobacco, lettuce, bean, pepper	*Botrytis cinerea*
T. asperellum T-203	Cucumber	*Pseudomonas syringae* pv. *lachrymans*
T. harzianum T-22; *T. atroviride* P1	Bean	*B. cinerea* and *Xanthomonas campestris* pv. *phaseoli*
T. harzianum T-1 and T22; *T. virens* T3	Cucumber	Green-mottle mosaic virus
T. harzianum T-22	Tomato	*Alternaria solani*
harzianum T-22	Maize	*Colletotrichum graminicola*
Trichoderma GT3-2	Cucumber	*C. orbiculare, P. syringae* pv. lachrymans
T. harzianum	Pepper	*Phytophthora capsici*
T. harzianum NF-9	Rice	*Magnaporthe grisea; Xanthomonas oryzae* pv. *oryzae*

Source: Adapted from Harman et al., 2004.

and 6-*n*-pentyl-6*H*-pyran-2-one, extracted from *Trichoderma* strains (*T. harzianum* strains T22, T39, and A6, and *T. atroviride* strain P1) exhibited induced systemic resistance in tomato and pea (Vinale et al., 2008a). 2-Phenylethanol (1) and tyrosol (2) obtained from *T. harzianum* are significantly inhibited the growth of *Armillaria mellea* at a 200 ppm, a fungus causing armillaria rot in tea plant (Tarus et al., 2003). Other metabolites such as 6-pentyl-α-pyrone and other α-pyrone analogues isolated from *T. harzianum* are also reduced the take all disease of wheat caused by fungus *Gaeumanmyces graminis* (Almassi et al., 1991; Ghisalberti and Rowland, 1993). Some secondary metabolites produced by *Trichoderma* spp. are given in Table 5.3.

5.5.4   BIOREMEDIATION

Trichoderma plays an important role in the bioremediation of soil that are contaminated with dangerous pesticides and herbicides and some pesticide

TABLE 5.3 Secondary Metabolites Produce by *Trichoderma* spp.

Secondary metabolites	Strain of *Trichoderma*	References
Mannitol	*T. hamatum*	Hussain et al. (1975)
2-Hydroxymalonic acid	*T. psuedokonigii*	Kamal et al. (1971)
Trichodermaol	*Trichoderma* spp.	Adachi et al. (1983)
p-Hydroxybenzyl alcohol	*T. koningii*	Huang et al. (1995)
1-Hydroxy-3-methylanthraquinone	*T. viride* PrL 2233	Slater et al. (1967)
Pencolide	*T. album*	Ren (1977)
Viridiofungin A	*T. viride*	Harris et al. (1993)
Trichodermene A	*T. psuedokonigii*	Kamal et al. (1971)
Carbolic acid	*Trichoderma* spp.	Turner and Aldridge (1983)
Dermadin methyl ester	*T. polysporum*	Jin and Jin (1989)
Isonitrin A	*T. harzianum* IMI 3198	Baldwina et al. (1991)
Spirolactone	*T. hamatum* HLX 1379	Baldwin et al. (1985)
Dermadin	*T. viride* 4875	Pyke and Dietz (1966)

are release volatile gases which pollute our environment for humankind. *Trichoderma* spp. has also ability to efficiently biodegrade toxic contaminants (Cao et al., 2008). *Trichoderma* decomposes and degrades wide range of insecticides: organochlorines, organophosphates, and carbonates. *Trichoderma* is compatible with organic manure and with biofertilizers like *Rhizobium*, *Azospirillum*, *Bacillus subtilis*, and *Phosphobacteria*. Trichoderma has wide range of applications such as in the industrial enzyme production, in food industry, and paper/pulp treatment (Singh and Singh, 2009). It is also useful to remediate the water and soil pollution, in preparation of biofilms in the nanotechnology (Ezzi and Lynch, 2005; Vahabi et al., 2011). Some species of *Trichoderma* are known to tolerate and accumulate heavy metals, namely, copper, cadmium, zinc, and arsenic (Errasquin and Vazquez, 2003; Zeng et al., 2010). In the root region, *Trichoderma* spp. enhances the nutrient uptake and other essential nutrients and also facilitates uptake of several metalloids and toxic metals, thus helping in phytoextraction activities (Cao et al., 2008). In *Brassica juncea*, *Trichoderma atroviridae* is very helpful in translocation and uptake of Zn, Ni, and Cd, while in *Salix fragilis*, *T. harzianum* is helpful in plant-growth promotion when plants are grown in metal-contaminated soil (Adams et al., 2007). In the diesel-contaminated compost, *Trichoderma* is reported as most dominant fungus, where it colonizes and degrades the diesel-contaminated soil (Hajieghrari, 2010). The efficacy of this fungus is due to extracellular enzyme system that catalyzes reactions that can degrade aromatic toxic compounds. Various chemical pesticide residues in soil like lindane, chlordane, and DDT (dichloro-diphenyl-trichloroethane) are also found to be degraded by *Trichoderma* spp. which makes them useful for the remediation of pesticide-contaminated sites (Zhou et al., 2007). In integrated crop management strategies, *Trichoderma* strains can be used with toxic chemicals and pesticides containing metal ions. Chlorpyrifos and photodieldrin are reported to degrade by *T. viride* (Mukherjee and Gopal, 1996). Dixit et al. (2011) reported the role of *Trichoderma* spp. in environmental bioremediation. They reported that glutathione transferase (roles in combating oxidative stresses induced by various heavy metals) gene transfers from *T. virens* to the host plant and enhance the tolerance of cadmium without increasing its accumulation in transgenic *Nicotiana tabacum*. Thus, through modulating the plant status at molecular, physiological, and phenotypic levels and alteration of gene expression, the role of *Trichoderma* in ameliorating abiotic stress tolerance is well elucidated (Harmosa et al., 2012).

5.5.5 DISEASE CONTROL

Plant diseases have been concerned with mankind since agriculture began and played a crucial role in the destruction of natural resources and contributing 13–20% losses in crop production worldwide (Pandya et al., 2011). *Trichoderma* is a potent biocontrol agent and used extensively for soilborne diseases. It has been used successfully against pathogenic fungi belonging to various genera, namely, *Fusarium*, *Phytophthara*, *Sclerotinia*, etc. In India, plant pathogens reduce the potential crop production by approximately 16% (Singh, 1996). Approximately 30,000 species of plant pathogens attack Indian crops, including 23,000 species of fungi (Pandotra, 1997) and 650 species of plant viruses (Patel and Patel, 1985). Chemical control of plant diseases can be very effective and risky but this is short-term measure and accumulation of these harmful chemical residues creates an ecological imbalances. In recent years, the huge and regular use of potentially hazardous pesticides for control of crop diseases has been resulted in growing concern of both environment and public health properties. Developing countries like India cannot afford such type of management strategies because these are costly and also harmful for mankind. Due to changes introduced in the farming, several plant pathogenic fungi are widely disseminated such as *Pythium*, *Phytophthora*, *Botrytis*, *Rhizoctonia*, and *Fusarium* during the last few decades. In addition, not only growing crops but also stored fruits prey to fungal infection (Chet et al., 1997). By contrast, biological control is risk-free, economical, and eco-friendly. Moreover, an integrated approach promotes a degree of disease control similar to that achieved with full fungicidal treatment. Weindling (1934) reported that *Trichoderma* spp. attacks other fungi to produce antibiotics that affect other microbes and to act as biocontrol microbes. Antagonists of phytopathogenic fungi have been used to control plant diseases and 90% of such applications have been carried out with different strains of *Trichoderma* (Monte and Llobell, 2003). The success of *Trichoderma* as biocontrol agents is due to their high reproductive capacity, ability to survive under very unfavorable conditions, efficiency in the utilization of nutrients, capacity to modify the rhizosphere, strong aggressiveness against phytopathogenic fungi, and efficiency in promoting plant growth and defense mechanisms. These properties have made *Trichoderma* a ubiquitous genus present in any habitat and at high population density (Misra and Prasad, 2003).

Trichoderma spp. along with many other fungal genera belongs to the soil is used against phytopathogenic fungi and nematodes for reducing disease incidence in most of the pulse crops, and among all, pigeon pea is found to be attacked by potential pathogens and having severe disease consequences (Brimner and Boland, 2003). *Trichoderma* spp. is evaluated against fungal pathogen specially *Fusarium udum* and also a wide range of soilborne pathogens (Bhatnagar, 1996; Somasekhara et al., 1998; Gundappagal and Bidari, 1997; Biswas and Das, 1999; Prasad et al., 2002; Roy and Sitansu, 2005). The review of Harman et al. (2004) reports that *Trichoderma* spp. is opportunistic, avirulent plant symbionts, as well as being parasites of other fungi with direct confrontation. Primarily, they are isolated from soil and decomposing organic matter (Monte and Llobell, 2003) but often considered as most frequently isolated soil fungi and presented nearly in all types of soils, namely, agriculture, forest, prairie, salt marsh, desert soils, etc. of temperate and tropical climatic zones (Danielson and Davey, 1973a,b; Domsch et al., 1980; Harman, 2006). *Trichoderma* spp. has been found to be effective against aerial, root, and soil pathogens (Weller, 1988;

Whipps et al., 1993; Elad et al., 1998; Elad, 2000; Chaube et al., 2002; Harman, 2006) and has long been recognized as biological agents by many researchers (Papavizas, 1985; Chet, 1987; Harman et al., 1989; Latunde-Dada, 1991; Shamim et al., 1997; Adekunle et al., 2001; Howell, 2003; Ranasingh et al., 2006; Saba et al., 2012) for the control of crop diseases due to their effective antagonists against the pathogens (Papavizas, 1985; Chet, 1987; Kumar and Mukerji, 1996; Ha, 2010), for their ability to increase plant root growth and development, crop productivity, abiotic stresses resistance, and uptake and use of nutrients (Monte and Llobell, 2003; Harman, 2006; Ranasingh et al., 2006; Saba et al., 2012) and much more indicated in Figure 5.2. Today, several biopesticides have been developed from several strains of *Trichoderma* spp., namely, *T. harzianum*, *T. viride*, *T. hamatum*, etc., and extensive studies on still progress to make these microbes versatile model organisms for research (Saba et al., 2012). *Trichoderma* spp. has been culminated in the commercial production in the United States, India, Israel, New Zealand, and Sweden for the protection and growth enhancement of a number of crops (McSpadden Gardener and Fravel, 2002) due to their well-known opportunistic plant symbionts nature and direct confrontation mechanisms with effective mycoparasitism against plant pathogenic agents (Papavizas, 1985; Harman and Kubicek, 1998; Howell, 2003; Saba et al., 2012). Bhatnagar (1996) studied the antifungal activity of three *Trichoderma* spp. as multiple action bioinoculants and to control variable pathogenesis against wilt pathogen at different pH, temperatures, and C/N ratios and found that all of them were equally efficient and showed maximum antagonistic properties at $35 \pm 2°C$ temperature and about of 6.5 pH. Apparently, Somasekhara et al. (1996) worked on two delivery systems (seed treatment and foliar application) by using six isolates of *Trichoderma* spp. and studied their efficacy which was found to be extreme on the 35 days of inoculation. Somasekhara et al. (1998) evaluated *Trichoderma* isolates and their antifungal extracts as potential biocontrol agents against pigeon pea wilt pathogen, *F. udum*. Biswas and Das (1999) performed in vitro experiments to reduce pathogenesis and tested five *Trichoderma* spp. *T. harzianum* was found to be most effective antagonist followed by *T. hamatum*, *T. longiconis*, and *T. koningii*. They also reported that by giving seed treatment of *T. harzianum* to pigeon pea, inoculants spores failed to reduce pathogen's growth, while soil amendment with *T. harzianum* in maize meal: sand applied at 40–60 g/kg soil resulted a significant reduction of wilt up to 90%. Khan and Khan (2002) confirmed differential behavior of multiple biocontrol agents including *Bacillus*, *Pseudomonas*, and *Trichoderma* which were controlling fusarium wilt and recorded 17–48% of decrease in disease incidence. The application of *Trichoderma* spp. for managing *Fusarium* wilt of pigeon pea has been recommended by several investigators (Mandhare and Suryawanshi, 2005; Jayalakshmi et al., 2003). The observation of the study suggested that the seeds of pigeon pea treated with *T. viride* followed by *T. harzianum* were found to be effective in reduction of the wilt disease, when compared with individual treatments. In 2006, differential efficacy of bioagents, namely, *T. viride*, *T. harzianum*, and *G. virens* were combined used by Dhar et al. (2006) against *F. udum* isolates and showed up to 35.5–57.3% of reduction in disease incidence in Fusarium wilt (FW) of pigeon pea. Gundappagal and Bidari (1997) used *T. viride* for seed treatment to resistant cultivar and became effective in integrated disease management of pigeon pea under dry land cultivation. Several strains of *Trichoderma* are commercially available to control plant disease in environment-friendly agriculture, for example, control of *S. rolfsii* in tobacco, *Nectria galligena* in apples, *R. solanii* in radish, bean and iris, strawberry, cucumber, potato, and tomato, *Chondrosterum purpureum* in stone-fruit, and other crops, and *B. cinerea* in apple (Cutler and Cutler, 1999). These species became popular biological agents due to their potential to protect crops against their pathogens all over the world (Howell, 2003; Ha, 2010) and are becoming widely used in agriculture and horticulture (Howell, 2003; Harman, 2006), since the first recognized application in the early of 1930, and have also been controlled and added many diseases to the list in subsequent years (Howell, 2003). Mycoparasitism describes the complex process in which several events are, namely, recognition of the host, attack, and subsequent penetration and killing are involved (Vinale et al., 2008b). *Trichoderma* spp. releases or produces a variety of chemical compounds that induce to systemic resistance responses in plants (Ranasingh et al., 2006) but have no harmful effects on humans, wildlife, and other beneficial organisms which contribute toward the control of plant pathogen, in addition of moderate effects on soil balance. They are really very effective and safe in both natural and controlled environments and never accumulate in the food chain (Saba et al., 2012).

Comparison of different biological control agents and their products showed reduced wilt incidence by *Trichoderma* spp., and seed treatment with its formulated cell mass at 8 g/kg seed recorded the lowest wilt incidence (Sawant et al., 2003). In 2002, many mutational and recombinant bioinoculants have been tried in this field to reduce the wilt incidence and found to be successful. Roy and Sitansu (2005) published that among the recombinant, *T. harzianum*, 50Th3II and 125Th4I are reduced the wilt disease in nonsterilized soil, while 75Th4IV in sterilized soil with a disease percentage of 36.51%, 33.86%, and 33.33%, respectively, were reported. *T. viride*, *T. harzianum*, and *G. virens* combinedly used by Ram and Pandey (2011) suggested that the combined use of *T. viride* and *P. fluorescens* is helpful in the reduction of *Fusarium* wilt. A summary on use of *Trichoderma* spp. in the management of diseases of crops has been summarized in Table 5.4.

5.6 CONCLUSION, SUMMARY, AND FUTURE RESEARCH OPPORTUNITIES

Crop protection includes several strategies used to reduce yield loss due to a variety of plant pathogens. Biocontrol agents are more sustainable alternatives to control the disease. Only by improving this method of disease control, India will be able to feed its growing population and eradicate extreme poverty and hunger. More than ever before sustainable economy

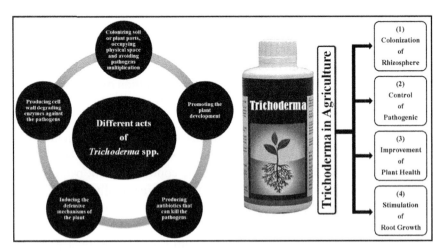

FIGURE 5.2 *Trichoderma* spp. as biological agents due to their different mechanisms (*Source*: Monte and Llobell, 2003) and advantages use in agriculture (*Source*: Adapted from Harman, 2006).

TABLE 5.4 *Trichoderma* spp. Protects Crops from Their Pathogens.

Name of crop	Name of disease	Plant pathogen
Lentil, pigeon pea	Wilt disease	*Fusarium* spp.
Ginger, turmeric, onion	Rhizome rot	*Pythium, Phytophthara, Fusarium*
Brinjal	Wilt disease, damping off and root rot	*Fusarium* spp., *Pythium, Phytophthora*
Clusterbean	Dry root rot	*Macrophomina phaseolina*
Banana, cotton	Wilt disease	*Fusarium* spp.
Mungbean	Dry root rot and wilt	*Macrophomina, Fusarium*
Tomato	Wilt disease and damping off	*Fusarium* spp., *Pythium, Phytophthora*
Chili	Damping off	*Pythium, Phytophthora*
Elephant foot yam	Collar rot	*Sclerotium rolfsii*
Chick pea	Dry root rot and Wilt disease	*Rhizoctonia* spp., *Fusarium* spp.
Beans, tomato, cotton	Collar rot, root rot	*Scelerotinia rolfsii, Rhizoctonia solani*
Wheat	Fusarium foot rot	*Fusarium* spp.
Maize	Stalk rot	Pythium *aphanidermatum, Fusarium graminearum*
Maize	Banded leaf and sheath blight	*Rhizoctonia solani* f. sp *sasakii*

Sources: Elad et al. (1998), Kumar et al. (1993), and Howell (2003).

and protection of our environment are dominant topics in our everyday life. Research on *Trichoderma* will help to mitigate the amount of chemicals being released into the environment. Investigation on the mechanism responsible for the biocontrol exerted by *Trichoderma* spp. is well studied. *Trichoderma* actively takes over a root zone and makes it difficult for pathogens to compete for space on the roots and for nutrients. It reduces growth, survival or infections caused by pathogens by different mechanisms like competition, antibiosis, mycoparasitism, hyphal interactions, and enzyme secretion. The discovery of new potential *Trichoderma* antagonists will lead to new challenges in research, development, and registration of biocontrol agents in a market where chemical pesticides dominate. Commercial use and application of biological disease control have been slow mainly due to their variable performances under different environmental conditions in the field. To overcome this problem and in order to take the biocontrol technology to the field and improve the commercialization of biocontrol, it is important to develop new formulations of biocontrol microorganisms with higher degree of stability and survival. In order to have more effective biological control strategies in the future, it is critical to carry out more research studies on some less developed aspects of biocontrol agents, mass production of biocontrol microorganisms and the use of biotechnology and nanotechnology in improvement of biocontrol mechanism strategies.

KEYWORDS

- *Trichoderma spp.*
- pathogenic control
- Rose bengal agar media
- biochemical elicitors
- bioremediation

REFERENCES

Adachi, T.; Aoki, H.; Osawa, T.; Namiki, M.; Yamane, T.; Ashida, T. Structure of Trichodermaol, Antibacterial Substance Produced in Combined Culture of *Trichoderma* sp. with *Fusarium oxysporum* or *Fusarium solani*. *Chem. Lett.* **1983**, *1*, 923–926.

Adams, P.; De-Leij, F. A. A. M.; Lynch, J. M. *Trichoderma harzianum* Rifai 1295–22 Mediates Growth Promotion of Crack Willow (*Salix fragilis*) Saplings in Both Clean and Metal-Contaminated Soil. *Microb. Ecol.* **2007**, *54*, 306–313.

Adekunle, A. T.; Cardwell, K. E.; Florini, D. A.; Ikotun, T. Seed Treatment with Trichoderma Species for Control of Damping-Off of Cowpea Caused by *Macrophomina phaseolina*. *Biocontrol Sci. Technol.* **2001**, *11*, 449–457.

Ahmed, A. S.; Sanchez, C. P.; Candela, M. E. Evaluation of Induction of Systemic Resistance in Pepper Plants (*Capsicum annuum*) to *Phytophthora capsici* Using *Trichoderma harzianum* and Its Relation with Capsidiol Accumulation. *Eur. J. Plant Pathol.* **2000**, *106*, 817–824.

Almassi, F.; Ghisalberti, E. L.; Narbey, M. J.; Sivasithamparam, K. New Antibiotics from Strains of *Trichoderma harzianum*. *J. Nat. Prod.* **1991**, *54*, 396–402.

Altomare, C.; Norvell, W. A.; Björkman, T.; Harman, G. E. Solubilization of Phosphates and Micronutrients by the Plant-Growth Promoting and Biocontrol Fungus *Trichoderma harzianum* Rifai 1295-22. *Appl. Environ. Microbiol.* **1999**, *65*, 2926–2933.

Anjaiah, V.; Cornelis, P.; Koedam, N. Effect of Genotype and Root Colonization in Biological Control of Fusarium Wilts in Pigeonpea and Chickpea by *Pseudomonas aeruginosa* PNA1. *Can. J. Microbiol.* **2003**, *49*, 85–91.

Asaduzzaman, M.; Alam, M. J.; Islam, M. M. Effect of *Trichoderma* on Seed Germination and Seedling Parameters of Chili. *J. Sci. Found.* **2010**, *8*(1/2), 141–150.

Baker, R. *Trichoderma* spp. as Plant Growth Stimulants. *CRC Crit. Rev. Biotechnol.* **1988,** *7,* 97–106.

Baldwin, J. E.; Adlington, R. M.; Chondrogianni, J.; Edenbrough, M. S.; Keeping, J. W.; Zielger, C. B. Structure and Synthesis of New Cyclopentyenyl Isonitriles from *Trichoderma hamatum* (Bon.) Bain. aggr. HLX 1379. *J. Chem. Soc., Chem. Commun.* **1985,** 816–817.

Baldwina, J. E.; Oneil, I. A.; Russell, A. T. Isonitrin A: Revision of the Structure and Total Synthesis in Racemic Form. *Synletters* **1991,** *1991*(8), 551–552.

Bhatnagar, H. Influence of Environmental Conditions on Antagonistic Activity of *Trichoderma* spp. Against *Fusarium udum*. *Indian J. Mycol. Plant Pathol.* **1996,** *26,* 58–63.

Bisset, J. A Revision of the Genus *Trichoderma*. II. Intrageneric Classification. *Can. J. Bot.* **1991,** *69,* 2357–2372.

Biswas, K. K.; Das, N. D. Biological Control of Pigeon Pea Wilt Caused by *Fusarium udum* with *Trichoderma* spp. *Ann. Plant Prot. Sci.* **1999,** *7,* 46–50.

Brimner, T. A.; Boland, G. J. A Review of the Non-Target Effects of Fungi Used to Biologically Control Plant Diseases. *Agric. Ecosyst. Environ.* **2003,** *100,* 3–16.

Cai, F.; Yu, G.; Wang, P.; Wei, W.; Fu, L.; Shen, Q.; Chen, W. Harzianolide, a Novel Plant Growth Regulator and Systemic Resistance Elicitor from *Trichoderma harzianum*. *Plant Physiol. Biochem.* **2013,** *73,* 106–113.

Cao, L.; Jiang, M.; Zeng, Z.; Du, A.; Tan, H.; Liu, Y. *Trichoderma atroviride* F6 Improves Phytoextraction Efficiency of Mustard [*Brassica juncea* (L.) Coss. var. Foliosa Bailey] in Cd, Ni Contaminated Soils. *Chemosphere* **2008,** *71,* 1769–1773.

Chang, Y. C.; Chang, Y. C.; Baker, R.; Kleifeld, O.; Chet, I. Increased Growth of Plants in the Presence of the Biological Control Agent *Trichoderma harzianum*. *Plant Dis.* **1986,** *70,* 145–148.

Chaube, H. S.; Mishra, D. S.; Varshney, S.; Singh, U. S. Biological Control of Plant Pathogens by Fungal Antagonistic: Historical Background, Present Status and Future Prospects. *Annu. Rev. Plant Pathol.* **2002,** *2,* 1–42.

Chet, I. *Trichoderma*—Application, Mode of Action and Potential as a Biocontrol Agent of Soil Borne Plant Pathogenic Fungi. In *Innovative Approaches to Plant Disease Control*; Chet, I., Ed.; John Wiley and Sons: New York, 1987; pp 137–160.

Chet, I.; Inbar, J.; Hadar, I. Fungal Antagonists and Mycoparasites. In *The Mycota IV: Environmental and Microbial Relationships*; Wicklow, D. T., Söderström, B., Eds.; Springer-Verlag: Berlin, 1997; pp 165–184.

Chowdappa, P.; Kumar, S. P. M.; Lakshmi, M. J.; Upreti, K. K. Growth Stimulation and Induction of Systemic Resistance in Tomato Against Early and Late Blight by *Bacillus subtilis* OTPB1 or *Trichoderma harzianum* OTPB3. *Biol. Control* **2013,** *65,* 109–117.

Collins, R. P.; Halim, A. F. Characterization of the Major Aroma Constituent of the Fungus *Trichoderma viride* (Pers.). *J. Agric. Food Chem.* **1972,** *20,* 437–438.

Cutler, S. J.; Cutler, H. G. *Biologically Active Natural Products: Pharmaceuticals*; CRC Press: New York; 1999.

Danielson, R. M.; Davey, C. B. The Abundance of *Trichoderma* Propagules and Distribution of Species in Forest Soils. *Soil Biol. Biochem.* **1973a,** *5,* 485–494.

Danielson, R. M.; Davey, C. B. Non-nutritional Factors Affecting the Growth of *Trichoderma* in Culture. *Soil Biol. Biochem.* **1973b,** *5,* 495–504.

Dhar, V.; Mishra, S.; Chaudhary, R. G. Differential Efficacy of Bioagents Against *Fusarium udum* Isolates. *Indian Phytopathol.* **2006,** *59,* 290–293.

Dixit, P.; Mukherjee, P. K.; Ramachandran, V.; Eapen, S. Glutathione Transferase from *Trichoderma virens* Enhances Cadmium Tolerance Without Enhancing Its Accumulation in Transgenic *Nicotiana tabacum*. *PLoS ONE* **2011,** *6*(1), e16360.

Dodd, S. L.; Hill, R. A.; Stewart, A. A Duplex-PCR Bioassay to Detect a *Trichoderma virens* Biocontrol Isolate in Non-Sterile Soil. *Soil Biol. Biochem.* **2004,** *36*, 1955–1965.

Domsch, K. H.; Gams, W.; Anderson, T. H. *Compendium of Soil Fungi, Vol. 1*; Academic Press: New York; 1980.

Doni, F.; Al-Shorgani, N. K. N.; Tibin, E. M. M.; Abuelhassan, N. N.; Anizan, I.; Che Radziah, C. M. Z.; Wan, W. Y. Mohtar, Microbial Involvement in Growth of Paddy. *Curr. Res. J. Biol. Sci.* **2013,** *5*(6), 285–290.

Doni, F.; Anizan, I.; Che Radziah, C. M. Z.; Salman, A. H. Enhancement of Rice Seed Germination and Vigour by Trichoderma spp. *Res. J. Appl. Sci. Eng. Technol.* **2014,** *7*(21), 4547–4552.

Doyle, J. J.; Doyle, J. L. Isolation of Plant DNA from Fresh Tissue. *Focus* **1990,** *12*(1), 13–15.

Elad, Y. Biological Control of Foliar Pathogens by Means of *Trichoderma harzianum* and Potential Modes of Action. *Crop Prot.* **2000,** *19*, 709–714.

Elad, Y.; David, D. R.; Levi, T.; Kapat, A.; Krishner, B.; Guvrin, E.; Levine, A. *Trichoderma harzianum*, Trichoderma-39 Mechanisms of Biocontrol of Foliar Pathogens. In *Modern Fungicides and Antifungal Compounds II*; Lyr, H., Ed.; Intercepts Ltd.: Andover, Hampshire, UK, 1998; pp 459–467.

Errasquin, E. L.; Vazquez, C. Tolerance and Uptake of Heavy Metals by *Trichoderma atroviride* Isolated from Sludge. *Chemosphere* **2003,** *50*, 137–143.

Ezzi, M. I.; Lynch, J. M. Biodegradation of Cyanide by *Trichoderma* spp. and *Fusarium* spp. *Enzyme Microb. Technol.* **2005,** *36*, 849–854.

Gams, W.; Bissett, J. Morphology and Identification of Trichoderma. In *Trichoderma and Gliocladium, Basic Biology, Taxonomy and Genetics*; Harman, G. E., Kubicek, C. P., Eds.; Taylor and Francis: London, UK, 1998; Vol. 1, pp 3–34.

Ghisalberti, E. L.; Rowland, C. Y. Antifungal Metabolites from *Trichoderma harzianum*. *J. Nat. Prod.* **1993,** *56*, 1799–1804.

Gundappagal, R. C.; Bidari, V. B. *Trichoderma viride* in the Integrated Management of Pigeon Pea Wilt Under Dryland Cultivation. *Adv. Agric. Res. India* **1997,** *7*, 65–69.

Ha, T. N. Using *Trichoderma* Species for Biological Control of Plant Pathogens in Vietnam. *J. ISSAAS.* **2010,** *16*, 17–21.

Hajieghrari, B. Effect of Some Metal-containing Compounds and Fertilizers on Mycoparasite *Trichoderma* Species Mycelia Growth Response. *Afr. J. Biotechnol.* **2010,** *9*, 4025–4033.

Hanania, U.; Avni, A. High-affinity Binding Site for Ethylene-Inducing Xylanase Elicitor on *Nicotiana tabacum* Membranes. *Plant J.* **1997,** *12*, 113–120.

Haran, S.; Schickler, H.; Oppenheim, A.; Chet, I. Differential Expression of *Trichoderma harzianum* Chitinase During Mycoparasitism. *Phytopathology* **1996,** *86*, 980–985.

Harman, G. E. Myths and Dogmas of Biocontrol. Changes in Perceptions Derived from Research on *Trichoderma harzianum* T-22. *Plant Dis.* **2000,** *84*, 377–393.

Harman, G. E. Overview of Mechanisms and Uses of *Trichoderma* spp. *Phytopathology* **2006,** *96*, 190–194.

Harman, G. E.; Björkman, T. Potential and Existing Uses of *Trichoderma* and *Gliocladium* for Plant Disease Control and Plant Growth Enhancement. In *Trichoderma and Gliocladium*; Kubicek, C. P., Harman, G. E., Eds.; Taylor and Francis: London, 1998; Vol. 2, pp 229–265.

Harman, G. E.; Kubicek, C. P. *Trichoderma and Gliocladium*; Taylor and Francis: London, 1998; pp 278.

Harman, G. E.; Shoresh, M. The Mechanisms and Applications of Opportunistic Plant Symbionts. In *Novel Biotechnologies for Biocontrol Agent Enhancement and Management*; Vurro, M., Gressel, J., Eds.; Springer: Amsterdam, The Netherlands, 2007; pp 131–153.

Harman, G. E.; Taylor, A. G.; Stasz, T. E. Combining Effective Strains of *Trichoderma harzianum* and Solid Matrix Priming to Improve Biological Seed Treatments. *Plant Dis.* **1989**, *73*, 631–637.

Harman, G. E.; Howell, C. R.; Viterbo, A.; Chet, I.; Lorto, M. *Trichoderma* Species— Opportunistic, Avirulent and Plant Symbionts. *Nat. Rev. Microbiol.* **2004**, *2*, 43–56.

Harmosa, R.; Viterbo, A.; Chet, I.; Monte, E. Plant Beneficial Effects of *Trichoderma* and its Genes. *Microbiology.* **2012**, *158*, 17–25.

Harris, G. H.; Jones, E. T. T.; Meinz, M.; Nallin-Omestead, M.; Helms, G. L.; Bills, G. F.; Zink, D.; Wilson, K. E. Isolation and Structure Elucidation of Viridiofungins A, B and C. *Tetrahedron Lett.* **1993**, *34*(33), 5235–5238.

Howell, C. R. Mechanisms Employed by *Trichoderma* Species in the Biological Control of Plant Diseases: The History and Evolution of Current Concepts. *Plant Dis.* **2003**, *87*, 4–10.

Huang, Q.; Tezuka, Y.; Kikuchi, T.; Nishi, A.; Tubaki, K.; Tanaka, K. Studies on Metabolites of Mycoparasitic Fungi. II. Metabolites of *Trichoderma koningii*. *Chem. Pharm. Bull.* **1995**, *43*(2), 223–239.

Hussain, S. A.; Noorani, R.; Qureshi, I. H. Microbiological Chemistry. Part I. Isolation and Characterization of Gliotoxin, Ergosterol, Palmitic Acid and Mannitol Metabolites of *Trichoderma hamatum* Bainier. *Pak. J. Sci. Ind. Res.* **1975**, *18*, 221–243.

Jayalakshmi, S. K.; Sreermula, K.; Benig, V. I. Efficacy of *Trichoderma* spp. Against Pigeon Pea Wilt Caused by *Fusarium udum*. *J. Biol. Control.* **2003**, *17*, 75–78.

Jiang, X.; Geng, A.; He, N.; Li, Q. New Isolates *Trichoderma viride* Strain for Enhanced Cellulolytic Enzyme Complex Production. *J. Biosci. Bioeng.* **2011**, *111*(2), 121–127.

Jin, L.; Jin, W. *Metabolites Produced by Fungus no. G-18. II. G-18-IV. an Immunosuppressive, Isolation and Identification*. Zhongguo Kangshengsu Zazhi. **1989**, *14*, 23–26. Chem. Abstr. 1991.115: 227914.

Kamal, A.; Akhtar, R.; Qureshi, A. A. Biochemistry of Microorganisms. XX. 2,5-Dimethoxybenzoquinone, Tartronic Acid, Itaconic Acid, Succinic Acid, Pyrocalciferol, Epifriedelinol, Lanosta-7,9(11), 24-triene-3ß,21-diol, Trichodermene A, Methyl 2,4,6-octatriene and Cordycepic Acid, *Trichoderma* Metabolites. *Pak. J. Sci. Ind. Res.* **1971**, *14*, 71–78.

Khan, J.; Ooka, J. J.; Miller, S. A.; Madden, L. V.; Hoitink, H. A. J. Systemic Resistance Induced by *Trichoderma hamatum* 382 in Cucumber Against *Phytophthora* Crown Rot and Leaf Blight. *Plant Dis.* **2004**, *88*, 280–286.

Khan, M. R.; Khan, S. M. Effect of Root-dip Treatment with Certain Phosphate Solubilizing Microorganisms on the *Fusarium* Wilt of Tomato. *Bioresour. Technol.* **2002**, *85*, 213–215.

Kikuchi, T.; Mimura, T.; Harimaya, K.; Yano, H.; Arimoto, T.; Masada, Y.; Inoue, T. Letter: Volatile Metabolite of Aquatic Fungi. Identification of 6-Pentyl-Alpha-Pyrone from *Trichoderma* and *Aspergillus* Species. *Chem. Pharm. Bull.* **1974,** *22,* 1946–1948.

Klein, D.; Eveleigh, D. E. Ecology of *Trichoderma.* In *Trichoderma and Gliocladium*; Kubicek, C. P., Harman, G. E., Eds. Taylor and Francis: London, 1998; Vol. 2, pp 57–69.

Kubicek, C. P.; Bissett, J.; Druzfinina, L.; Kulling, G.; Szakacs, G. Genetic and Metabolic Diversity of *Trichoderma*: A Case Study on Southeast Asian Isolates. *Fungal Genet. Biol.* **2002,** *38,* 310–319.

Kumar, C. P. C.; Garibova, I. V.; Vellickanov, L. L.; Durinina, E. P. Biocontrol of Wheat Root Rots Using Mixed Cultures of *Trichoderma viride* and *Glomus epigaeus. Indian J. Plant Prot.* **1993,** *21*(2), 145–148.

Kumar, R. N.; Mukerji, K. G. Integrated Disease Management Future Perspectives. In *Advances in Botany*; Mukerji, K. G., Mathur, B., Chamala, B. P., Chitralekha, C., Eds.; APH Publishing Corporation: New Delhi, 1996; pp 335–347.

Lalitha, P.; Srujana; Arunalakshmi, K. Effect of *Trichoderma viride* on Germination of Mustard and Survival of Mustard Seedlings. *Int. J. Life Sci. Biotechnol. Pharm. Res.* **2012,** *1*(1), 137–140.

Latunde-Dada, A. O. The Use of *Trichoderma koningii* in the Control of Web Blight Disease Caused by *Rhizoctonia solani* in the Foliage of Cowpea (*Vigna unguiculata*). *J. Phytopathol.* **1991,** *133,* 247–254.

Lynch, J. M.; Wilson, K. L.; Ousley, M. A.; Whipps, J. M. Response of Lettuce to *Trichoderma* treatment. *Lett. Appl. Microbiol.* **1991a,** *12,* 56–61.

Lynch, J. M.; Lumsden, R. D.; Atkey, P. T.; Ousley, M. A. Prospects for Control of *Pythium* Damping-off of Lettuce with *Trichoderma, Gliocladium,* and *Enterobacter* spp. *Biol. Fertil. Soils.* **1991b,** *12,* 95–99.

Maisuria, V. B.; Gohel, V.; Mehta, A. N.; Patel, R. R.; Chhatpar, H. S. Biological Control of *Fusarium* Wilt of Pigeonpea by *Pantoea dispersa,* a Field Assessment. *Ann. Microbiol.* **2008,** *58*(3), 411–419.

Mandhare, V. K.; Suryawanshi, A. V. Application of *Trichoderma* Species Against Pigeonpea Wilt. *JNKVV Res. J.* **2005,** *32*(2), 99–100.

Martínez-Medina, A.; Roldán, A.; Albacete, A.; Pascual, J. A. The Interaction with Arbuscular Mycorrhizal Fungi or *Trichoderma harzianum* Alters the Shoot Hormonal Profile in Melon Plants. *Phytochemistry* **2011,** *72,* 223–229.

McSpadden Gardener, B. B.; Fravel, D. R. Biological Control of Plant Pathogens: Research, Commercialization, and Application in the USA. *Plant Health Progr.* (Online) 2002. http://naldc.nal.usda.gov/download/12141/PDF (accessed April 25, 2016).

Ming, Q.; Han, T.; Li, W.; Zhang, Q.; Zhang, H.; Zheng, C.; Huang, F.; Rahman, K.; Qin, L. Tanshinone IIA and Tanshinone I Production by *Trichoderma atroviride* D16, an Endophytic Fungus in *Salvia miltiorrhiza. Phytomedicine* **2012,** *19,* 330–333.

Mishra, A.; Salokhe, V. M. Rice Growth and Physiological Responses to SRI Water Management and Implications for Crop Productivity. *Paddy Water Environ.* **2011,** *9,* 41–52.

Misra, A. K.; Prasad, B. *Trichoderma*—A Genus for Biocontrol. In *Biopesticide and Bioagents in Integrated Pest Management of Agriculture Crops*; Shrivastava, R. P., Ed.; International Book Distribution Co.: Lucknow, 2003; pp 811–833.

Monte, E.; Llobell, A. In *Trichoderma in Organic Agriculture*, Proceedings V World Avocado Congress (Actas V Congreso Mundial del Aguacate), Dec 10–14, 2003; pp 725–733.

Morsy, E. M.; Abdel-Kawi, K. A.; Khalil, M. N. A. Efficiency of *Trichoderma viride* and *Bacillus subtilis* as Biocontrol Agents Against *Fusarium solani* on Tomato Plants. *Egypt. J. Phytopathol.* **2009,** *37*(1), 47–57.

Moss, M. O.; Jackson, J. M.; Rogers, D. The Characterization of 6-(pent-1-enyl)-alpha-pyrone from *Trichoderma viride*. *Phytochemistry* **1975,** *14*, 2706–2708.

Mukherjee, I.; Gopal, M. Degradation of Chlorpyrifos by Two Soil Fungi *Aspergillus niger* and *Trichoderma viride*. *Toxicol. Environ. Chem.* **1996,** *57*, 145–151.

Oerke, E. C.; Dehne, H. W. Safeguarding Production-losses in Major Crops and the Role of Crop Protection. *Crop Prot.* **2004,** *23*, 275–285.

Okoth, S. A.; Otadoh, J. A.; Ochanda, J. O. Improved Seedling Emergence and Growth of Maize and Beans by *Trichoderma harziunum*. *Trop. Subtrop. Agroecosyst.* **2011,** *13*, 65–71.

Pandey, K. K.; Upadhyay, J. P. Selection of Potential Bio-control Agents based on Production of Volatile and Non-volatile Antibiotics. *Veg. Sci.* **1997a,** *24*, 144–146.

Pandey, K. K.; Upadhyay, J. P. Effect of Different Nutrient Media and pH on Growth and Sporulation of Different Isolates of *Trichoderma* spp. and *Gliocladium virens*. *Veg. Sci.* **1997b,** *24*(2), 140–143.

Pandotra, V. R. *Illustrated Fungi of North India with Special Reference to J & K State*; International Book Distributors: Dehradun, India, 1997.

Pandya, J. R.; Sabalpara, A. N.; Chawda, S. K. *Trichoderma*: A Particular Weapon for Biological Control of Phytopathogens. *J. Agric. Sci. Technol.* **2011,** *7*(5), 1187–1191.

Papavizas, G. C. *Trichoderma* and *Gliocladium*: Biology, Ecology, and Potential for Biocontrol. *Annu. Rev. Phytopathol.* **1985,** *23*, 23–54.

Patel, B. N.; Patel, G. J. In *Virus Diseases of Tobacco and Their Management*, Integrated Pest and Diseases Management. Proceedings of the National Seminar; Jayaraj, S., Ed.; Tamil Nadu Agricultural University: Coimbatore, India, 1985; pp 261–266.

Persoon, C. H. Neuer Veersuch Einer Systematischen Eintheilung der Schwamme (Dispositio methodica fungorum). *Romer's News Mag. Bot.* **1794,** *1*, 63–128.

Prasad, R. D.; Rangeshwaran, R.; Hegde, S. V.; Anuroop, C. P. Effect of Soil and Seed Application of *Trichoderma haezianum* on Pigeonpea WILT Caused by *Fusarium udum* Under Field Conditions. *Crop Prot.* **2002,** *21*, 293–297.

Pyke, T. H.; Dietz, A. U-21,9263, a New Antibiotic. I. Discovery and Biological Activity. *Appl. Microbiol.* **1966,** *14*, 506–510.

Ram, H.; Pandey, R. N. Efficacy of Bio-control Agents and Fungicides in the Management of wilt of Pigeon Pea. *Indian Phytopathol.* **2011,** *64*(3), 269–271.

Ranasingh, N.; Saurabh, A.; Nedunchezhiyan, M. Use of *Trichoderma* in Disease Management. *Orissa Rev.* **2006,** *September–October,* 68–70.

Ratnakumari, R. Y.; Nagamani, A.; Sarojini, C. K.; Adinarayana, G. Effect of *Trichoderma* Species on Yield of *Mentha arvensis* L. *Int. J. Adv. Res.* **2014,** *2*(7), 864–867.

Ren, W. Y. Synthesis and Stereochemistry of Pencolide and Structure and Synthesis of a New Cyclopentanone Derivative from *Trichoderma album*. *Diss. Abstr. Int. B 1978.* **1977,** *39*, 12193 (Chem.Abstr.1979, 91: 175119).

Rifai, M. A Revision of Genus *Trichoderma*. *Mycol. Pap.* **1969,** *116*, 1–56.

Roiger, D. J.; Jeffers, S. N.; Caldwell, R. W. Occurrence of *Trichoderma* Species in Apple Orchard and Woodland Soils. *Soil Biol. Biochem.* **1991**, *23*, 353–359.

Roy, A.; Sitansu, P. Biological Control Potential of Some Mutants of *Trichoderma harzianum* and *Gliocladium virens* Against wilt of Pigeon Pea. *J. Interacademicia.* **2005**, *9*(4), 494–497.

Saba, H.; Vibhash, D.; Manisha, M.; Prashant, K. S.; Farhan, H.; Tauseef, A. *Trichoderma*—A Promising Plant Growth Stimulator and Biocontrol Agent. *Mycosphere* **2012**, *3*, 524–531.

Samuels, G. J. Trichoderma: Systematics, the Sexual State, and Ecology. *Phytopathology* **2006**, *96*, 195–206.

Sawant, D. M.; Kolase, S. V.; Bachkar, C. B. Efficacy of Bio-Control Agents Against wilt of Pigeon Pea. *J. Maharashtra Agric. Univ.* **2003**, *28*, 303–304.

Shamim, S.; Ahmad, N.; Rahaman, A.; Haque, S. E.; Chaffer, A. Efficacy of *Pseudomonas aeruginosa* and Other Biocontrol Agents in the Control of Root Rot Infection in Cotton. *Acta Agrobotanica.* **1997**, *50*, 5–10.

Sharma, G.; Pandey, R. R. Influence of Culture Media on Growth, Colony Character and Sporulation of Fungi Isolated from Decaying Vegetable Wastes. *J. Yeast Fungal Res.* **2011**, *1*(8), 157–164.

Sharma, P.; Patel, A. N.; Saini, M. K.; Deep, S. Field Demonstration of *Trichoderma harzianum* as a Plant Growth Promoter in Wheat (*Triticum aestivum* L). *J. Agric. Sci.* **2012**, *4*(8), 65–73.

Singh, H. B.; Singh, D. P. From Biological Control to Bioactive Metabolites: Prospects with *Trichoderma* for Safe Human Food. *Pertanika J. Trop. Agric. Sci.* **2009**, *32*, 99–110.

Singh, S. P. Biological Control. In *50 Years of Crop Science Research in India*; Paroda, R. S., Chadha, K. L., Eds.; Indian Council of Agricultural Research: New Delhi, 1996; pp 88–116.

Slater, G. P.; Haskins, R. H.; Hagge, L. R.; Nesbitt, L. R. Metabolic Products from a *Trichoderma viride* Pers. ex Fries. *Can. J. Chem.* **1967**, *45*, 92–96.

Somasekhara, Y. M.; Anilkumar, T. B.; Siddaramaiah, A. L. Bio-control of Pigeon Pea [*Cajanus cajan* (L.) Millsp.] Wilt (*Fusarium udum* Butler). *Mysore J. Agric. Sci.* **1996**, *30*, 159–163.

Somasekhara, Y. M.; Siddaramaiah, A. L.; Anilkumar, T. B. Evaluation of *Trichoderma* Isolates and Their Antifungal Extracts as Potential Bio-control Agents Against Pigeon Pea Wilt Pathogen, *Fusarium udum* Butler. *Curr. Res.* **1998**, *27*, 158–160.

Sriram, S.; Manasa, S. B.; Savitha, M. J. Potential Use of Elicitors from *Trichoderma* in Induced Systemic Resistance for the Management of *Phytophthora capsici* in Red Pepper. *J. Biol. Control.* **2009**, *23*(4), 449–456.

Tančić, S.; Skrobonja, J.; Lalošević, M.; Jevtić, R.; Vidić, M. Impact of *Trichoderma* spp. on Soybean Seed Germination and Potential Antagonistic Effect on *Sclerotinia sclerotiorum*. *Pestic. Phytomed.* **2013**, *28*(3), 181–185.

Tarus, K.; Lang'at-Thoruwa, C. C.; Wanyonyi, A. W.; Chhabra, S. C. Bioactive Metabolites from *Trichoderma harzianum* and *Trichoderma longibrachiatum*. *Bull. Chem. Soc. Ethiop.* **2003**, *17*, 185–190.

Thompson, J. D.; Higgins, D. G.; Gibson, T. J. CLUSTAL W: Improving the Sensitivity of Progressive Multiple Sequence Alignment Through Sequence Weighting,

Position-Specific Gap Penalties and Weight Matrix Choice. *Nucleic Acids Res.* **1994,** *22*(22), 4673–4680.

Tucci, M.; Ruocco, M.; De Masi, L.; De Palma, M.; Lorito, M. The Beneficial Effect of *Trichoderma* spp. on Tomato Is Modulated by the Plant Genotype. *Mol. Plant Pathol.* **2011,** *12,* 341–354.

Turner, W. B.; Aldridge, D. C. *Fungal Metabolites II*; Turner, W. B., Aldridge, D. C., Eds.; Academic Press: London, 1983.

Upadhyay, R. S.; Rai, B. Effect of Cultural Practices and Soil Treatments on Incidence of wilt Disease of Pigeon Pea. *Plant Soil* **1981,** *62*, 309–312.

Vahabi, K.; Mansoori, G. A.; Karimi, S. Biosynthesis of Silver Nanoparticles by Fungus Trichoderma Reesei: A Route for Largescale Production of AgNPs. *Insciences J.* **2011,** *1*(1), 65–79.

Vinale, F.; Sivasithamparam, K.; Ghisalberti, E. L.; Marra, R.; Barbetti, M. J.; Li, H.; Woo, S. L.; Lorito, M. A Novel Role for *Trichoderma* Secondary Metabolites in the Interactions with Plants, *Physiol. Mol. Plant Pathol.* **2008a,** *72*, 80–86.

Vinale, F.; Sivasithamparam, K.; Ghisalberti, E. L.; Woo, S. L.; Nigro, M.; Marra, R.; Lombardi, N.; Pascale, A.; Ruocco, M.; Lanzuise, S.; Manganiello, G.; Lorito, M. *Trichoderma* Secondary Metabolites Active on Plants and Fungal Pathogens. *Open Mycol. J.* **2014,** *8*(Suppl-1, M5), 127–139.

Vinale, F.; Sivasithamparamb, K.; Ghisalbertic, E. L.; Marraa, R.; Wooa, S. L.; Moritoa, L. *Trichoderma*–plant–pathogen Interactions. *Soil Biol. Biochem.* **2008b,** *40*, 1–10.

Vipul, K.; Shahid, M.; Singh, A.; Srivastava, M.; Biswas, S. K. RAPD Analysis of *T. longibrachiatum* Isolated from Pigeon Pea Fields of Uttar Pradesh. *Indian J. Agric. Biochem.* **2011,** *24*(1), 80–82.

Wang, Y.; Kausch, A. P.; Chandlee, J. M.; Luo, H.; Ruemmele, B. A.; Browning, M.; Jackson, N.; Goldsmith, M. R. Co-transfer and Expression of Chitinase, Glucanase and *Bar* Genes in Creeping Bentgrass for Conferring Fungal Disease Resistance. *Plant Sci.* **2003,** *165*(3), 497–506.

Wardle, D. A.; Parkinson, D.; Waller, J. E. Interspecific Competitive Interactions between Pairs of Fungal Species in Natural Substrates. *Oecologia* **1993,** *94*, 165–172.

Weindling, R. Studies on Lethal Principle Effective in the Parasitic action of *Trichoderma lignorum* on *Rhizoctonia solani* and Other Soil Fungi. *Phytopathology* **1934,** 24.

Weller, D. M. Biological Control of Soil Borne Plant Pathogens in the Rhizosphere with Bacteria. *Annu. Rev. Phytopathol.* **1988,** *26*, 379–469.

Whipps, J. M.; McQquilken, M. P.; Budge, S. P. Use of Fungal Antagonists for Biocontrol of Damping-off and Sclerotinia Disease. *Pestic. Sci.* **1993,** *37*, 309–313.

White, T. J.; Bruns, T.; Lee, S.; Taylor, J. Amplification and Direct Sequencing of Fungal Ribosomal DNA for Phylogenetics. In *PCR Protocols: A Guide to Methods and Applications*; Innes, M. A., Gelfand, D. H., Sninsky, J. J., White, T. J., Eds.; Academic Press, Inc.: San Diego, California, 1990; pp 315–322.

Williams, J. G. K.; Kubelik, A. R.; Livak, K. J.; Rafalski, J. A.; Tingey, S. V. DNA Polymorphisms Amplified by Arbitrary Primers Are Useful as Genetic Markers. *Nucleic Acid Res.* **1990,** *18*, 6531–6535.

Woo, S. L.; Scala, F.; Ruocco, M.; Lorito, M. The Molecular Biology of the Interactions Between *Trichoderma* spp., Phytopathogenic Fungi, and Plants. *Phytopathology* **2006,** *96,* 181–185.

Yedidia, I.; Benhamou, N.; Chet, I. Induction of Defense Responses in Cucumber Plants (*Cucumis sativus* L.) by the Biocontrol Agent *Trichoderma harzianum. Appl. Environ. Microbiol.* **1999,** *65,* 1061–1070.

Yedidia, I.; Benhamou, N.; Kapulnik, Y.; Chet, I. Induction and Accumulation of PR Proteins Activity During Early Stages of Root Colonization by the Mycoparasite *Trichoderma harzianum* strain T-203. *Plant Physiol. Biochem.* **2000,** *38,* 863–873.

Yedidia, I.; Srivastva, A. K.; Kapulnik, Y.; Chet, I. Effect of *Trichoderma harzianum* on Microelement Concentrations and Increased Growth of Cucumber Plants. *Plant Soil* **2001,** *235,* 235–242.

Zadoks, J. C.; Waibel, H. From Pesticides to Genetically Modified Plants: History, Economics and Politics. *Neth. J. Agric. Sci.* **1999,** *48,* 125–149.

Zeng, X.; Su, S.; Jiang, X.; Li, L.; Bai, L.; Zhang, Y. Capability of Pentavalent Arsenic Bioaccumulation and Biovolatilization of Three Fungal Strains Under Laboratory Conditions. *Clean—Soil Air Water* **2010,** *38,* 238–241.

Zhang, Z.; Schwartz, S.; Wagner, L.; Miller, W. A Greedy Algorithm for Aligning DNA Sequences. *J. Comput. Biol.* **2000,** *7,* 203–214.

Zheng, Z.; Shetty, K. Enhancement of Pea (*Pisum sativum*) Seedling Vigour and Associated Phenolic Content by Extracts of Apple Pomace Fermented with *Trichoderma* spp. *Process Biochem.* **2000,** *36,* 79–84.

Zhou, X.; Xu, S.; Liu, L.; Chen, J. Degradation of Cyanide by *Trichoderma* Mutants Constructed by Restriction Enzyme Mediated Integration (REMI). *Bioresour. Technol.* **2007,** *98,* 2958–2962.

PART 3

Microbiology for Soil Health and Crop Productivity Improvement

CHAPTER 6

NITROGENASE: A KEY ENZYME IN MICROBIAL NITROGEN FIXATION FOR SOIL HEALTH

DEEPAK KUMAR VERMA[1*], BALRAJ KAUR[2], ABHAY K. PANDEY[3*], and BAVITA ASTHIR[2*]

[1]*Agricultural and Food Engineering Department, Indian Institute of Technology, Kharagpur 721302, West Bengal, India*

[2]*Department of Biochemistry, College of Basic Sciences and Humanities, Punjab Agriculture University, Ludhiana 141004, Punjab, India*

[3]*PHM-Division, National Institute of Plant Health Management, Ministry of Agriculture and Farmers Welfare, Government of India, Rajendranagar 500030 Hyderabad, Telangana, India*

Corresponding authors. E-mail: deepak.verma@agfe.iitkgp.ernet.in/ rajadkv@rediffmail.com; abhaykumarpandey.ku@gmail.com; b.asthir@rediffmail.com

6.1 INTRODUCTION

Biological nitrogen fixation (BNF) is accountable for the major reduction of dinitrogen to the ammonia which is utilized by plants followed by animals indirectly. Gaseous nitrogen (N_2) is bound by one of the strongest covalent bonds in nature. Haber–Bosch process is another way to break this bond and convert it into ammonia, requiring about 500°C and 450 bar. Fertilizer production by this process results in significant release of the "greenhouse gases," particularly carbon dioxide (CO_2). Nature, in contrast, converts dinitrogen to ammonia at ambient conditions by employing the bacterial enzyme nitrogenase. BNF is considered as major route ammonia production (Seefeldt et al., 2009). However, in diverse taxonomic group, diazotrophs are the prokaryotes, in which this process is limited (Hartmann and Barnum, 2010), may be due to difficulty in detection at experimental level. One of the three sub types (i.e., *Nif, Vnf,* and *Anf*) of nitrogenase are mostly found in diazotrophs. The similarity was found with respect to structure, phylogeny as well as mechanism in these nitrogenase sub types. But they differ in metal content, whereas at their catalytic site, two distinct proteins was found, namely, dinitrogenase comprising component proteins D and K and dinitrogenase reductase comprising protein H (Seefeldt et al., 2009; Hartmann and Barnum, 2010).

6.2   HISTORICAL RESUME OF MICROBIAL N$_2$ FIXATION

In 1886, the German scientists named Hellriegel and Wilfarth who were first time reported nitrogen fixation in legume (root nodules) (Hellriegel and Wilfarth, 1988). Then during 1888, Dutch microbiologist named as Beijerinck worked on bacterial isolation and isolated *Rhizobium leguminosarum* from root nodules. Further, *Azotobacter* spp. was isolated in 1901 by Beijerinck and again in 1903 by Lipman; during 1901 *Clostridium pasteurianum* was isolated by Winodgradsky (Stewart, 1969). Then later on nitrogen fixation phenomenon was found in blue green algae (Stewart, 1969). By 1960, the nitrogen fixation phenomenon was found only in dozen of genera of free-living soil bacteria (Henson et al., 2004; Lindström and Martínez-Romero, 2007). In short, during old time, due to technical limitation as well as lack of tools for taxonomy and phylogeny, the detection of new N$_2$-fixing genera/species has been obstructed for long.

BNF has been recommended to be an ancient and possibly even primordial process (Fani et al., 2000). This concept is based on simulations of Archaean atmospheric chemistry that compete the decrease of CO$_2$ that led to abiotic N$_2$ oxidation to NO led to a nitrogen crises at ~3.5 Ga (Kasting and Walker, 1981). This time frame corresponds to a period of earth history where inferred fixed N levels are thought to have become warning, the concentration of O$_2$ started to increase, and the concentration of dissolved molybdenum (Mo) started to increase (Navarro-González et al., 2001; Anbar et al., 2007; Anbar, 2008). Various geochemical changes linked with this period of geological history are possible a consequence of oxygen production by proliferating the populations of oxygenic phototrophs (Anbar and Knoll, 2002; Anbar, 2008). The production of oxygen may have opened up new ecological niches, allowing the global biome to diversify and radiate into new environmental realms. This expansion of the biosphere would have created additional demand on the bioavailable N$_2$ pool and may have increased the selective pressure to evolve a biological mechanism to increase the local bioavailable N$_2$ pool.

6.3   NITROGENASE: KEY ENZYME

Nitrogenase is a unique enzyme used by some group of organisms to produce ammonium ion (NH_4^+) through the reduction of molecular dinitrogen (N_2) as given in the following equation:

$$N_2 + nMgATP + 8e^- + 10H^+ \longrightarrow 2NH_4^+ + nMgADP + H_2 \ldots\ldots[I]$$
$$(n > 2 \text{ per electron}) \qquad (6.1)$$

The breakage of N≡N bond (225 kcal/mol) requires high-activation energy and results in overall exothermic reaction. There is no report regarding the existence of nitrogenase activity in eukaryotic system; therefore, it may conclude that nitrogenase enzymes activity is mainly linked to bacteria and the Archaea (Young, 1992). The legumes have the ability to form symbiotic relationship with certain nitrogen-fixing bacteria and thus nitrogen fixation named as plant process. However, the main role is played nitrogenase enzyme complex in these symbioses, which is synthesized and localized in the endophyte of these symbioses as named as endosymbiont. These nitrogenases found in various organisms have very similar properties. The nitrogenase synthesizing organisms include free-living diazotrophs (viz., green sulfur bacteria, cyanobacteria, azotobacteraceae, etc.) and symbiotic diazotrophs (viz., *Rhizobia*, *Frankia*, etc.).

Instead of producing ammonium ion from dinitrogen, there are other reactions also catalyzed by nitrogenase enzyme:

$HC{\equiv}CH \rightarrow H_2C{=}CH_2$

$N^-{=}N^+{=}O \rightarrow N_2 + H_2O$

$N{=}N{=}N^- \rightarrow N_2 + NH_3$

$C{\equiv}N^- \rightarrow CH_4, NH_3, H_3C{-}CH_3, H_2C{=}CH_2 (CH_3NH_2)$

$N{\equiv}C{-}R \rightarrow RCH_3 + NH_3$

$C{\equiv}N{-}R \rightarrow CH_4, H_3C{-}CH_3, H_2C{=}CH_2, C_3H_8, C_3H_6, RNH_2$

$O{=}C{=}S \rightarrow CO + H_2S$

$O{=}C{=}O \rightarrow CO + H_2O$

$S{=}C{=}N^- \rightarrow H_2S + HCN$

$O{=}C{=}N^- \rightarrow H_2O + HCN, CO + NH_3$

Basically, to date, vanadium (VFe), molybdenum (MoFe), and iron-only (FeFe) are categorized as three main nitrogenase systems that have been found in nature (Table 6.1). Out of these, extensively studied nitrogenase system is MoFe, but the alternative systems, that is, vanadium-dependent (*Vnf*) and iron-only (*Anf*) systems are less well characterized, with respect to elements used for their biosynthesis as well as activity. The diazotrophs hold a common core of *nitrogen fixation (Nif) genes which is* necessary for nitrogen fixation in various organisms, whereas vanadium nitrogen fixation (*Vnf*) is a gene specific for vanadium-dependent diazotrophy or alternative nitrogen fixation (*Anf*) for iron-dependent diazotrophy.

6.3.1   Nif GENE

The *Nif* genes encode the enzyme which catalyzed the conversion of atmospheric nitrogen into that form which existing to living organisms. The Gary Ruvkun and Sharon R. Long were the first one who cloned Rhizobium genes by the 1980s (Spaink, 1998), named as "*Nif*" for nitrogen fixation and "nod" for nodulation in the laboratory of Frederick M. Ausubel's. The nitrogenase complex is the primary enzyme encoded by the *Nif* genes which converts atmospheric N_2 to ammonia. Besides this enzyme complex, the transcript of *Nif* genes also encodes a number of regulatory proteins that are necessary for its expression and activity. Therefore, in various plants, *Nif* genes are found to be associated with both the free-living nitrogen-fixing and symbiotic bacteria. Basically, in the host plant, the activity of *Nif* genes is induced at low N and oxygen concentrations at the periphery environment of root.

Nif genes transcription is regulated by *Nif*A protein, which is also nitrogen sensitive. At low nitrogen level available for organisms' use, *Ntr*C is activated which triggers *Nif*A protein expression, results in *Nif* genes cascade activation and produce nitrogenase enzyme. Whereas on the other hand, when nitrogen is present in excess, then it activates *Nif*L protein which inactivates *Nif*A protein and no nitrogenase is formed. The *Nif*L protein activity is further controlled by the *GlnD* and *GlnK* gene product. The *Nif* genes are localized on other genes for example, *nod* gene or on bacterial plasmids/chromosomes. In *Azotobacter vinelandii,* Walmsley et al. (1994) examined the role of *Nif*A regulatory gene in expression of three nitrogenases. It was found that expression of *Nif*A gene cause active

TABLE 6.1 Alternate Nitrogenase.

Type	Metals of dinitrogen reductase	H$_2$/NH$_3$ Mol ratio	Activity (NH$_3$ formed nmol/min/ mg enzyme)	Organisms	Gene code
Mo-nitrogenase	Fe, Mo	2:1	1000	All	*Nif*
V-nitrogenase	Fe, V	5:1–2:1	1000–300	*Clostridium* spp., *Azotobacter* spp. and *Anabaena* spp.	*Vnf*
Nitrogenase-3	Fe	10:1	<50	*Azotobacter* spp., *Rhodobacter* spp., and *Rhodopseudomonas* spp. (photosynthetic)	*Anf*

expression of *NifM*. The following are the examples of *Nif* gene in various diazotrophs.

6.3.1.1 Klebsiella pneumoniae

Klebsiella pneumoniae is a free-living anaerobic nitrogen-fixing bacterium. It contains 20 *Nif* genes that are located in a 24-Kb region on the chromosome. Nitrogenase subunit is encoded by *Nif*H, *Nif*D, and *Nif*K, whereas proteins required for nitrogenase subunit assembly and iron and molybdenum atoms incorporation into the subunit is encoded by other genes, namely, *Nif*N, *Nif*E, *Nif*U, *Nif*V, *Nif*S, *Nif*W, *Nif*B, *Nif*X, and *Nif*Q. Instead of this, *Nif*F and *Nif*J encode proteins that help in reduction process by electron transfer. The other regulatory proteins like *Nif*A and *Nif*L regulate the expression of the other *Nif* genes.

In *K. pneumonia*, *Nif* gene plays a task in the maturation of MoFe protein (Orme-Johnson, 1985); however, mutant *Nif*V lacks the Fe–Mo cofactor and the defect is now known to be in the enzyme homocitrate synthase (Hoover et al., 1987) which is essential for the synthesis of the Fe–Mo cofactor. No major role of *Nif*W and *Nif*Z genes has been studied by DNA sequence analysis (Arnold et al., 1988). Howard et al. (1986) found that *Nif*M cooperates a major role in the creation of active Fe protein. *Nif*Q product is thought to be involved in the assimilation of molybdenum into nitrogenase (Imperial et al., 1984). *Nif*B mutants are defective in an early stage of Fe–Mo cofactor synthesis and the product is mostly membrane-bound in *K. pneumoniae* extracts and appears to be required in stoichiometric amounts in an in vitro system that makes Fe–Mo cofactor (Shah et al., 1988).

6.3.1.2   *Rhodospirillum rubrum*

Rhodospirillum rubrum is a free-living anaerobic photosynthetic bacterium and shows transcriptional controls as described earlier. It also regulates *Nif* genes activity through ADP-ribosylation, which is caused by *Dra*G and *Dra*T at specific arginine residue in reversible way and thereby inhibits the electron transfer to nitrogenase complex (Merrick and Edwards, 1995). In *R. rubrum*, *Nif*A (transcriptional activator for the *Nif* genes) is activated posttranslationally by the uridylylated form of GlnB, one of the three P(II) homologs in the organism (Zhu et al., 2006). In *R. rubrum*, Zhu et al. (2006) identified the GlnB variants. GlnB variants are altered in their capability to interrelate with different targets in the response to nitrogen status signals. The role of uridylylation of GlnB is principally to shift the stability of GlnB from a "nitrogen-sufficient" form to a "nitrogen-deficient" form, each of which interacts with different but overlapping receptor proteins in the cell. These GlnB variants apparently shift that equilibrium through direct structural changes.

Nitrogen fixation in *R. rubrum* is tightly regulated at two levels, that is, transcriptional regulation of *Nif* expression and posttranslational regulation of dinitrogenase reductase by reversible ADP ribosylation. Zhang et al. (2000) studied the functional characterization and mutagenesis of the glnB, glnA, and *Nif*A genes from the photosynthetic bacterium *R. rubrum*. The reaction is catalyzed by the DRAT-DRAG (dinitrogenase reductase ADP-ribosyltransferase–dinitrogenase reductase-activating glycohydrolase) system. The adenylylation of GS plays no noteworthy function in *Nif* expression or the ADP-ribosylation of dinitrogenase reductase, since a mutant expressing GS-Y398F showed normal nitrogenase action and normal alteration of dinitrogenase reductase in response to NH_4 1 and darkness treatments.

6.3.1.3 Rhodobacter capsulatus

Rhodobacter capsulatus is a free-living anaerobic phototroph. It also regulates *Nif* gene expression through the action of *NifA* protein as described before, but in little bit different manner. In this case, activator known as *NtrC* is required for *NifA* protein expression (Merrick and Edwards, 1995). The *Nif* genes expression in *R. capsulatus* depends on rpoN and *Nif*A regulatory genes, encoding a transcriptional activator and alternative sigma factor of RNA polymerase both are *Nif*-specific, respectively. The expression of *Nif* gene in this bacterium has also been studied by Hubner et al. (1991) by using NtrC-independent repression by high ammonium ion concentrations. The two *Nif*A genes were found to be induced first, followed by *Nif*H and finally by rpoN upon weak, medium, and strong nitrogen starvation, respectively. *Nif*A moderately expressed in the presence of ammonia or oxygen, whereas *Nif*H and *Nif*R4 do not express under these conditions. In *R. capsulatus*, expression of *Nif* genes is compared and discussed with the situation of another free-living diazotroph, that is, *K. pneumoniae* (Hubner et al., 1991). Similarly in the study of Paschen et al. (2001), ammonium was found to be repressed the nitrogen fixation genes expression in *R. capsulatus* at different regulatory levels including an NtrC-independent mechanism controlling *Nif*A activity.

6.3.1.4 Rhizobium spp.

Rhizobium spp. is a genus of Gram-negatives soil bacteria which involved in establishing and maintaining a symbiotic relationship with legume species. By the 1980s, Gary Ruvkun and Sharon R. Long were first time cloned the *Nif* and nod genes in *Rhizobium* for nitrogen fixation and for nodulation (Spaink, 1998). Some researchers found that these genes are also associated with host specificity and are clustered tightly on a transmissible plasmid in *R. leguminosarum viciae* with *Nif* genes *parallel to* situation in *R. leguminosarum trifolii.* For diverse *Rhizobium* species, the rule of genome organization is different. In some cases, the symbiosis genes are found to be clustered, and in some cases, they are dispersed. In few cases, these genes are on plasmids and can spread by conjugation at high frequency, while these genes are spread among many plasmids and chromosomes in others. The "nodulation" genes are activated in bacteria in response to signals produce by plant roots and in return bacteria synthesize signals that induce a nodule meristem formation and thereby enable the entry of bacteria via an infection thread. Plant products like glycoproteins and glycolipids are synthesized de novo within the nodule and infection thread (Merrick and Edwards, 1995). Preliminary stages of nodule formation require expression of precise nodulation (nod) genes by rhizobia as stated above. The gene product of *nodABCFELMN* is involved in the synthesis of a cluster of signal molecules also known as Nod factors which induce nodule morphogenesis. These signals are acylated derivatives of an oligo *N*-acetyl-glucosamine polymer. *Nod* F and *Nod* E determine the type of *N*-acyl group (C18:4 in *R. leguminosarum* biovar *viciae*) on the first glucosamine on the oligomer, whilst NodL carries out an *O*-acetylation of the first glucosamine residue. In parallel, bacteria secrete "*NodO*" Ca_2^+ binding protein that stimulate the infection procedure through its interaction with root cell membranes. Then infection threads raise to the nodule and release the rhizobia in these cells by infecting its inner tissue, where they differentiate morphologically into bacteroids ultimately fix nitrogen from the atmospheric/elemental nitrogen into ammonium, that is, utilized form of plant ($NH_3 + H^+ \rightarrow NH_4^+$), using the enzyme nitrogenase. In return, the plant provides carbohydrates, proteins, oxygen, etc. to the bacteria. Plant protein named as leghemoglobin helps in keeping the low oxygen concentration so as not to reduce nitrogenase activity.

6.3.2 *Vnf* GENE

Vanadium nitrogenase (V-nitrogenases) is also a nitrogen-fixing enzyme that is used as an alternative to molybdenum nitrogenase (Mo-nitrogenases) which is expressed only under molybdenum deficient conditions (Rehder, 2000). It is an important component of the nitrogen cycle, by converting nitrogen gas to ammonia, thereby building otherwise inaccessible N available to plants (Eady, 1989). V-nitrogenases are also participating in reduction of carbon monoxide (CO) to ethane, ethylene and propane. Basically, V-nitrogenase consists of Fe protein and *VnfDGK* act as two component system and encoded vanadium–iron (VFe) protein. Out of these two proteins, Fe_4S_4 cluster-homodimer was found in Fe protein similar to the Fe protein of Mo-nitrogenase. Whereas VFe protein contains $\alpha_2\beta_2$-tetrameric protein but they are different from MoFe protein.

In addition, it consists of δ-subunit encoded by *VnfG* genes along with the α- (encoded by *VnfD)* and β-subunits encoded by *VnfK*.

Vanadium nitrogenases are active under anaerobic conditions and are found in the bacterial genus like *Azotobacter*, also found in other species like *Rhodopseudomonas palustris* and *Anabaena variabilis* (Larimer et al., 2004; Rehder, 2000). Like Mo-nitrogenases, competitive inhibitor of nitrogen fixation is dihydrogen and noncompetitive is carbon monoxide (CO) (Janas and Sobota, 2005). An $\alpha_2\beta_2'Y_2$ subunit structure is found in V-nitrogenases and $\alpha_2\beta_2$ in Mo-nitrogenases, although the structural genes encoding V-nitrogenases demonstrate only 15% conservation with Mo-nitrogenases. Same types of redox center are found in these two nitrogenases complex. At room temperature, efficiency of V-nitrogenase is with respect to nitrogen-fixing than Mo-nitrogenases because it results in side reaction that converts more H^+ to H_2 (Eady, 1989). However, the effect of temperature on V-nitrogenases activity is more than the Mo type which is found to be 10 times more below 5°C (Miller and Eady, 1988; Rehder, 2000).

The reaction mechanism of nitrogen fixation which is catalyzed by V-nitrogenases is summarized as follows:

$$N_2 + 12e^- + 14H^+ + 24MgATP \rightarrow 2NH_4^+ + 3H_2 + 24MgADP + 24HPO_4^{2-}$$

Woodley et al. (1996) recognized the important sequence of *Vnf* genes in *A. vinelandii* by the *Vnf*A transcriptional activator (*Vnf*E-, *Vnf*H-, or *Vnf*D-lacZ) to examine regulation of the vanadium-dependent nitrogenase. They concluded that *Vnf*A (active form) is perhaps interacting dimers, a tetramer, or a higher order oligomer since two regions of dyad symmetry are necessary for its interaction with the DNA.

6.3.3 Anf GENE

The *Anf* genes encode for Fe-alternative nitrogenase protein. It reduces N_2 and C_2H_2 at lower rate but produce more H_2 than Mo-nitrogenase. It also reduces C_2H_2 beyond C_2H_2 to C_2H_6. Thus, the formation of C_2H_6 indicates alternative nitrogenase expression in an organism. Schüddekopf et al. (1993) characterized the presence of *Anf* genes specific for the nitrogenase in *R. capsulatus*. The recognition of several Tn5 insertions mapping outside of *Anf* region A showed that at least 10 genes specific for the alternative nitrogenase are present in *R. capsulatus*.

6.4 COMBINING PHOTOSYNTHESIS WITH NITROGENASE

Since all fixed carbon and nitrogen are basic requirement of various organisms, therefore, it must combine photosynthesis with nitrogenase for an organism, to be self-sufficient. There is problem related to linking photosynthesis with nitrogenase activity, since oxygen is produced by photosynthesis to which nitrogenase is sensitive. Legumes are group of organisms (pea and soybean) that photosynthesize like all plants, but with the help of nitrogen-fixing bacteria also produce fixed nitrogen called rhizobia that forms symbiotic relationship with legumes and are being traditionally used in crop rotation in order to increase the amount of fixed nitrogen in the soil. However, *Rhizobia* form cysts or nodules in the roots of legumes to fix C produced by the plant. Due to leghemoglobin a protein the root nodules exhibit a low oxygen concentration. This protein protects the nitrogenase enzyme by binding with oxygen.

In some cases like cyanobacteria, for example, *Anabaena sperica* and *Nostoc punctiforme* fixed both carbon and nitrogen but in separate compartments. So, each specialized cells called heterocyst carry out either photosynthesis or nitrogen-fixation and then exchange the products. In heterocyst, nitrogenase enzyme must be synthesized, but they must also halt photosynthesis. In order to do this, some group of enzymes must be degraded that play important role to make up photosystem II (part of the photosynthetic machinery that produces oxygen). Another way to protect them from oxygen in heterocyst is to produce additional cell walls, which is still produced by neighboring cells. The leghemoglobin (oxygen binding protein) is also produced by heterocyst. Therefore, to cope and to meet with the increasing energy demands of nitrogen-fixation, glycolysis rate also increased in heterocyst that produces energy from glucose. In *Trichodesmium*, a marine cyanobacteria, photosynthesis and nitrogenase action has also been well studied (Capone et al., 1997). During the daylight hours, this bacterium evolves oxygen and fixes nitrogen simultaneously within the colonies without isolated intracellular structures or specific regions of nitrogen fixation. Due to both properties this bacterium is also involved in the carbon cycle in two ways, they are as follows: (1) it makes direct role to carbon cycling as a primary producer and (2) indirectly by fixing atmospheric nitrogen to encourage growth of surrounding phytoplankton species (Orchard et al., 2009).

6.5 STRUCTURE AND COMPOSITION

The two important component proteins of nitrogenase enzyme complex are dinitrogenase and dinitrogenase reductase. Dinitrogenase reductase acts as electron donor to dinitrogenase. Then, dinitrogenase catalyzes the reduction of important substrate (N_2) like biological and agronomic. The primary sequence similarity among and across kingdoms is at least 70% identical in the case of dinitrogenase reductase in contrast to dinitrogenase subunit proteins sequence which is highly conserved. Therefore, functional nitrogenase can be formed by mixing component proteins from two different organisms, that is, dinitrogenase reductase from one organism and a dinitrogenase from another. If mixed components failed to show nitrogenase activity, it might be due to tight packing of different component proteins rather than inability to interact with each. Further studies regarding complex formation between different components was done. It was found that dinitrogenase reductase undergoes various conformational changes upon binding to dinitrogenase protein, but the structure of dinitrogenase protein remains relatively unchanged (Schindelin et al., 1997). Instead of it, it also undergoes number of posttranslational modifications. The ribosylation process results in binding of ADP-ribose at Fe_4S_4 site of dinitrogenase reductase protein, which prevents the formation of functional nitrogenase.

6.5.1 NITROGENASE COMPLEX: TWO COMPONENT PROTEINS

The structure of nitrogenase complex consists of two components: (1) Fe-protein/dinitrogenase reductase and (2) MoFe-protein/dinitrogenase, each containing FeS cluster(s). During the process of substrate reduction, these proteins are responsible for the electron flow. These are a family of metalloenzymes, and in global nitrogen cycle, it catalyzes a key step: the adenosine triphosphate (ATP)-dependent reduction of dinitrogen (N_2) to ammonia (NH_3). The best characterized member of this enzyme family is molybdenum (Mo)-nitrogenase. Some key concepts are associated with nitrogenase complex given as following:

- The complex consists of two Fe protein dimers and one MoFe protein tetramer; each component harbors metallocluster(s) that within the complex mediate electron flow to the active center.
- The active center of Mo-nitrogenase, designated the FeMoco or the M-cluster, is a [$MoFe_7S_9C$-homocitrate] cluster.
- Where substrate reduction occurs, electrons are transferred from the [Fe_4S_4] cluster in the Fe protein to the P-cluster and then the M-cluster in the MoFe protein.
- Under ambient temperature and pressure nitrogenase catalyzes the ATP-dependent reduction of dinitrogen (N_2) to ammonia (NH_3).
- Nitrogenase has been found to catalyze the reduction of alternative substrates, such as H^+, N_3^-, CN^-, C_2H_2, and CO.

6.5.1.1   Fe PROTEIN

The Fe protein is a dimer of identical subunits having one [Fe_4S_4] cluster and weighs about 60–64 kDa. Fe protein transport electrons from a reducing agent, that is, ferredoxin or flavodoxin to the MoFe protein. This electron transfer requires an effort of chemical energy which comes from the binding and hydrolysis of ATP. Hydrolysis of ATP causes conformational changes within the nitrogenase complex, and for easier electron transfer bringing both proteins that is, Fe and MoFe protein closer together. Fe protein is essential for nitrogenase catalysis and biosynthesis (*Burgess and Lowe, 1996*).

6.5.1.1.1   Fe Protein Cycle

In Fe protein cycle, during the hydrolysis of two MgATP molecules, P cluster acts as bridge and transfer electrons from nitrogenase Fe protein to MoFe protein. The structure of Fe protein is homodimer having (MgATP/MgADP)-binding sites within each subunit and consists of single [4Fe–4S] cluster between two subunits (Fig. 6.1) (Howard and Rees, 1994; Seefeldt and Dean, 1997; Georgiadis et al., 1992).

Still the details regarding the mechanism of Fe protein cycle is not clear (Seefeldt and Dean, 1997). There are so many questions, like, how nucleotide binding sites in the interface between two subunits, that is, Fe protein and MoFe protein interact with each other. Before its attachment to MoFe protein complex, the Fe protein shows undetectable rate of MgATP hydrolysis. Binding of Fe protein to the MoFe protein accompanied by the hydrolysis of MgATP. The MgATP binding sites located >10 Å away from the MoFe protein docking interface within the Fe protein (Georgiadis et al., 1992); as a result, it causes protein conformational changes within the Fe protein that activate MgATP hydrolysis (Howard and Rees, 1994; Seefeldt and Dean, 1997). There is a stretch of amino acid within the Fe protein which is specific for docking interface and MgATP-binding sites and lead to the movement of catalytic residues as well as nucleotide hydrolysis. But details mechanism is still not clear. Other reviews provide a detailed historical perspective on the role of the Fe protein in nitrogenase catalysis (Burgess and Lowe, 1996; Seefeldt and Dean, 1997; Lowe et al., 1995; Rees and Howard, 2000).

6.5.1.2 *MoFe PROTEIN*

MoFe protein is a heterotetramer having two α-subunits and two β-subunits about 240–250 kDa weigh. This protein also contains two iron–sulfur clusters P-clusters (located at the interface between the α and β subunits) and within the α subunits two FeMo cofactors (Fig. 6.3).

- The core (Fe_8S_7) of the P-cluster takes the form of two $[Fe_4S_3]$ cubes linked by a central sulfur atom. Each P-cluster is covalently linked to the MoFe protein by six cysteine residues.
- Each FeMo cofactor (Fe_7MoS_9C) consists of two nonidentical clusters: $[Fe_4S_3]$ and $[MoFe_3S_3]$, which are linked by three sulfide ions. Each FeMo cofactor is covalently linked to the α subunit of the protein by one cysteine residue and one histidine residue.

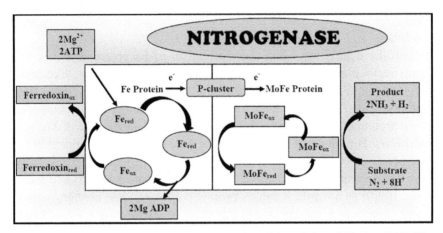

FIGURE 6.1 Nitrogenase complex. (*Source:* Adapted from Taiz and Ziegler, 1998). The role of Fe protein in nitrogenase mechanism can also described in the form of three-state cycle named as Fe protein cycle as given in Fig. 6.2. In this cycle, the reduced form of Fe protein having [4Fe–4S] cluster (1+oxidation state) bound to two MgATP molecules which then associate with the MoFe protein in this state (Hageman and Burris, 1978). As a result, two MgATP molecules hydrolyzed to two MgADP molecules and the electron is transferred from the Fe protein [4Fe–4S] cluster toward the other protein (MoFe). The Fe protein [4Fe–4S] cluster oxidized to Fe protein ([4Fe–4S]$^{2+}$) having two bound MgADP molecules, which dissociates from the MoFe protein. This is the over-all rate-limiting step for this reaction (Thorneley and Lowe, 1985). Then, the reaction occur in reverse direction, MgADP molecules are replaced by MgATP and the [4Fe–4S]$^{2+}$ cluster is reduced to the 1$^+$ oxidation state results in regeneration of Fe protein (Mortenson, 1964).

6.5.1.2.1 FeMo Cofactor

The MoFe protein is an$\alpha 2\beta 2$-heterotetramer, consists of P cluster, that is, [8Fe–7S] and FeMo cofactor, that is, [7Fe–Mo–9Shomocitrate-X]. One of each type of metal cluster is contained in an$\alpha\beta$-unit, and thus, each MoFe protein consists of two catalytic units (Fig. 6.1) (Kim and Rees, 1992). There is a substrate binding site (known as *R*-homocitrate) at FeMo cofactor which is connected through its 2-hydroxy and 2-carboxyl groups to the Mo atom (Hoover et al., 1987) (Fig. 6.4). There is X atom which is connected at one corner to the cofactor A [4Fe–3S] subcluster to [3Fe–Mo–3S] subcluster and also to three bridging inorganic sulfides. The presence of the X atom was detected through high-resolution (1.16 Å) structure analysis of the MoFe protein (Einsle et al., 2002) and electron density for X indicated that it is most consistent with N, C, or O elements

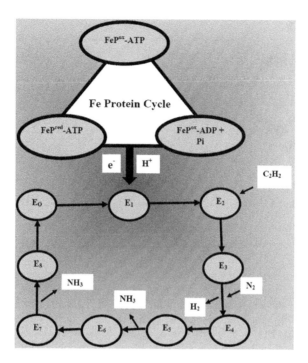

FIGURE 6.2 **(See color insert.)** Fe and MoFe protein cycle (three state cycle Fe protein at top) and (an eight state cycle MoFe protein) at bottom. Fe protein (abbreviated FeP) can exist in the 1^+ reduced state (Red) or the 2^+ oxidized state (Ox) and two MgATP molecules bound as ATP or two MgADP with two Pi as ADP+Pi. When FeP gets associated to MoFe protein, then the exchange of an electron occurs. In the MoFe protein cycle, the MoFe protein is successively reduced by one electron, with reduced states represented by En, where n is the total number of electrons donated by the Fe protein. Acetylene (C_2H_2) binds to E_2 and N_2 binds to E_3 and E_4. The N_2 binding results in displacement of H_2 and two ammonia molecules being liberated from later E states. (*Source*: Adapted from Seefeldt et al. 2009)

but not yet identified (Lee et al., 2003). The FeMo cofactor is connected by α-275Cys and α-442His to MoFe protein at one end to Fe atom and to the Mo atom at the other end, respectively. A MoFe protein that citrate substitutes for homocitrate results in structure alteration, thereby exhibits altered catalytic properties (Imperial et al., 1989; Madden et al., 1990). Similarly, various mutant strains that are deficient in biosynthesis of FeMo cofactor are inactive but on addition of FeMo cofactor activates it again (Christiansen et al. 1998; Pollock et al., 1995; Christie et al. 1996; Lee et al., 1997).

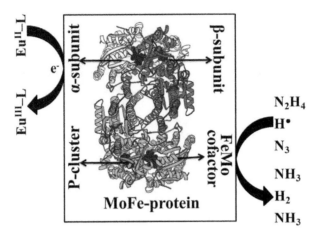

FIGURE 6.3   Phenomena of nitrogenase catalyzed substrates reduction in which electrons are delivered by Fe protein (nucleotide-dependent) to MoFe protein where FeMo cofactor is the active site. (*Source*: Adapted from Rees et al., 2005; Danyal et al. 2015; *Note*: for more details on the phenomena illustrated by figure, reader should refer to the paper of Danyal et al., 2015)

6.5.1.3   P CLUSTER

The MoFe proteins contain one P cluster in each $\alpha\beta$-unit which is basically [8Fe–7S] clusters and transfer electron between Fe protein and FeMo cofactor (Fig. 6.5). The P cluster consists of common bridging μ6-sulfide at one corner to link two [4Fe–4S] subclusters (S1 in Fig. 6.5). Along with typical cysteinate-S ligands, the [Fe–S] cluster contains serinate-O (β-188Ser) and amide-N (α-88Cys) ligands. P clusters results in the displacement of both the serinate-O and amide-N ligands from PN on rearrangement from resting state (dithionite reduced, PN) to oxidation state (Fig. 6.5) (Kim and Rees, 1992). Since the mechanism of nitrogenase is not fully understood with respect to structural changes, induction of different oxidation states of the P cluster occurs after addition of electron transfer agents to the MoFe protein. The Mossbauer spectroscopy analysis showed that in the P^N state, all of the Fe atoms are in the ferrous oxidation state (eq 6.2).

$$P^N \rightleftarrows P^{1+} \rightleftarrows P^{2+} \rightleftarrows P^{3+} \ldots II \qquad (6.2)$$

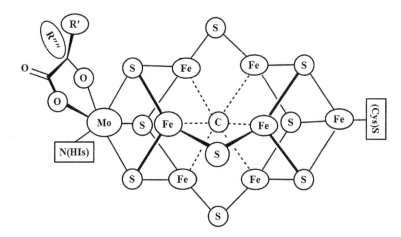

FIGURE 6.4   Structure of the iron–molybdenum (FeMo) cofactor of nitrogenase.

During catalysis, PN get oxidized by three electrons to P^{1+}, P^{2+}/POX, and P^{3+}oxidation states (eq 6.2). The $P^{3+/2+}$ redox couple does not exist in vivo which is also supported by the fact that this couple is irreversible. During catalysis, in order to explain electron transfer through P cluster to FeMo cofactor, a model was suggested to explain $P^{1+/N}$ and $P^{2+/1+}$ couples that involve P cluster electron transfer (Lindahl et al., 1988; Yoo et al., 2000).

6.5.2 MoFe PROTEIN CYCLE

In the MoFe protein cycle, on reduction of N_2, the FeMo accepts eight electrons as given in eq 6.1 which can be represented by a nomenclature of MoFe protein having different catalytic states (Fig. 6.2) (Thorneley and Lowe, 1985). The n is the number of electrons added to the MoFe protein (E) (Fig. 6.2) having different states denoted by E_0 to E_8 before it returns to the ground state (E_0).

The number of studies was done regarding the mechanism of nitrogenase, and it was found that on binding of N_2 three or four electrons get accumulate within the MoFe protein (Thorneley and Lowe, 1985), which results in released of H_2 (Liang and Burris, 1988). Further its reduction by nitrogenase required to attempts to accumulate an En state for study. But under N_2 absence conditions, protein would only access low En states. The reduction of alkyne acetylene to ethylene is the most commonly used

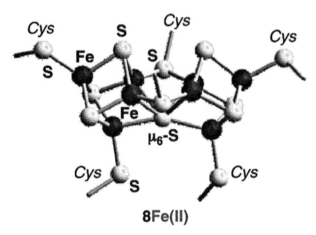

8Fe(II)

FIGURE 6.5   Structure of P cluster. P^N and P^{OX}-reduced and oxidized form of P cluster [8Fe–7S], α-88^{Cys} and β-188^{Ser}—MoFe protein amino acid ligands, S1-central S atom, Fe atoms 5 and 6. The PBD files used were 3MIN for the PN state and 2MIN for the POX state. (*Source* Adapted from Ohki et al., 2009 and Seefeldt et al., 2009).

method to study nitrogenase activity is (Burgess and Lowe, 1996), since N_2 binds to the higher states (E_3, E_4) and acetylene binds to a less-reduced E state (E_2). Therefore, it may be concluded that acetylene appears to be a noncompetitive inhibitor of N_2 reduction, even both binds to the same site on the FeMo cofactor when both acetylene and N_2 are present (Dos Santos et al., 2005).

Reduction of dinitrogen (N_2) to two ammonia molecules (NH_3) is catalyzed by nitrogen-fixing bacteria which have major role in fixation of atmospheric nitrogen in the biogeochemical nitrogen cycle. The most widely studied nitrogenase is the Mo-dependent enzyme. The N_2 reduction by this enzyme involves the transitory interaction of two component proteins, that is, Fe protein and the MoFe protein, and minimally requires sixteen MgATP, eight protons, and eight electrons. How these proteins and small molecules together effect the reduction of N_2 to ammonia is reviewed recently. The roles of the Fe protein and MgATP hydrolysis, information on the roles of the two metal clusters enclosed in the MoFe protein in catalysis, insights gained from recent achievement in trapping substrates and inhibitors at the active site metal cluster FeMo-cofactor, and at last, considerations of the mechanism of N_2 reduction catalyzed by nitrogenase (Fig. 6.6) (Seefeldt et al., 2009).

6.6   MICROORGANISMS ASSOCIATED WITH N₂-FIXING

Nearly 80% of N_2 gas constitutes the atmosphere, but this form N is not used or viable to the plants. Several groups of soil bacteria/microorganisms living in the soil converts nitrogen gas into ammonium (NH_4^+) which plants easily used. Nitrogenase enzyme catalyzed this transformation. There are few microbes/ free living bacteria in soil which activate the nitrogenase complex through their slack associations with root surfaces or by symbiotic associations with plants. The various examples of N fixing microbes are as given in Table 6.2. These nitrogen-fixing organisms are prokaryotes (bacteria) and live independently of other organisms so-called free-living nitrogen-fixing bacteria. Others live symbiotically with plants or with other organisms (e.g., protozoa). Examples are shown in Table 6.3.

6.7   MECHANISM OF MICROBIAL N₂ FIXATION BY NITROGENASE ACTION

BNF is defined as the conversion of atmospheric nitrogen into ammonia by the action of nitrogenase enzyme.

$$N_2 + 6\,H^+ + 6\,e^- \rightarrow 2\,NH_3$$

In this procedure, 16 molecules of ATP get hydrolyzed which is accompanied by the conformation of one molecule of H_2 as given in Figure 6.7. The nitrogenase generated ammonium which is assimilated into glutamate

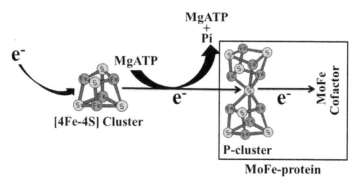

FIGURE 6.6   Roles of Fe and Mg metals in MoFe protein.

through the glutamine synthetase/glutamate synthase cycle in free-living diazotrophs, and results in synthesis of other amino acids/proteins. But this enzyme is very vulnerable to breakage by oxygen. Only in anaerobic situation many nitrogen-fixing organisms exist, respiring to draw down oxygen levels, or binding the oxygen with a protein such as leghemoglobin.

In N_2 fixation, the conversion of dinitrogen gas (N_2) to ammonia (NH_3) occurs. We know that ammonia is essential for the creation of biologically important, nitrogen-containing compounds; a fixed nitrogen source is needed to sustain life on earth. Furthermore, the ammonia is essential to support biosynthetic reactions is continually sequestered into sediments or reconverted to N_2 through the biological processes of nitrification and denitrification (Fig. 6.8). For the maintenance of the diversity of life on earth, nitrogen fixation is also needed. There are three ways through which nitrogen fixation occurs in the biosphere are lightning and other natural combustion processes, the industrial Haber–Bosch process, and BNF. Among these, the most significant contributor is BNF, accounts for about 65% of the total. Nitrogen fixation has also agronomic significance because the availability of fixed nitrogen—commonly referred to as fertilizer nitrogen—usually limits to crop production.

6.7.1   NITRIFICATION

The conversion of ammonium to nitrate is called nitrification, carried out by the nitrifying bacteria. These bacteria are specialized to gain their energy by oxidizing ammonium, while using CO_2 as their source of carbon to synthesize organic compounds. These bacteria are found in soils and waters of moderate pH. Some of them include *Nitrosomonas* species which are specialized to convert ammonium to nitrite (NO_2^-) and others are *Nitrobacter* species which convert nitrite to nitrate (NO_3). In fact, nitrite accumulation inhibits *Nitrosomonas*, so it depends on *Nitrobacter* to convert this to nitrate, whereas *Nitrobacter* depends on *Nitrosomonas* to generate nitrite.

6.7.2   DENITRIFICATION

In this process, nitrate is converted to nitric oxide, nitrous oxide and N_2 by microorganisms. The intermediate step is the production of nitrite

TABLE 6.2   N$_2$-Fixing Microbes.

S. No.	Microorganisms associated with N$_2$-fixing
1.	*Actinomyces* spp.
2.	*Anabaena* spp.
3.	*Anabaenopsis* spp., e.g., *A. circularis*
4.	*Azomonas* spp., e.g., *A. macrocytogenes*
5.	*Azospirillum* spp.
6.	*Azotobacter* spp., e.g., *A. beijerinckii, A. chroococcum, A. paspali, A. vinelandii*
7.	*Bacillus* spp., e.g., *B. macerans, B. polymyxa*
8.	*Beijerinckia* spp., e.g., *B. fluminensis, B. indica*
9.	*Calothrix* spp.
10.	*Calothrix* spp., e.g., *C. brevissima*
11.	*Chlorobium* spp., e.g., *C. thiosulfatophilu*
12.	*Chlorogloea* spp., e.g., *C. fritschii*
13.	*Chromatium* spp., e.g., *C. vinosum*
14.	*Clostridium* spp., e.g., *C. butylicum, C. butyricum, C. pasteurianum*
15.	*Corynebacterium* spp., e.g., *C. autotrophicum*
16.	*Cylindrospermum* spp.
17.	*Ectothiorhodospira* spp., e.g., *E. shaposnikovii*
18.	*Enterobacter* spp., e.g., *E. aerogenes, E. cloacae*
19.	*Escherichia* spp., e.g., *E. intermedia*
20.	*Klebsiella* spp., e.g., *K. aerogenes, K. pneumoniae*
21.	*Methylobacter* spp.
22.	*Methylococcus* spp.
23.	*Methylocystis* spp.
24.	*Methylomonas* spp.
25.	*Methylosinus* spp.
26.	*Mycobacterium* spp., e.g., *M. flavum*
27.	*Nostoc* spp., e.g., *N. muscorum*
28.	*Oscillatoria* spp., e.g., *O. limnetica*
29.	*Phormidium* spp.
30.	*Rhizobium* spp.
31.	*Rhodopseudomonas* spp., e.g., *R. capsulatus, R. gelatinosa, R. palustris, R. spheroids, R. rubrum*
32.	*Scytonema* spp.
34.	*Thiobacillus* spp., e.g., *T. ferrooxidans*
35.	*Thiocapsa* spp.
36.	*Thiopedia* spp.
37.	*Tolypothrix* spp., e.g., *T. tenuis*

TABLE 6.3   The Nitrogen-fixing Organisms.

Symbiotic with plants		Free living	
Legumes	**Other plants**	**Anaerobic**	**Aerobic**
Rhizobium	*Frankia, Azospirillum*	*Clostridium* (some), *Desulfovibrio*, purple sulfur bacteria[a], purple nonsulfur bacteria[a], green sulfur bacteria[a]	*Azotobacter, Beijerinckia, Klebsiella* (some), Cyanobacteria (some)[a]

[a]Nitrogen-fixing bacteria that is denoted as photosynthetic bacterium.

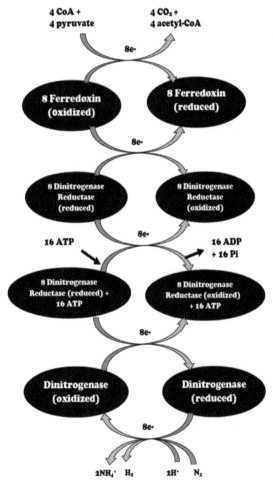

FIGURE 6.7   Mechanism of nitrogenase. (*Source* Adapted from Nelson and Cox, 2005).

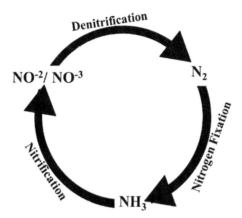

FIGURE 6.8 Process of nitrogen fixation.

(NO_2^-). In anaerobic conditions, several bacteria carry out this change when growing on organic matter. Due to scarcity of oxygen for normal aerobic respiration, they utilize nitrate in place of oxygen as the terminal electron acceptor, known as anaerobic respiration and can be defined by the following equation:

$$C_6H_{12}O_6 + 6O_2 = 6CO_2 + 6H_2O + energy$$

When oxygen is absence, nitrate (NO_3^-) a reducible substance could serve the same role and be reduced to nitrite, nitric oxide, nitrous oxide or N_2. Some species of *Alkaligenes*, *Bacillus*, and *Pseudomonas* are the common denitrifying bacteria. Their actions result in considerable losses of nitrogen into the atmosphere, roughly balancing the amount of nitrogen fixation that occurs each year.

6.8 ENERGETICS OF MICROBIAL N_2 FIXATION

BNF is displayed by the equation given below, in which 2 moles of ammonia are formed from one mole of nitrogen gas, at the expense of 16 moles of ATP and a supply of electrons and protons (hydrogen ions):

$$N_2 + 8H^+ + 8e^- + 16ATP = 2NH_3 + H_2 + 16ADP + 16Pi$$

The reaction is carried out by prokaryotic microorganisms, using nitrogenase enzyme complex. The reactions take place while N_2 is bound to the nitrogenase enzyme complex (Fig. 6.9). The Fe protein is first reduced by electrons donated by ferredoxin. Then ATP bind to the reduced Fe protein and reduces the MoFe protein, which donates electrons to N_2, producing HN=NH. In two further cycles of this process (each requiring electrons donated by ferredoxin), HN=NH is reduced to $H_2N–NH_2$, and this in turn is reduced to $2NH_3$. The reduced ferredoxin which supplies electrons for this process depending on the type of microorganisms it is generated by photosynthesis, respiration or fermentation. Table 6.4 summarizes some estimates of the amount of nitrogen fixed on a global scale.

6.9   FACTORS AFFECTING NITROGENASE ACTIVITY IN N_2-FIXATION

Several studies have addressed the effect of different environmental factors on nitrogenase activity in nitrogen fixation. Because the past conditions influence current fixation activity through the amount of ATP, reductants, nitrogenase enzymes, and N compounds present in the most of the cyanobacteria.

6.9.1   OXYGEN SENSITIVITY

The nitrogenase activity is quite sensitive to oxygen in which, out of two proteins, Fe protein is irreversibly damaged by it. In order to defend enzyme activity, two defensive mechanisms named as conformational defense and respiratory defense. In the conformational defense mechanism, as name suggests, conformational change occur in enzyme which results in its insensitivity with respect to oxygen and not capable to catalyze dinitrogen fixation. Whereas in the mechanism of respiratory defense, there is an oxygen scavenging process, which results to high respiratory activity and maintains enzyme in catalytically active form. *Azospirillum* species are aerobic bacteria which require low limited pressure of oxygen (pO_2) for expression of hydrogenase activity. The fixation of nitrogen is not possible unless nitrogenase is protected under fully aerobic conditions. For nitrogen fixation studies, *Azospirillum chroococcum* and *A. vinelandii* have been used under conditions of high pO_2. *Azotobacter* can carry out

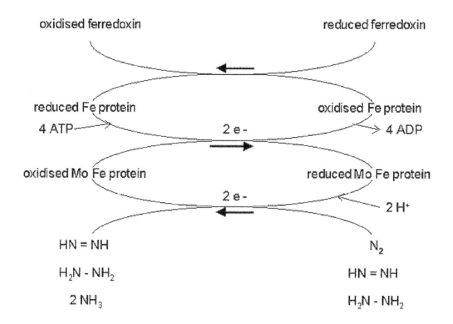

FIGURE 6.9 Energetics of microbial N_2 fixation. (*Source:* Reprinted from Deacon, 2016)

TABLE 6.4 Amount of Nitrogen Fixed (1012 g per Year, or 106 Metric Tons per Year) on a Global Scale.

Type of fixation	N_2 fixed
Biological	
Agricultural land	About 90
Forest and nonagricultural land	About 50
Sea	About 35
Total	**About 175**
Nonbiological	
Industrial	About 50
Combustion	About 20
Lightning	About 10
Total	**About 80**

Source: Bezdicek and Kennedy (1998).

nitrogen fixation in air because both protection mechanisms operate in nitrogenase and also enzyme locates within the cell. Nitrogen fixation takes place in *Azospirillum*, at optimal rate between 0.005 and 0.007 atmospheres (0.507 and 0.709 kPa) (Seefeldt et al., 2004).

6.9.2   TEMPERATURE AND pH

In *Azospirillum*, nitrogenase activity is sensitive to temperatures below 15°C, because optimum temperature is 32–40°C for H_2-depentent growth, whereas for *Azospirillum*, the optimal pH is 6.8–7.8 for N_2 dependent growth. Therefore, these requirements are met within the cells of roots or at the surface. When soil pH is low, roots give optimal pH conditions to bacteria (Schrauzer, 2003). Prévost et al. (1987) studied the effect of low temperatures by arctic rhizobia on nitrogenase activity in sainfoin (*Onobrychis viciifolia*) nodulated. They found that nitrogenase activity measured at aforesaid temperature with detached nodules/with whole plants is higher than that of temperate rhizobia. At 5°C and 10°C, nitrogenase activity of arctic rhizobia showed 12% and 33% of those measured at 20°C, while lower values of 3.7% and 22.4% were observed with temperate rhizobia. Drouin et al. (2000) investigated that in strain of *R. leguminosarum* bv. *viciae,* the physiological adaptation to low temperatures associated with *Lathyrus* spp., namely, *Lathyrus japonicas* and *Lathyrus pratensis*. In symbiosis cold adaptation trait was slightly reflected with the agronomic legume, *Lathyrus sativus*, with which slightly higher nitrogenase activity was observed in one cold-adapted strain and shoot dry matter yield than a commercial strain under a sub-optimal temperature regime. In soybean, also the relation between low root temperature and nitrogenase activity has been investigated (Layzell et al., 1984), and it is observed increase in relative efficiency with lower temperatures showed that, during the study of symbioses, evolution of H_2 may have provided a buffer which permitted the upholding of high levels of N_2 fixation during short-term or diurnal fluctuations in soil temperature.

6.9.3   SOIL MOISTURE

A correlation between nitrogensase activity and soil moisture has been reported in recent years, as high nitrogenase activity was found at wet areas in the soil. The acetylene rate reduction is also found to be increased as soil moisture content increased. This concluded that change in pO_2 also affects nitrogenase activity. These observations give support to the fact that wet soil increased nitrogen fixation in nonnodular plants. Similarly, in wetland grasses, nitrogenase activity has been recorded to be high than mesic or dry soils growing plants. For nitrogen fixation, high levels of combined nitrogen are as denitrification and leaching of nitrates, the available combined nitrogen content of wet soils is low (Postgate, 1982, 1987, 1998). In the study of Roper (1983), nitrogenase activity improved with mean daily soil temperature (up to at least 30°C) and reduced as the soil dried from field capacity. He concluded that in soils fixation of nitrogen by free-living microorganisms amended with straw may supply to the nitrogen status of the soil and thus, reduce the need for nitrogen fertilizers. Later, Roper (1985) on nitrogenase activity also investigated effect of soil moisture (C_2H_2 reduction). He reported that with added glucose nitrogenase activity occurred at much lower soil moisture contents in the Gunnedah soil (0.5–1.75 times—10 kPa water content) than in the Cowra soil (1.0–2.5 times—10 kPa water content). Albrecht et al. (1984) in soybeans grown in the field studied the association of nitrogenase activity to plant water stress. They observed that leaf conductance to water vapor, leaf water potential, apparent carbon exchange, nodule moisture content, and nitrogenase activity declined as the duration of the stress period increased.

6.10 CONCLUSION

Nitrogen fixation is an old topic but its study at gene level revealing detailed mechanism is still unknown. It is a phenomenon happening in all known ecosystems. The described chapter revealed that nitrogenase is the key component play very significant function in the nitrogen fixation. This fixation is also achieved by role of various microorganisms as described understory. The number of experiments proved that unique species of plants bears nitrogenase that performs nitrogen fixation. Instead of this, the use of biofertilizers is increasing day by day which has number of environmental as well as health hazards. Therefore, at biotechnology/ plant breeding level, crops are being engineered by introducing particular genes which show association/symbiosis with microbes or nitrogenase is directly introduces into the plant, so as to reduce the application of fertilizers. But only partial experimental approaches were carried out due to the complexity of the nature of the BNF process.

6.11 SUMMARY

Nitrogenase is the known family of enzymes which carry out the reduction of atmospheric nitrogen into ammonia. The nitrogen molecule consists of strong N≡N bond, and therefore, it requires high energy to break. So using nitrogenase nitrogen fixation requires rather large inputs of energy to drive the process. The nitrogenase complex is also known as two protein component systems and consists of dinitrogenase (MoFe–protein) and dinitrogenase reductase (Fe protein) components. The MoFe cofactor consists of iron (Fe) and molybdenum (Mo) in dinitrogenase (MoFe–protein). The nitrogenase enzyme is active under anaerobic environment since it denatured by exposure to oxygen.

KEYWORDS

- **biological nitrogen fixation**
- **nitrogenase**
- *Klebsiella pneumoniae*
- *Rhodospirillum rubrum*
- *Rhodobacter capsulatus*

REFERENCE

Albrecht, S. L.; Bennett, J. M.; Boote, K. J. Relationship of Nitrogenase Activity to Plant Water Stress in Field-grown Soybeans. *Field Crops Res.* **1984**, *8*, 61–71.

Anbar, A. D. Elements and Evolution. *Science* **2008**, *322*, 1481–1483.

Anbar, A. D.; Knoll, A. H. Proterozoic Ocean Chemistry and Evolution: A Bioinorganic Bridge. *Science* **2002**, *297*, 1137–1142.

Anbar, A. D.; Duan, Y.; Lyons, T. W.; Arnold, G. L.; Kendall, B.; Creaser, R. A.; et al. A Whiff of Oxygen Before the Great Oxidation Event. *Science* **2007**, *317*, 1903–1906.

Arnold, W.; Rump, A.; Klipp, W.; Priefer, U. B.; Puhler, A. Nucleotide Sequence of a 24,206-Base-Pair DNA Fragment Carrying the Entire Nitrogen Fixation Gene Cluster of *Klebsiella pneumonia*. *J. Mol. Biol.* **1988**, *203*, 715–738.

Bezdicek, D. F.; Kennedy, D. F. *Microorganisms in Action;* Lynch, J. M., Hobbie, J.E., Eds.; Blackwell Scientific Publications: Oxford, United Kingdom, 1998.

Burgess, B. K.; Lowe, D. J. Mechanism of Molybdenum Nitrogenase. Chem. Rev. **1996,** *96, 2983–3011.*

Capone, D. G.; Zehr, J. P.; Paerl, H. W.; Bergman, B.; Carpenter, E. J. *Trichodesmium*, a Globally Significant Marine Cyanobacterium. *Science* **1997**, *276*, 1221–1229.

Christiansen, J.; Goodwin, P. J.; Lanzilotta, W. N.; Seefeldt, L. C.; Dean D. R. Catalytic and Biophysical Properties of a Nitrogenase apo-MoFe Protein Produced by a *nifB*-Deletion Mutant of *Azotobacter vinelandii*. *Biochemistry* **1998**, *37*, 12611–12623.

Christie, P. D.; Lee, H. I.; Cameron, L. M.; Hales, B. J.; Orme-Johnson, W. H.; Hoffman, B. M. Identification of the CO-Binding Cluster in Nitrogenase MoFe Protein by ENDOR of 57Fe Isotopomers. *J. Am. Chem. Soc.* **1996**, *118*, 8707–8709.

Danyal, K.; Rasmussen, A. J.; Keable, S. M.; Inglet, B. S.; Shaw, S.; Zadvornyy, O. A.; Duval, S.; Dean, D. R.; Raugei, S.; Peters, J. W.; Seefeldt, L. C. Fe Protein-Independent Substrate Reduction by Nitrogenase MoFe Protein Variants. *Biochemistry* **2015**, *54*, 2456–2462.

Deacon, J. *The Microbial World: The Nitrogen cycle and Nitrogen fixation Produced*; Institute of Cell and Molecular Biology, The University of Edinburgh, Scotland, 2016. http://archive.bio.ed.ac.uk/jdeacon/microbes/nitrogen.htm (accessed Oct 8, 2016).

Dos Santos, P. C.; Igarashi, R. Y.; Lee, H. I.; Hoffman, B. M.; Seefeldt, L. C.; Dean, D. R. Substrate Interactions with the Nitrogenase Active Site. *Acc. Chem. Res.* **2005**, *38*, 208–214.

Drouin, P.; Prévost, D.; Antoun, H. Physiological Adaptation to Low Temperatures of Strains of *Rhizobium leguminosarum* bv. Viciae Associated with *Lathyrus* spp. *FEMS Microbiol Ecol.* **2000**. DOI: http://dx.doi.org/10.1111/j.1574-6941.2000.tb00705.x.111-120.

Eady, R. R. The Vanadium Nitrogenase of *Azotobacter*. *Polyhedron* **1989**, *8*, 1695–1700.

Einsle, O.; Tezcan, F. A.; Andrade, S. L. A.; Schmid, B.; Yoshida, M.; Howard, J. B.; Rees, D. C. Nitrogenase MoFe–Protein at 1.16 Å Resolution: A Central Ligand in the FeMo-Cofactor. *Science* **2002**, *297*, 1696–1700.

Fani, R.; Gallo, R.; Liò, P. Molecular Evolution of Nitrogen Fixation: The Evolutionary History of the *nifD, nifK, nifE*, and *nifN* Genes. *J. Mol. Evol.* **2000**, *51*, 1–11.

Georgiadis, M. M.; Komiya, H.; Chakrabarti, P.; Woo, D.; Kornuc, J. J.; Rees, D. C. Crystallographic Structure of the Nitrogenase Iron Protein from *Azotobacter vinelandii*. *Science* **1992**, *257*, 1653–1659.

Hageman, R. V.; Burris, R. H. Nitrogenase and Nitrogenase Reductase Associate and Dissociate With Each Catalytic Cycle. *Proc. Natl. Acad. Sci. U.S.A.* **1978**, *75*, 2699–2702.

Hartmann, L. S.; Barnum S. R. Inferring the Evolutionary History of Mo-Dependent Nitrogen Fixation from Phylogenetic Studies of nifK and nifDK. *J. Mol. Evol.* **2010**, *71*, 70–85.

Hellriegel, H.; Wilfarth, H. Studies on the Nitrogen Nutrition of Gramineae and Legumes. *Supplement to the Journal of the Association for Rubenzucker Industry German Reichs;* Post: Kayssler & Co.: Berlin, Germany, 1888; p 234.

Henson, B. J.; Watson, L. E.; Barnum S. R. The Evolutionary History of Nitrogen Fixation, as Assessed by nifD. *J. Mol. Evol.* **2004**, *58*, 390–399.

Hoover, T. R.; Robertson, A. D.; Cerny, R. L.; Hayes, R. N.; Imperial, J.; Shah, V. K.; Ludden, P. W. Identification of the V Factor Needed for the Synthesis of the Iron Molybdenum Cofactor of Nitrogenase as Homocitrate. *Nature* **1987**, *329*, 855–857.

Howard, J. B.; Rees, D. C. Nitrogenase: A Nucleotide-dependent Molecular Switch. Annu. *Rev. Biochem.* **1994**, *63*, 235–264.

Howard, K. S.; McLean, P. A.; Hansen, F. B.; Lemley, P. V.; Koblan, K. S.; Orme-Johnson, W. H. *Klebsiella pneumoniae* nifM Gene Product Is Required for Stabilization and Activation of Nitrogenase Iron Protein in *Escherichia coli. J. Biol. Chem.* **1986**, *261*, 772–778.

Hubner, P.; Willison, J. C.; Vignais, P. M.; Bickle, T. A. Expression of Regulatory nif Genes in *Rhodobacter capsulatus. J. Bacteriol.* **1991**, *173*, 2993–2999.

Imperial, J.; Hoover, T. R.; Madden, M. S.; Ludden, P. W.; Shah, V. K. Substrate Reduction Properties of Dinitrogenase Activated In Vitro Are Dependent Upon the Presence of Homocitrate or Its Analogue During Iron–Molybdenum Cofactor Synthesis. *Biochemistry* **1989**, *28*, 7796–7799.

Imperial, J.; Ugalde, R. A.; Shah, V. K.; Brill, W. J. Role of the nifQ Gene Product in the Incorporation of Molybdenum into Nitrogenase in *Klebsiella pneumoniae. J. Bacteriol.* **1984**, *158*, 187–194.

Janas, Z.; Sobota P. Aryloxo and Thiolato Vanadium Next Term Complexes as Chemical Models of the Active Site of Nitrogenase. *Coord. Chem. Rev.* **2005**, *249*(21–22), 2144–2155.

Kasting, J. F.; Walker, J. C. G. Limits on Oxygen Concentrations in the Prebiological Atmosphere and the Rate of Abiotic Nitrogen Fixation. *J. Geophys. Res.* **1981**, *86*, 1147–1158.

Kim, J.; Rees, D. C. Structural Models for the Metal Centers in the Nitrogenase Molybdenum-Iron Protein. *Science* **1992**, *257*, 1677–1682.

Larimer, F. W.; Chain, P.; Hauser, L.; Lamerdin, J.; Malfatti, S.; Do, L.; Land, M. L.; Pelletier, D. A.; Beatty, J. T.; Lang, A. S.; Tabita, F. R.; Gibson, J. L.; Hanson, T. E.; Bobst, C.; Torres, J. L. T. Y.; Peres, C.; Harrison, F. H.; Jane Gibson, J.; Harwood, C. S. Complete Genome Sequence of the Metabolically Versatile Photosynthetic Bacterium *Rhodopseudomonas palustris. Nat. Biotechnol.* **2004**, *22*, 55–61.

Layzell, D. B.; Rochman, P.; Canvin, D. T. Low Root Temperatures and Nitrogenase Activity in Soybean. *Canadian J. Bot.* **1984**, *62*, 965–971.

Lee, H. I.; Cameron, L. M.; Hales, B. J.; Hoffman, B. M. CO Binding to the FeMo Cofactor of CO-Inhibited Nitrogenase: 13CO and 1H Q-Band ENDOR Investigation. *J. Am. Chem. Soc.* **1997**, *119*, 10121–10126.

Lee, H. I.; Benton, P. M.; Laryukhin, M.; Igarashi, R. Y.; Dean, D. R.; Seefeldt, L. C.; Hoffman, B. M. The Interstitial Atom of the Nitrogenase FeMo-Cofactor: ENDOR and ESEEM Show It Is Not an Exchangeable Nitrogen. *J. Am. Oil Chem. Soc.* **2003**, *125*(19), 5604–5605.

Liang, J.; Burris, R. H. Hydrogen Burst Associated with Nitrogenase-catalyzed Reactions. *Proc. Natl. Acad. Sci. U.S.A.*, **1988**, *85*, 9446–9450.

Lindahl, P. A.; Papaefthymiou, V.; Orme-Johnson, W. H.; Münck, E. Mössbauer Studies of Solid Thionin Oxidized MoFe Protein of Nitrogenase. *J. Biol. Chem.* **1988,** *263,* 19412–19418.

Lindström, K.; Martínez-Romero, E. International Committee on Systematics of Prokaryotes Subcommittee on the Taxonomy of Agrobacterium and Rhizobium: Minutes of the Meeting, 23–24 July 2006, Århus, Denmark. *Int. J. Syst. Evol. Microbiol.* **2007,** *57,* 1365–1366.

Lowe, D. J.; Ashby, G. A.; Brune, M.; Knights, H.; Webb, M. R.; Thorneley R. N. F. ATP Hydrolysis and Energy Transduction by Nitrogenase. In *Nitrogen Fixation: Fundamentals and Applications*; Tikhonovich, I. A., Provorov, N. A., Romanov, V. I., Newton, W.E., ed.; Boston: Kluwer, 1995, pp 103–108.

Madden, M. S.; Kindon, N. D.; Ludden, P. W.; Shah, V. K. Diastereomer-Dependent Substrate Reduction Properties of A Dinitrogenase Containing 1-Fluorohomocitrate in the Iron–Molybdenum Cofactor. *Proc. Natl. Acad. Sci. U.S.A.* **1990,** *87,* 6517–6521.

Merrick, M. J.; Edwards R. A. Nitrogen Control in Bacteria. *Microbiol. Rev.* **1995,** *59*(4), 604–622.

Miller, R. W.; Eady, R. R. Molybdenum and Vanadium Nitrogenases of *Azotobacter chroococcum.* Low Temperature Favours N2 Reduction by Vanadium Nitrogenase. *Biochem. J.* **1988,** *256*(2), 429–432.

Mortenson, L. E. Ferredoxin Requirement for Nitrogen Fixation by Extracts of *Clostridium pasteurianum. Biochim. Biophys. Acta* **1964,** *81,* 473–478.

Navarro-González, R.; McKay, C. P.; Mvondo, D. N. A Possible Nitrogen Crisis for Archaean Life Due To Reduced Nitrogen Fixation by Lightning. *Nature* **2001,** *412,* 61–64.

Nelson, D. L.; Cox, M. M. *Biosynthesis of Amino Acids, Nucleotides, and Related Molecules.* In: *Lehninger Principles of Biochemistry*; 4th Ed. Freeman, W. H. and Company: New York, 2005.

Ohki, Y.; Imada, M.; Murata, A.; Sunada, Y.; Ohta, S.; Honda, M.; Sasamor, T.; Tokitoh, N.; Katada, M.; Tatsumi, K. Synthesis, Structures, and Electronic Properties of [8Fe-7S] Cluster Complexes Modeling the Nitrogenase P-Cluster. *J. Am. Chem. Soc.* **2009,** *131*(36), 13168–13178.

Orchard, E. O.; Webb, E. A.; Dyhrman, S.T. Molecular Analysis of the Phosphorus Starvation Response in *Trichodesmium* spp. *Environ. Microbiol.* **2009,** *11,* 2400–2411.

Orme-Johnson, W. H. Molecular Basis of Biological Nitrogen Fixation. *Annu. Rev. Biophys. Biophys. Chem.* **1985,** *14,* 419–459.

Paschen, A.; Drepper, T.; Masepohl, B.; Klipp, W. *Rhodobacter capsulatus* nifA Mutants Mediating nif Gene Expression in the Presence of Ammonium. *FEMS Microbiol. Lett.* **2001,** *200,* 207–213.

Pollock, C. R.; Lee, H.-I.; Cameron, L. M.; DeRose, V. J.; Hales, B. J.; Orme-Johnson, W. H.; Hoffman, B. M. Investigation of CO Bound to Inhibited Forms of Nitrogenase MoFe Protein by 13C ENDOR. *J. Am. Oil Chem. Soc.* **1995,** *117,* 8686–8687.

Postgate, J. *Nitrogen Fixation.* 3rd ed.; Cambridge University Press: Cambridge, UK, 1998.

Postgate, J. R. *The Fundamentals of Nitrogen Fixation*; Cambridge University Press: Cambridge, England, 1982.

Postgate, J. R. *Nitrogen Fixation.* 2nd ed.; Studies in Biology, no. 92, Edward Arnold: London, 1987.

Prévost, D.; Antoun, H.; Bordeleau, L. M. Effects of Low Temperatures on Nitrogenase Activity in Sainfoin (*Onobrychis viciifolia*) Nodulated by Arctic Rhizobia. *FEMS Microbiol Ecol.* **1987**, 205–210. DOI: http://dx.doi.org/10.1111/j.1574-6968.1987.tb02357.x.

Rees, D. C.; Howard, J. B. Nitrogenase: Standing at the Crossroads. *Curr. Opin. Chem. Biol.* **2000**, *4*, 559–566.

Rees, D. C.; Tezcan, F. A.; Haynes, C. A.; Walton, M. Y.; Andrade, S.; Einsle, O.; Howard, J. B. Structural Basis of Biological Nitrogen Fixation. *Philos. Trans. R. Soc. A: Math. Phys. Eng. Sci.* **2005**, *363*(1829), 971–984.

Rehder, D. Vanadium Nitrogenase. *J. Inorg. Biochem.* **2000**, *80*(1–2), 133–136.

Roper, M. M. Field Measurements of Nitrogenase Activity in Soils Amended with Wheat Straw. *Austr. J. Agric. Res.* **1983**, *34*, 725–739.

Roper, M. M. Straw Decomposition and Nitrogenase Activity (C2H2 Reduction): Effects of Soil Moisture and Temperature. *Soil Biol. Biochem.* **1985**, *17*, 65–71.

Schindelin, H.; Kisker, C.; Schlessman, J. L.; Howard, J. B.; Rees, D. C. Structure of ADP-AIF42-Stabilized Nitrogenase Complex and Its Implications for Signal Transduction. *Nature* **1997**, *387*, 370–376.

Schrauzer, G. N. The Nutritional Significance, Metabolism and Toxicology of Selenomethionine. *Adv. Food Nutr. Res.* **2003**, *47*, 73–112.

Schüddekopf, K.; Hennecke, S.; Liese, U.; Kutsche, M.; Klipp, W. Characterization of anf Genes Specific for the Alternative Nitrogenase and Identification of nif Genes Required for Both Nitrogenases in *Rhodobacter capsulatus*. *Mol. Microbiol.* **1993**, *8*(4), 673–684.

Seefeldt, L. C.; Dean, D. R. Role of Nucleotides in Nitrogenase Catalysis. *Acc. Chem. Res.* **1997**, *30*, 260–266.

Seefeldt, L. C.; Dance, I. G.; Dean, D. R. Fe Versus Mo. *Biochemistry* **2004**, *43*(6), 1401–1409.

Seefeldt, L. C.; Hoffman, B. M.; Dean, D. R. Mechanism of Mo-Dependent Nitrogenase. *Annu. Rev. Biochem.* **2009**, *78*, 701–722.

Shah, V. K.; Hoover, T. R.; Imperial, J.; Paustian, T. D.; Roberts, G. P.; Ludden, P. W. In *Nitrogen Fixation: A Hundred Years After*; Bothe, H., de Bruijn, F., Newton, W., Eds.; Gustav Fischer: New York, 1988, pp 115–120.

Spaink, H. P.; Kondorosi, A.; Hooykaas, P. *The Rhizobiaceae: Molecular Biology of Model Plant-Associated Bacteria*; Spaink, H. P., Kondorosi, A., Hooykaas, P., Eds.; Springer: Netherlands, 1998, pp. 402. DOI:10.1007/978-94-011-5060-6.

Stewart, W. D. P. Biological and Ecological Aspects of Nitrogen Fixation by Free-Living Microorganisms. *Proc. R. Soc. B (Lond.)* **1969**, *172*, 367–388.

Taiz, L.; Zeiger, E. *Plant Physiology;* 2nd Ed. Sinauer Associates, Inc.: Sunderland, Massachusetts, 1998.

Thorneley, R. N. F.; Lowe, D. J. *Kinetics and Mechanisms of the Nitrogenase Enzyme System.* In *Molybdenum Enzymes*; Spiro, T. G., Ed.; New York: Wiley, 1985, pp 221–284.

Walmsley, J.; Toukdarian. A.; Kennedy, C. The Role of Regulatory Genes nifA. vnfA, anfA, nfrX, ntrC and rpoN in Expression of Genes Encoding the Three Nitrogenases of *Axtobocler rinelandii.* *Arch. Microbial.* **1994**, *162*, 422–429.

Woodley, P.; Buck, M.; Kennedy, C. Identification of Sequences Important for Recognition of vnf Genes by the VnfA Transcriptional Activator in *Azotobacter vinelandii*. *FEMS Microbiol. Lett.* **1996**, *135*, 213–221.

Yoo, S. J.; Angove, H. C.; Papaefthymiou, V.; Burgess, B. K.; Münck, E. Mössbauer Study of the MoFe Protein of Nitrogenase from *Azotobacter vinelandii* Using Selective 57Fe Enrichment of the M-Centers. *J. Am. Oil Chem. Soc.* **2000,** *122,* 4926–4936.

Young, J. P. W. *Phylogenetic Classification of Nitrogen-Fixing Organisms.* In: *Biological Nitrogen Fixation*; Stacey, G., Burris, R.H., Evans, H.J. Eds., Chapman and Hall: New York, 1992, pp 43–86.

Zhang, Y.; Pohlmann, E. L.; Ludden, P. W.; Roberts, G. P. Mutagenesis and Functional Characterization of the glnB, glnA, and nifA Genes from the Photosynthetic Bacterium *Rhodospirillum rubrum*. *J. Bacteriol.* **2000,** 182, 983–992.

Zhu, Y.; Conrad, M. C.; Zhang, Y.; Roberts, G. P. Identification of *Rhodospirillum rubrum* GlnB Variants That Are Altered in Their Ability to Interact with Different Targets in Response to Nitrogen-Status Signals. *J. Bacteriol.* **2006,** *188,* 1866–1874.

PART 4

Microbiology for Environmental Security and Pollution Control

CHAPTER 7

MICROBIOLOGY OF AQUEOUS ENVIRONMENTS: INTERACTIONS, EFFECTS, AND HOMEOSTASIS

DHIRAJ PAUL[1]*, SATISH KUMAR[2], SUNIL BANSKAR[1], BALARAM MOHAPATRA[3], and YOGESH S. SHOUCHE[1]*

[1]*Microbial Culture Collection, National Centre for Cell Science, Savitribai Phule University of Pune Campus, Ganeshkhind, Pune 411007, Maharashtra, India*

[2]*ICAR-National Institute of Abiotic Stress Management, Baramati, Pune 411007, Maharashtra, India*

[3]*Department of Biotechnology, Indian Institute of Technology, Kharagpur 721302, West Bengal, India*

Corresponding authors: E-mail: pauldhiraj09@gmail.com; yogesh@nccs.res.in

7.1 INTRODUCTION

Aquatic system covers more than 70% of the Earth's surface and offers diverse goods and amenities for human civilizations, signifying great economic value (Costanza, 1998). Major portions of this occur between continents, as oceans and the remaining occur within the continental boundaries, as groundwater and polar ice. Aquatic microbiology is the study of microscopic organisms, namely, bacteria, archaea, viruses, and fungi, and their relation to other organisms in this entire aquatic environment, that is, organisms living in fresh or saline water ecosystems. Microorganisms, mainly bacteria, archaea, virus, and eukarya (protists and fungi), are predominantly present in these ecosystems in terms of abundance and biomass. Approximately 10^6 eukaryotic cells and $\sim 10^8$ prokaryotic cells (Whitman et al., 1998) are found in 1 L of sea water. It is noted that a large and diverse pool of microbial species are present in aquatic ecosystems, for instances, approximately 10^6 bacterial species are estimated within the global ocean (Curtis et al., 2002; Šlapeta et al., 2005; Wilhelm and Matteson, 2008; Auguet et al., 2010). Probably due to terrestrial existence, living in the aquatic environment (freshwater and marine) has numerous advantages that include physical support (buoyancy), passive movement of water current, easily access of available soluble organic and inorganic nutrient elements, dispersal of motile elements (gametes, genetic materials) in water, etc. (Covich et al., 2004; Sigee, 2005).

Microorganisms present in such diverse aquatic environment plays a crucial role in maintaining global C, N, and S biogeochemical cycle by conducting a vast array of metabolic function in different redox zone/strata of various aquatic ecosystems (Cotner and Biddanda 2002; Venter et al., 2004; Rusch et al., 2007; Falkowski et al., 2008). In aquatic environment, namely, ocean and eutrophic lakes, autotrophic microbes play a major role in the primary production and produce nearly 50% of the net primary production of the world (Field et al., 1998), whereas heterotrophic microorganisms and protists together formed microbial loop which contribute predominantly to organic matter and nutrient recycling (Azam et al., 1983; Pernthaler 2005; Pomeroy et al., 2007). Eukaryotic organisms present in aquatic and other systems are metabolically less capable, that is, able to aerobic respiration of organic carbon, oxygenic photosynthesis (CO_2 fixation), and carbon cycling between inorganic and organic forms producing and consuming oxygen, compared to prokaryotes, whereas bacteria and archaea show an enormous range of metabolic capacities, including energy generating which is possible for moving electrons from oxidized and reduced forms of, for example, Fe, N_2, S, etc., including the aerobic respiration and oxygenic photosynthesis that are characteristic of eukarya. Therefore, organisms especially bacteria and archaea are exclusively driving all other biogeochemical cycles including carbon and oxygen. As reported in oceans and other ecosystems (Behrenfeld, 2011), primary productivity is likely to be affected by global warming and other anthropogenic contaminants (heavy metal, hydrocarbons, etc.), which may affect the indigenous microbial community structure and food web and thus threaten water resources both freshwater and marine water (Dudgeon et al., 2006; Worm et al., 2006; Nogales et al., 2011). Therefore, a detailed study on microbial diversity of the aquatic ecosystems is essential for monitoring and anticipating their major activities and sustainability (Duffy and Stachowicz, 2006).

In the following sections, microbial community structure and composition of the diverse aquatic ecosystem and the effect of contamination on the aquatic microbiota are presented. A broad overview of groundwater, estuaries and lakes (freshwater and hypersaline), drinking and wastewater microbial structure are presented in the following section.

7.2   TYPES OF AQUATIC ECOSYSTEMS

The aquatic ecosystem is the most diverse ecosystem on the Earth. Water covers about 70% of the Earth of which more than 95% exists in gigantic oceans. A very less amount of water is present in the rivers (0.00015%) and lakes (0.01%), which include the most valuable fresh/drinking water resources. Based on the salinity, the global aquatic ecosystems are broadly classified into saltwater and freshwater ecosystems. The saltwater ecosystem contains a high concentration of salt (averaging about 3.5%), whereas freshwater ecosystems have low concentrations of salts (<500 mg L^{-1}). Saltwater ecosystem is mainly marine water and some hypersaline lake water systems, while freshwater ecosystems are, namely, river water, stream water, and lake water. Beside this groundwater also constitute a significant part of available fresh water on Earth. In the following section, we will limit our discussion of major aquatic systems such as freshwater (rivers, lakes, and estuaries), groundwater, and saline water (hypersaline lakes or soda lakes), on their microbial composition and interactions.

7.3   FRESHWATER ENVIRONMENT AND MICROBIOME

The freshwater environment is broadly classified as lentic (standing water) and lotic systems (flowing water). Flowing water is rivers, estuaries, and canals, while standing water are mostly ponds, lakes and other closed water bodies.

7.3.1   GEOCHEMICAL FACTORS OF FRESHWATER

Natural water bodies are in constant contact with rocks and sediments having different composition of mineral and compounds; therefore, the water inevitably gains a specific composition. Additionally, an anthropogenic activity alters the natural composition of water and also introduces the effects of pollution. Geochemical studies of river water provide significant information regarding the chemical weathering of rock and sediments, isotopic, and chemical compositions of drainage and on the elements cycled in the continent–river–ocean system (Reeder et al., 1972; Stallard and Edmond, 1983; Goldstein and Jacobsen, 1987; Elderfield et al., 1990; Zhang et al., 1995). The geochemical properties of river water or any fresh water determine its utility for municipal, industrial, agricultural and portability purpose. Simeonov et al. (2003) have analyzed a large number of river water samples in Northern Greece for geochemical analysis. They have analyzed 27 geochemical parameters of streams, ditches, tributaries, and the major river systems (Aliakmon, Axios, Gallikos, Loudias, and Strymon). Six major, namely, factors which affect river water quality includes weathering, organic, nutrient, physicochemical, soil-leaching, and toxic-anthropogenic factors. Geochemical analysis of the river Damodar is investigated by Singh et al. (2008). It is found that Ca, Na, Mg, HCO_3, and Cl ions are dominated, except samples collected from mining areas where SO_4^{2-} concentration is higher in the samples. Another study is conducted in river Tsurumi River, Yokohama, Japan, where mostly trace metal concentrations are measured. It is found that lead and molybdenum concentration is higher than standard values, while manganese and iron is lower than that of surface water standards.

7.3.2 FRESHWATER CONTAMINATION

Freshwater contamination is caused by the both geogenic/natural and anthropogenic activities. Continuously increasing industrial activities, mostly mining has initiated pollution or contamination (organic, inorganic compound, heavy metal, etc.) into the river systems or any other surface water systems (Silveira et al., 2006; Farkas et al., 2007; Widmeyer and Bendell-Young, 2008). A lot of studies reported that different sites of river water are contaminated by a large number of hydrocarbon compounds like chlorinated aliphatic hydrocarbons (CAHs), crude oils containing more than 17,000 compound, etc. The major sources of contamination include natural sources, for example, hydrocarbons produced by biological activities, seepage from natural sources, such as the bitumen deposits in Canada, northeastern Alberta, and Deep Ocean deposits. The anthropogenic sources of environmental hydrocarbons include all activities associated with the production, transportation, exploitation, and use of petroleum hydrocarbons. Chemical pollutants, namely, heavy metal are one of the major contaminates of most of the river system. Many compounds (toxic metals/metalloids) from industrial/mining activities are released into fresh water/river water without knowing the environmental risk associated with this. Metals concentration in rivers is usually very low, and it is mainly received from weathering of rock and soil (Reza and Singh, 2010). The major sources of heavy metal contamination in fresh water bodies are mining and smelting activities, dumping of untreated and partially treated effluents, chelates from diverse industries and pesticides in agricultural fields (Nouri et al., 2008; Reza and Singh, 2010). In complex aqueous environments, the formation of different ion species of metals plays an important role in its bioavailability. According to the Free Ion Activity Model, the toxicity of metals to aquatic organisms is related to the free metal ion activity. Different types of heavy metal (Cr, As, F, Fe, etc.) contamination in freshwater is reported. Within this, arsenic contamination is a major pollution in all aquatic systems. It is noted that As^{3+} and As^{5+} species are mostly found in aquatic systems, and between these two, As^{3+} is 100 more toxic than As^{5+}. Therefore, according to the availability of speciation form of the metals in the aquatic system, aquatic microorganisms are responded (Paul et al., 2015).

7.3.3 MICROBIAL DIVERSITY, INTERACTION, AND FUNCTION

Microorganisms present in the freshwater ecosystem are categorized into two major groups, that is, autotrophic and heterotrophic. Autotrophic organisms, including algae and autotrophic bacteria, can fix CO_2 into complex compounds, and during CO_2 fixation, inorganic compounds are used as a source of nitrogen and phosphorus by them (primary producers). Heterotrophic organism includes most of the microorganisms, namely, bacteria, protozoa, and fungi and can utilize organic compounds, which are synthesized by the primary producer or other dissolved organic compounds (e.g., leaf litter). In the case of fresh water ecosystem, prokaryotes play an important role in biogeochemical cycling (Pernthaler, 2013).

Flowing water ecosystems like rivers and streams are extremely hetero-geneous and sporadic input of inorganic and organic materials and various arrangements of habitat patchiness are their unique features (Proia et al., 2012). In flowing water (lotic ecosystem), microbial loop is dependent on the allochthonous source of carbons particularly leaf litter. It serves as a substrate for bacterial and fungal growth and a source of dissolved organic carbon. The core bacterial groups presented in most or all fresh-water ecosystems are the members of *Proteobacteria* (*Alpha-* and *Beta-*subdivisions), *Cytophaga–Flavobacterium–Bacteroides*, *Actinobacteria*, and *Verrucomicrobia*. Except *Verrucomicrobia*, the widespread occur-rence of all above-mentioned bacterial groups in the freshwater ecosystem is also validated by in situ hybridization studies, where probes specifically for these groups are used (Glöckner et al., 1999). In that study, 34 phylo-genetic clusters of closely related sequences that are either restricted to freshwater or dominated by freshwater sequences are observed (Zwart et al., 2002; Mueller-Spitz et al., 2009; Newton et al., 2011).

Previously revealed general dominance of the phyla *Proteobacteria* and *Actinobacteria* and more specifically the dominance of the phylum *Acti-nobacteria* in the Amazon River corroborate well with metagenomic study (Ghai et al., 2011)—even their predominance is also confirmed in three recent high-throughput sequencing studies in riverine bacterioplankton on the Upper Mississippi River (USA); Yenisei River (Russia); and River Thames (UK) (Staley et al., 2013; Kolmakova et al., 2014; Read et al., 2015). Read et al. (2015) have investigated the bacterioplankton community structure at 23 different sites of Thames basin. They observed a shift from a *Bacteroidetes* dominated community in the upstream to an *Actinobacteria* dominated community in the downstream, reaches near the river mouth. They interpret the patterns as evidence of ecological succession along the river continuum. Although these studies are mostly limited on relatively small river basins and/or a small number of sampling sites very recently, Domenico et al. (2015) have presented a broad dataset detailing the microbial diversity along with the midstream of the Danube River and its tributaries. In this study, they cover a total of 2600 km of the river Danube, and more than 100 samples are used for microbial diversity analysis purpose (Savio et al., 2015). 16S rRNA gene amplicon sequencing analysis reveals abundance of bacterial members *Actinobacteria*, *Proteobacteria*, *Bacteroidetes*, *Verrucomicrobia*, and *Candi-date_division_OD1*. Sequence reads assign to *Bacteroidetes* decreased significantly along the river in the free-living fraction.

Anthropogenic activities likely large-scale mining, dumping of the waste materials, and other activities from various industries, located near the rivers and other surface water sites, resulted contamination of aquatic environments around the world. Changes in the geochemical and microbiological characteristics of the freshwater due to heavy metal contaminated environments are well documented. Heavy metal contamination can decrease water quality, and therefore, it is toxic/harmful for numerous microorganisms. Effect of chromium contamination on microbial community structure of river Alviela and its two tributaries is investigated using both culture-dependent and -independent approaches (Branco et al., 2005). Bacterial phylum *Actinobacteria*, *Firmicutes*, *Bacteroidetes*, and *Proteobacteria* are predominantly detected in the culture-dependent study, whereas *Acidobacteria* and *Deltaproteobacteria* are abundant in culture-independent study. It is observed that within *Proteobacteria*, *Gammaproteobacteria* is particularly abundant in chromium contaminated sites of the Alviela River (Branco et al., 2005). In another study, the relationships between microbial community structure and heavy-metal contamination in six different sites of Clark Fork River (USA) are investigated. No positive correlation between total bacterial abundance and heavy-metal contaminations is observed (Feris et al., 2004). Only the correlations between heavy-metal contamination and the abundance of specific bacterial groups, that is, *Alpha-* and *Gammaproteobacteria* and *Cyanobacteria* are found apparent during the fall and early winter. Microbial diversity study using high-throughput sequencing from the specific sites of Shenzhen River, that is, polluted with volatile sulfide compounds and nitrate is investigated (Chen et al., 2013). Bacterial members *Beta-*, *Gamma-*, and *Epsilonproteobacteria* are predominantly present in nitrate-rich sites, whereas *Firmicutes*, *Chloroflexi*, and *Deltaproteobacteria* are abundant in nitrate unexposed sites. Using PhyloChip and clone library-based approach, microbial community structure of heavy metal contaminated Coeur d'Alene River is explored. PhyloChip and clone libraries based analyses capture metabolically versatile indigenous microbial populations that include ammonia oxidizers, iron-reducers and -oxidizers, methanogens, and sulfate-reducers bacterial populations that belong to *Proteobacteria*, *Firmicutes*, and *Actinobacteria*. Microbial diversity of the Zenne River, contaminated with CAHs discharged (Vilvoorde, Belgium), is investigated using denaturing gradient gel electrophoresis (DGGE) analysis (Hamonts et al., 2014). The predominance of bacterial members CAH-respiring

Dehalococcoides followed by methanogens and sulfate-reducing bacteria is observed. Culture-independent shotgun metagenomics (C-iSM) (pyro-sequencing) approach is employed to investigate microbial diversity in river Patancheru (Hyderabad, India) polluted with wastewater from the production of antibiotics. A wide range of antibiotic resistance genes and the elements of horizontal gene transfer, including integrons, transposons, and plasmids, are identified. Two uncharacterized resistance plasmids are identified (Kristiansson et al., 2011). Therefore, the study highlights that the antibiotic contaminations play a crucial role for promoting the resistance genes and subsequently their mobilization from indigenous microbes to other species and finally to human pathogens.

7.4 ESTUARIES ENVIRONMENT AND MICROBIOME

An estuary principally is a semienclosed system of freshwater that is directly connected with sea where the sea water is diluted by freshwater from land drainage (Abreu et al., 2001). Estuarine ecosystems offer the utmost physicochemical variations in composition, that is, biological and chemical gradients which is due to the mixing of seawater and river/freshwater and simultaneous effect of indigenous biological activity. Factors, namely, salinity, nutrient availability, organic matter compositions, are thought to impact the structure of indigenous microbial communities. Compared to open sea ecosystems, estuaries' ecosystems are 20 times more productive. Along with the marine floor and water bodies and cold deep sediment, estuary ecosystem proves to be the second most extreme choice for exploring microbial diversity due to a different ecological behavior, that is, intermediate zone between terrestrial and marine ecosystems and thought to have a significantly different bacterial community than marine ecosystem (Orcutt et al., 2011).

7.4.1 GEOCHEMICAL FACTORS OF ESTUARIES ECOSYSTEM

The most important geochemical factors of the estuary ecosystems are dissolved oxygen (DO_2), salt concentration, and sediment chemistry. The massive spatial variation is noted incase saline, that is, from near 0% to 3.4%. The variation occurs over time interval and seasons which creates very harsh environment for organisms including microorganisms. DO_2 which is responsible for oxic–anoxic transition creates problems for life forms in this ecosystem (Abreu et al., 2001; Basak et al., 2015a).

7.4.2 MICROBIAL DIVERSITY, INTERACTION, AND FUNCTION

Microbial community structure of the estuarine ecosystem is expected to be high due to continuous mixing of marine and river/freshwater and the resuspension of sediments and particles from many sources, including benthic and sea grass beds. With respect to salinity gradient, bacterial community composition and dynamics of the Parker River estuary and Plum Island Sound, in northeastern Massachusetts, are investigated using culture-independent DGGE techniques. The abundance of bacterial phyla *Alpha-*, *Beta-*, *Gamma-*, and *Delta-Proteobacteria*, *Actinobacteria*, *Cyanobacteria*, and *Bacteroidetes* is observed (Zwart et al., 2002). The effect of spatial and temporal disparities on the bacterial community structure of the unique coastal mangrove of Sundarbans is observed using high-throughput pyrosequencing. A total of 24 bacterial phyla are noted. The predominance of bacterial phyla *Proteobacteria* is observed followed by *Alphaproteobacteria*, *Deltaproteobacteria*, and *Gammaproteobacteria*. Along with *Proteobacteria*, sequences affiliated *Actinobacteria*, *Bacteroidetes*, *Planctomycetes*, *Acidobacteria*, *Chloroflexi*, *Cyanobacteria*, *Nitrospira*, and *Firmicutes* are also found. Furthermore, presence of microaerophilic and anaerobic microbial population is noted, indicating anaerobic and reducing environment of the sediments in Sundarbans (Basak et al., 2015a). In another study, bacterial community structure at different depth of Sundarbans estuaries (India) is investigated. The taxonomic analysis indicates the presence of 2746 bacterial species that belong to 33 different phyla revealing the dominance of *Proteobacteria*, *Firmicutes*, *Chloroflexi*, *Bacteroidetes*, *Acidobacteria*, *Nitrospirae*, and *Actinobacteria*, respectively (Basak et al., 2015b). Crump et al. (1999) are investigated free-living and particle associated bacterial community from Columbia River estuary. It is noted particle-attached bacteria play a major role in the estuarine food web. Most of the particle-associated bacterial populations (75%) are represented the sequence identity with the phyla *Cytophaga* and *Alpha-*, *Gamma-*, and *Deltaproteobacteria*. It is observed,

48% of the free-living estuarine bacterial clones showed close similarity with the river or the coastal ocean (Crump et al., 1999). The microbial composition and its relation to changes with environmental conditions along an estuarine gradient of Tillamook Bay, Oregon are explored. It is found, bacterial phyla *Gammaproteobacteria*, *Betaproteobacteria*, and *Bacteroidetes* are dominated in freshwater samples, while *Alphaproteobacteria* and *Cyanobacteria* are predominant in marine samples. Changes in microbial community composition are strongly correlated with salinity, dissolved silicon, and precipitation (Bernhard et al., 2005). Ammonia-oxidizing bacteria (AOB) play a major role in maintaining nitrogen biogeochemical cycle in estuary systems, but a very limited study has been conducted about the distribution and community structure of AOB organisms in relation to geochemical changes encountered in estuary ecosystems. Bernhard et al. investigated AOB bacterial diversity and distribution using T-RFLP (Terminal-Restriction Fragment Length Polymorphism) analysis of different sites of the Plum Island Sound Estuary (located in north-east Massachusetts). At the high-salinity sites, AOB are closely related to *Nitrosospira* spp., while at the mid- and low-salinity, sites are mostly dominated by the sequences related to *Nitrosomonas ureae/oligotropha* and *Nitrosomonas* sp. Nm143. Therefore, the study indicates the salinity/salt concentration is a major role on AOB community's structure and distribution in this estuary ecosystem (Bernhard et al., 2005).

7.5   LAKE WATER ENVIRONMENT AND MICROBIOME

Lakes are large water filled areas in the deep depressions of the lands away from the sea and oceans. Excluding the inlet and outlet of river, rivulets, or the stream, these are surrounded by the land areas from all sides. The sources of water for these water bodies can be a smaller river, rivulets, or the ground water. The lakes are rich in aquatic plants, zooplanktons, phytoplanktons, and bacterial communities. Broadly, lakes can be categorized into two major groups, that is, freshwater lake and extreme soda lakes.

Freshwater lakes are characterized by their neutral pH and salinity (≤ 1 g L^{-1}). These are among the largest source of the freshwater around the globe carrying more than 125,000 km^3 of the water (Sigee, 2005). The freshwater lakes usually have three major stages that is, oligotrophic lake:

these are deeper and relatively colder, and there is less availability of nutrients, therefore, have limited vegetation into it and has less productivity. These types of lakes are ideal for swimming and bathing. Hence, the water is clean and is characterized by a relatively lower load of bacteria. Mesotrophic lake: With time, oligotrophic lakes progresses toward the mesotrophic lake. These types of lakes have specific characteristics, that is, higher productivity, relatively warm, shallow, and have higher vegetation. It has higher bacterial load than the oligotrophic stage. Eutrophic lake: Lake naturally progresses toward the eutrophic stage, but the anthropogenic activities and the pollution, fertilizers wash off speedups the whole process. Hence, high-productivity rate, higher algal blooms, and higher abundance of the aquatic and other plants are the characteristics of eutrophic lake. The water quality is very much deteriorated and unsuitable for the household usage. Eutrophic lakes have the highest bacterial load in its water.

7.5.1 BACTERIAL COMMUNITIES OF FRESHWATER LAKES

Different lakes are in constant touch with the surrounding soil, vegetation and hence greatly affected by these factors; therefore, the bacterial communities of different lakes vary greatly (Newton et al., 2011). Most aquatic bacterial communities are heterotrophic and constitute mainly the Gram-negative bacteria and relatively fewer Gram-positive ones (e.g., *Bacillus pituitans*). The bacteria are free-living or symbiotic, even parasitic with the other aquatic organisms in the lake water. An important example of bacteria which is parasitic as well as free living at different phases of life in water is *Bdellovibrio* life cycle, includes both free-living (nonnutritive) and parasitic phases (Sigee, 2005).

In general, *Betaproteobacteria* and *Actinobacteria* have been found to be the most dominating groups in freshwater lake (Methé et al., 1998; Glöckner et al., 2000). As previously reported, most of the oligotrophic lake is dominated by the *Actinobacteria* and *Bacteroidetes*, whereas the mesotrophic and eutrophic lakes show the dominance of the *Cyanobacteria* (Van der Gucht et al., 2005). The eutrophic and hypertrophic lakes mostly dominated by *Alpha-* and *Betaproteobacteria*, *Bacteroidetes* are followed by *Gammaproteobacteria* and *Planctomycetes* (Zwart et al., 2002). ACK-M1, an *Actinobacteria* cluster, has been reported to present in most of the freshwater lakes among 81 studied by the Zwart et al. (2003). Similarly, *Rhodoferax* spp. Bal 47 and *P. necessaries* clusters of *Betaproteobacteria* are represented in 7–8 lakes, out of 12 lakes studied (Zwart et al., 2002).

7.5.2   ROLE OF BACTERIA IN LAKE ENVIRONMENT

Bacteria play an important role in the detritus cycle to decompose the organic matters in environments. Similarly, in the aquatic environment, they play an important role in the breakdown of the organic components entered into the water body from its shore or the dead aquatic plants and algae, etc. Additionally, bacteria perform an important role in the productivity in the aquatic environment by living heterotrophic (use organic compounds as the source of energy) and autotrophic (fix CO_2 to obtain energy in the presence of sunlight, i.e., *photoautotroph*, or in the presence of inorganic substrate, i.e., *lithoautotroph*) mode of life. Hence, aquatic bacteria play a crucial role in the cycling of the compounds. These bacteria also play an important role in the synthesis of organic mass in the lake system (Pernthaler, 2013).

7.5.3   EFFECT OF POLLUTION IN THE LAKE MICROBIAL COMMUNITIES

The waste produced as the results of various anthropogenic activities including agricultural, industrial, municipal, and household are readily disposed into the water including the lake water. The native microbial communities with the flora and fauna of aquatic system are negatively affected by the toxic wastes. Since, the bacterial or the microbial degradation is the key to the recycling of the waste material; the waste provides the unwanted selective pressure for the overgrowth of waste degrading microbial population. Additionally, the changes caused by physiochemical properties of the water also affect the native bacterial communities. For example, Gough and Stahl (2011) demonstrate the increase abundance of *Crenarchaeota* in the prolonged exposure of heavy metal in the sediment samples of Lake DePue, Illinois, USA. Further, metabolic profiling studies on the sediment sample collected from Lake DePue indicate the enrichment of the sulfate-reducing bacterial communities and heavy-metal resistance gene carrying bacteria. Additionally, the tolerant group of bacteria also increases significantly in contaminated lake water; for example, enhancement of the relative abundance of spore-forming bacteria *Firmicutes* has been demonstrated in the highly contaminated sites of Lake Geneva (Sauvain et al., 2014).

7.5.4   MICROBIOLOGY OF SODA LAKES

Soda lake or the extreme hypersaline and hyperakaline lakes are characterized by the high minerals, salt concentration, and high pH (9.0–12.0) (Melack and Kilham, 1974; Grant, 2006). These lakes have been reported throughout the geological period. Soda lakes are usually formed when there is no water outlet from the lake and continuous evaporation of the water leaves the salts and minerals behind causes the increase in total salt concentration of the water. Similarly, high inflow, low outflow of the water and a high evaporation rate also lead to the creation of soda lakes. Despite extremities of alkalinity and salts, soda lakes reported to have high productivity rate (>10 g cm^{-2} per day) (Melack and Kilham, 1974).

7.5.5   GEOCHEMICAL PROPERTIES OF THE SODA LAKES

Despite high salinity and high alkalinity, different soda lakes have been reported from various parts of the world to contain different heavy metals in their water contents, for example, Shar Burdiin Lake of Mongolia has high concentrations of Uranium (62.5 mM) (Linhoff et al., 2011), the Mono Lake and Searles Lake of America have high concentrations, that is, 3000 and 200 mM of dissolved arsenic levels, respectively (Oremland et al., 2005). Rift valley soda lake and Mono Lake reported to have higher concentrations of dissolved sodium, sulfates, chlorides, and potassium, whereas the Lonar lake (India) has been reported to contain a higher nitrates and organic contents (Antony et al., 2010). Additionally, Mono Lake is reported having high amounts of radioactive Carbon-14, small detectable amounts of Uranium, Thorium, and Plutonium (www.mono-lake.org). The comparison of geochemical properties of various soda lakes around the world are presented in Table 7.1.

7.5.6   MICROBIAL COMMUNITIES OF SODA LAKE

The major bacterial communities of soda lakes ecosystems are halophilic and alkaliphilic which can tolerate the high salt and high pH conditions. From all the soda lakes around the world, a moderate load of halophilic bacteria has been reported from water and sediment samples. For example, bacterial counts were reported in the range of about 10^2–10^4 per mL of water and 10^2–10^6 CFU g^{-1} of sediments from the Lonar Lake, India (Joshi et al., 2008). Similarly, from the African soda lakes, about 10^5–10^6 CFU mL^{-1} of water has been reported (Grant et al., 1990; Jones et al., 1998). The major bacterial communities of Lonar Lake reported belongs to *Firmicutes, Actinobacteria, Alphaproteobacteria, Betaproteobacteria,* and *Gammaproteobacteria* phylum and major bacterial genera includes the *Halomonas, Alkalimonas, Bacillus, Paracoccus,* and *Mythylophaga,* etc. (Antony et al., 2012). Bacterial members, namely, *Halomonas, Bacillus, Arthrobacter,* and alkaliphilic archaea related to *Natronococcus* and *Natronobacterium* have been reported from East African soda lakes (Duckworth et al., 1996). It is noted that *Arthrospira* spp. along with seasonal *Cyanobacterial* species related to *Chroococcus, Synechococcus,* and *Cyanospira* are the major contributors to the primary productivity of the East African soda lakes (Jones et al., 1998). Further, a higher abundance of *Actinobacteria* from the surface and *Firmicutes* from the bottom part of the Mono Lake is reported (Humayoun et al., 2003).

7.5.7   BACTERIAL ROLE IN SODA LAKE

Cyanobacteria and many other autotrophic bacteria play an important role in the primary productivity of the lake ecosystem. Additionally, bacteria have also been reported to play an important role in heavy-metal cycling in soda lakes. For example, arsenate-respiring heterotrophs, for example, *Bacillus arsenicoselenatis* and *B. selenitireducens,* and chemoautotrophs including *Alkalilimnicola ehrlichii* and *Halarsenatibacter silvermanii* have been reported from the Mono Lake and Searles Lake (Switzer Blum et al., 1998; Hoeft et al., 2004; Oremland et al., 2005; Hoeft et al., 2007). Recently, haloarchaea from the arsenic-rich Diamante lake (Argentina) have been reported to use arsenic as the bioenergetics substrate (Rascovan et al., 2016); indicating hypersaline soda lakes could be the important

TABLE 7.1   Properties of Different Soda Lakes.

Physicochemical Characteristic	Lonar Lake	Lake Bogoria	Lake Nakuru	Lake Elmenteita	Mono Lake
pH	9.8, 10[a]	9.93	10.35	9.90	9.8
Temp. (°C)	26	27.9	24.1	23.1	20
Salinity (%)	1.1	4.9	1.8	2.6	8.4
Alkalinity (M) (CO_3 &/or HCO_3)	0.03	0.11–1.16	0.11–1.16	0.11–1.16	0.5–1.0
Dissolved oxygen (mg L^{-1})	13.3	13	17	8	3.8
Dissolved organic nitrogen (mg L^{-1})	53.95	2.33	11.38	8.42	ND
Total phosphorus (mg L^{-1})	2.8, 31.6[ab]	6.2	2.4	1.2	400
Nitrates (mg L^{-1})	2.4[a]	0.3	1.4	0.3	ND
Cl^- (meq L^{-1})	71.9, 87.4[a]	176	67.3	215	500
SO_4^{2-} (meq L^{-1})	2.3, 1.0[a]	8.2	5	14.1	260
Na^+ (meq L^{-1})	136.1, 159.1[a]	1348.7	413.1	718.8	1300
K^+ (meq L^{-1})	0.4, 0.2[a]	19.2	8.5	15.9	30.7
$F-$ (meq L^{-1})	–	19.2	8.5	15.9	ND

Source: Modified from Antony et al. (2013).
ND = no data available; mean values are shown.
[a]Antony et al. unpublished data.
[b]Phosphate (μm).

sites of heavy-metal recycling. Sulfur-reducing bacteria *Thioalkalimicrobium* spp. have been reported from the Soap lake (Washington) (Sorokin et al., 2007). The sulfate-reducing bacteria of different genera including *Desulfonatronospira*, *Desulfonatronum*, and *Desulfonatronovibrio* other heterotrophic sulfate-reducing bacteria belonging to *Desulfobacteraceae*, *Desulfobotulus alkaliphilus*, *Desulfobulbus alkaliphilus*, and *Synthrophobacteraceae* from soda lakes in Kulunda Steppe are also reported (Sorokin et al., 2011). From all the soda lakes methanogens, methanotrophs and methylotrophs have been reported. Methanotrophs like *Methylobacter*, *Methylomicrobium*, *Methylothermus*, and *Methylocystis* spp. are reported from Mono lake (Lin et al., 2005). Single isotope probing (SIP) analysis performs to identify the active living bacteria at the Lonar Lake led to the identification of novel methylotroph, that is, *Methylophaga lonarensis*. Additionally, different heterotrophs including *Georgenia satyanarayanai*, *Cecembia lonarensis*, *Nitritalea halakaliphila*, and *Indibacter alkaliphilus* have been reported from the Lonar Lake (Cristobal et al., 2006).

Similar SIP experiments performed on Lake Suduntuiskii Torom and Lake Gorbunka identified the methanotrophs majority of affiliated to *Methylobacter*, *Methylomicrobium*, *Methylothermus*, and *Methylomonas* spp. (Lin et al., 2004). The attempts to identify the eukaryotic communities have identified the dominance of *Oxytrichia* (ciliated protozoan) and fungi like *Candida* spp. (Antony et al., 2013). Hence, the bacteria are reported to play a significant role in biogeochemical cycle of the soda lake ecosystems.

7.6 GROUNDWATER ENVIRONMENT AND MICROBIOME

Groundwater is used as the most important water source for drinking and irrigation purpose throughout the world. Many places of the world, mostly villages, are solely depending on the groundwater for drinking and irrigation purpose. One survey in the United States by United States Geological Survey observed that >50% of the US population and almost 99% rural population of US depend on the groundwater for drinking purpose, which indicates its significance as drinking water in human life. Groundwater is present in many naturally occurring open spaces like fractures or pore spaces within grains in sediments/soils and different types of rocks (like basalts or fractured bedrock). It starts as rainfall and soaks into the sediment and preserves as groundwater beneath the earth's surface which is known as aquifers. Due to less pathogenic organism's contamination compared to surface water, groundwater uses as one of the best sources of drinking water throughout the world.

7.6.1 GEOCHEMICAL FACTORS OF GROUNDWATER

Groundwater/aquifer ecosystems are lacking of photosynthesis and deficient of fresh and easily accessible/labile organic/inorganic matters that create a significant difference between the surface water and groundwater microbial community composition. Groundwater/aquifer ecosystems are characterized by hydrological, chemical, and geological heterogeneity (Madsen and Ghiorse, 1993). The vertical stratification of the aquifers is unique and very complex for each groundwater systems. However, within the different zones/layer, environmental parameters are very stable. Due to lack of light, limited nutrient accessibility, and relatively low temperature, the aquifer systems are known as "extreme" ecosystems (Danielopol et al., 2000). But interestingly aquifers microbes are well adapted to these conditions and to survive and maintain the important biogeochemical cycle within this extreme ecosystem (Paul et al., 2015). Groundwater/ aquifer ecosystems are varying in size and complexity ranging from small ecosystems (a few kilometers) to large regional aquifers (more than 100 km) (Danielopol et al., 2003). Groundwater contains nearly neutral pH, low DO_2 (0.1–1 mg L^{-1}) and negative oxidation reduction potential depending upon the aquifer depth and environment. The temperature of the water is within the range of 20–30°C. Dissolve organic content is low (1–5 mg L^{-1}) (Dhar et al., 2011; Paul et al., 2015). Groundwater containing less than 500 mg L^{-1} of the total dissolved solids is acceptable for the domestic and industrial purpose. Among the anions and cations, the major compounds are bicarbonate, chloride, and sulfate within the anions and sodium, calcium, magnesium, and potassium within the cations. Depending upon the concentration of these ions groundwater called Ca–HCO_3 type or Ca–Mg–HCO_3 type water (Dhar et al., 2011; Paul et al., 2015). Besides this depending upon the geological environment, many times it also contains a high concentration of arsenic, iron, and/or nitrate, which are varied depending upon the depth and redox conditions of the aquifers.

7.6.2   GROUNDWATER CONTAMINATION

Groundwater or aquifers are contaminated mostly by geogenic and anthropogenic activities. Groundwater percolating through soils picks up naturally occurring minerals, salts, and organic compounds. During water migration toward downwards, the concentration of soluble minerals and salts typically increase and many times the percolating water accumulates high concentration of minerals that groundwater no longer can be used as a drinking water supply or irrigation purpose without treatments. During flow through the deposited sedimentary rocks and soils, groundwater may retrieve a large number of compounds, mostly metals such as Mg, Ca, and Cl. Some aquifers have a high natural concentration of dissolved constituents such as H_2S, As, Cr, Br, F, Se, radon, etc. (Watanabe et al., 2000; Holmes et al., 2002; Sarkar et al., 2013; Paul et al., 2015). Therefore, the quality of groundwater depends on the contaminant types and its concentrations present in the aquifers. A number of contamination occurs in aquifers such as hydrogen sulfide originated for the decomposition of organic matter present in the aquifers, radioactive gas radon occurred from the natural decay of uranium presented as minerals in the different rocks and sediments. Besides heavy-metals contamination, arsenic contamination in groundwater knows as the worst mass poisoning in the world. It is observed more than 100 million people and more than 70 countries of the world are affected by arsenic contamination in groundwater. Within this Bengal delta plain (BDP) covering most of the part of West Bengal (India) and Bangladesh is mostly affected by arsenic. Today, more than 85 million of the Bangladesh people are taking arsenic-contaminated groundwater for drinking and irrigation purpose. Many aquifers are also contaminated by nitrate and polyaromatic hydrocarbons (Paul et al., 2015). Among the anthropogenic activities, household chemicals and cleaning products, nitrogen-containing fertilizers, industrial waste materials, and fuels, chemical spills or railroad accidents, pesticides, petroleum/hydrocarbon-containing products or chemicals, and waste disposal sites or dumps, etc. are the most important activities which created contamination in groundwater/aquifers. The presence of nitrate and nitrite in water results due to mixing of agricultural runoff containing excessive fertilizers and adversely affect bottle-fed infants by causing methemoglobinaemia, also called blue baby syndrome. Elevated levels of lead which often come from solders were used in pipe fittings or industrial discharge, result in severe neurological disorders. Another important mineral, the fluoride is naturally present in water, and its concentration in naturally occurring in water depends on the geographical factors of the area, with some areas having relatively very high concentrations in groundwater. The compounds of fluorine, like sodium fluoride and fluorosilicates, can undergo dissolution and becomes part of groundwater when it transports through cracks or spaces between rocks. It also can enter drinking water by agricultural runoff or aluminum factories. The EPA has prescribed the maximum contaminant level goals for fluoride level of 4.0 mg L^{-1} in water. Lower amount of fluoride in drinking water provide protection against dental caries while a higher concentration of the fluoride may result in dental fluorosis and adverse effects on skeletal tissues (Power and Schepers, 1989; Kundu et al., 2001; Cerejeira et al., 2003).

7.6.3 MICROBIAL DIVERSITY, INTERACTION, AND FUNCTION IN GROUNDWATER

Before 1970, people mostly depended on surface water for drinking purposes, but due to pathogenic microbial contamination, namely, diarrhea and other diseases related organisms, people have shifted from surface water to groundwater for getting pathogen-free drinking water. According to coli forming counting method, the groundwater is portable for drinking purpose, but in the advancement of technology mainly next generation sequencing (NGS) and other culture-independent methods, it is found that groundwater contains different type microflora including pathogenic organisms. Now it is well known that aquifer microorganisms play a crucial role in biogeochemical cycles in this extreme ecosystem. But our current understanding about the groundwater microbial community structure and composition, which is involved in and responsible for ecosystem functioning, is still inadequate. How anthropogenic or geogenic contamination affect the microbial diversity and how community shifting will affect biogeochemical processes in individual aquifers are urgently required.

Due to low DO_2 and nutrient limited condition, indigenous microbial community or activity is dominated by the populations that can able to use ferric iron, sulfate, or carbon dioxide as electron acceptor(s). So, microbial communities within the aquifers are largely consisting with heterotrophic organisms, which grow and survive well into the nutrient limited and oligotrophic environments (Ghiorse and Wilson, 1988; Madsen and Ghiorse, 1993). Lithoautotrophs, which fix CO_2 and meet their energy requirements by oxidizing inorganic electron donors, are another major component of aquifer microbial communities (Stevens and McKinley, 1995; Kotelnikova and Pedersen, 1997). Previously, most of the studies of groundwater microbiology are carried out by culture-dependent approach, where bacterial count varied from 10^2 to 10^6 cell cm^{-3} of groundwater and $10^4–10^8$ cell per cm^{-3} of sediment. Culture-based analyses of both shallow and deep groundwater are showed presence of heterotrophic bacteria, namely, *Proteobacteria, Bacteroidetes, Actinobacteria*, and *Firmicutes* (Boivin-Jahns et al., 1995; Zlatkin et al., 1996). At lower taxonomy, members of the genera *Rhodococcus, Microbacterium, Arthrobacter* (*Actinobacteria*), *Pseudomonas, Aeromonas* (*Gammaproteobacteria*), *Brevundimonas, Methylobacterium* (*Alphaproteobacteria*), *Flavobacterium, Chryseobacterium*

(*Bacteroidetes*), *Burkholderia*, and *Acidovorax* (*Betaproteobacteria*) are mostly detected in groundwater samples of various part of the Earth (Ultee et al., 2004). Most of these microorganisms are well reported as nitrate-, iron-, manganese-, and sulfate-reducing, acetogenic and C1 or methane-utilizing bacteria (Ultee et al., 2004). In a culture-dependent study, it is also found that methanogenic bacteria increase with the depth of the aquifers, whereas sulfur-reducing bacterial abundance is decreased.

In culture-independent techniques (like clone library, DGGE or NGS), a large number of microorganisms that are uncaptured by culturable methods are documented (Amann et al., 1995). Phylogenetic analysis indicates the bacterial phyla *Firmicutes* and class *Deltaproteobacteria* are prevalent in the groundwater. The abundance of hydrogen utilizing bacteria, sulfate reducers, and homoacetogens indicated that H_2 and CO_2 are the major energy and carbon sources in this nutrient deprived oligotrophic environment. Several metal-reducing bacteria including sulfate-reducing bacteria, for example, *Geobacter*, *Anaeromyxobacter*, *Desulfovibrio*, *Desulfitobacterium*, etc. have been reported from the various aquifers (Petrie et al., 2003). Members of the bacterial genera *Acinetobacter*, *Aquamonas*, *Aquabacterium*, *Aquaspirillum*, *Brevundimonas*, *Herbaspirillum*, *Hydrogenophaga*, *Methyloversatilis*, *Methylophilus*, *Pseudomonas*, and *Rhizobium* are abundantly detected in groundwater of different part of the world including South Asia countries. Besides, bacterial groups *Achromobacter*, *Burkholderiales*, *Herbauspirillum*, *Ideonella*, *Nitrosomonas*, *Propionivibrio*, *Ralstonia*, *Sideroxydans*, *Undibacterium*, *Vogesella*, and so on are reported from the groundwater of BDP which are detected as relatively minor population (Gault et al., 2005; Sutton et al., 2009; Paul et al., 2015). Bacterial genera *Achromobacter*, *Herbaspirillum*, and *Burkholderiales* are known for their role in many biogeochemical cycles including arsenic in the aquifers ecosystem (Sarkar et al., 2013; Paul et al., 2015). In another study, microbial community filterable by 0.2 μm pore size filters from aquifer samples of Mediterranean region is investigated and member of bacterial groups *Alphaproteobacteria*, *Gammaproteobacteria*, and *Cytophaga–Flavobacterium–Bacterioides* are detected (Miyoshi et al., 2005). These bacterial groups are previously reported for their metabolic versatility including the ability to utilize multiple carbon source (organic/inorganic), hydrocarbons, and even HCO_3 during their autotrophic/heterotrophic mode of growth which clearly demonstrates the metabolic flexibility to sustain life under different nutritional limited condition. It also demonstrates their potentiality to sustain at the interface of oxic and anoxic environment and their crucial role in maintaining biogeochemical cycles in this extreme environment (Paul et al., 2015).

Microbial community structure is investigated within arsenic contaminated groundwater from different parts of Asia, Europe and America. The predominance of *Alpha-*, *Beta-*, and *Gamma-proteobacteria*, *Actinobacteria*, and *Firmicutes* that are well known as aerobic/facultative anaerobic organisms is observed. Presence of bacterial genera *Thiobacillus*, *Herminimonas*, *Hydrogenophaga* (members of *Betaproteobacteria*); *Pseudomonas*, *Methylophaga*, *Actinobacter* (members of *Gammaproteobacteria*); *Planococcus*, *Bacillus* (members of *Firmicutes*) along with *Planctomyces*, *Acidobacteria*, *Verrucomicrobiales*, and iron-reducing *Geobacter* (Kinegam et al., 2008; Price et al., 2013; Jiang et al., 2014; Paul et al., 2014, 2015) are detected. These are well reported as highly sulfide/thiosulfate-oxidizing, denitrifying, and heavy-metal-resistant bacteria. The aquifer-containing higher concentrations of Fe, As, methane and low concentration of SO_4^{2-} and NO_3^- are dominated by *Acinetobacter*, *Geobacter*, *Thermoprotei*, and *Methanosaeta*, whereas *Pseudomonas* and *Nitrosophaera* are abundant, and *Thermoprotei* and methanogens are absent in higher concentrations of SO_4^{2-} and NO_3^- (Kinegam et al., 2008; Jiang et al., 2014). Hemme et al. (2010) have done a metagenomic analysis of high concentrations of heavy metal, nitric acid and organic solvent containing groundwater. Microbial community analysis reveals that prolonged exposure (approximately 50 years), the groundwater samples are dominated by *Beta-* and *Gammaproteobacterial* populations. In microcosm-based experiment where the groundwater is exposed to uranium for a long time, the abundance of bacterial sequences related to sulfate reducers, that is, *Desulfobacter* and *Desulfocapsa* spp. are observed (Holmes et al., 2002). Microbial community structure of petroleum-contaminated groundwater samples of Kuji, Japan is investigated (Watanabe et al., 2000). The dominance of bacterial phyla *Epsilon-*, *Beta-*, and *Deltaproteobacteria* are noted.

7.7   WASTEWATER ENVIRONMENT AND MICROBIOME

With the constant increase in urban populations in the developing countries, huge amounts of freshwaters are diverted to meet out the increased demand for domestic, commercial, and industrial sectors that, in turn results in generation of greater volumes of wastewater (Asano et al., 2007). Wastewater is the general term that refers to the "adversely effected used water" produced as a result of domestic, industrial, agricultural, or any other anthropogenic activity. Wastewater often contains elevated levels of undesirable substances. These undesirable substances may be classified into physical [total dissolved solids and suspended solids (SS)], chemical (hazardous heavy metals, pesticides, herbicides), and biological (pathogenic microorganisms) impurities. The household generated wastewater, the sullage (also called gray water), municipal wastewater, sewage (also called black water), and industrial effluents are well known and most often encountered wastewater systems in the present civil settings. In the absence of good quality water, the marginal waters (quality compromised water) are often used for diverse purpose like crop farming and aquaculture. The proper planning for the reuse of municipal wastewater is emerging as an important dimension of water resources planning in many countries across the globe. Complying with these priorities, some countries like Hashemite Kingdom of Jordan and the Kingdom of Saudi Arabia have evolved a national policy to reuse all treated wastewater effluents (FAO, 2015). The treatment of wastewater is of paramount importance, in order to prevent spreading the diseases and illness caused by the sewage and also to address the issues related to the overall environmental health and water pollution. Public exposure to untreated sewage often occurs through contamination of nearby drinking water reservoirs or consumption of foods harvested directly from polluted water bodies. The untreated wastewater is not only the risk for human health, but it also adversely affects the aquatic life because it contains abundant nitrogen and phosphorus (N, P) and leads to the phenomena of eutrophication, growth of huge amounts of algae (algal blooms) and other aquatic plants that deplete local waters of oxygen ultimately leading to the deterioration of the water quality (Conley et al., 2009). Sewage derived spread of pollution to surface water, groundwater, exposed soils can be restricted by improved policies, institutional dialogues, and financial mechanisms for proper disposal and recycling of the wastewater, before releasing it into the environment. The effective management of wastewater is also essential for nutrient recycling and for maintaining ecosystem integrity.

7.7.1   GEOCHEMICAL FACTORS OF WASTEWATER

The composition of wastewater varies greatly depending on the source of the origin, but broadly, it contains 99.9% water and 0.1% solids (Mason et al., 2013). The composition of the wastewater is effected by the interplay of natural geochemistry, geology and mineralogy, of the area and anthropogenic interference in natural processes. Understanding the characteristics of geomorphological traits of local catchment of a wastewater source is important in order to identify, understand, and manage the scenarios that could lead to water pollution. The quality of the raw water received at a wastewater treatment plant is governed by multiple factors including flora and fauna of the area, climate, topography, geology, vegetation, and the human interventions.

The physicochemical characteristics of wastewater, like color, odor, temperature, turbidity, viscosity, pH, DO_2, biological oxygen demand (BOD), chemical oxygen demand, concentration of nitrogenous compounds (ammonia, nitrites, nitrates), phosphorus content, sulfur content, concentration of heavy metals (lead, mercury, chromium, arsenic), and other toxic compounds like silver, boron, cyanides, pesticides and fertilizers, etc., are basically the manifestation of prevailing geochemical and anthropogenic factors. The particulate matter in sewage includes SS, dissolved solids, colloidal solids. Some gases and living organisms (mainly microorganisms) do also constitute a significant portion of this 0.1% nonwater part of sewage (Wang et al., 2007). Some of the physicochemical parameters give indirect indications about the stage of decomposition of the sewage. For instance, the color of the sewage indicates the freshness of sewage. The freshly generated sewage appears grayish brown or yellowish in color, which turns blackish with the passage of time due to the onset of putrefaction (Singh and John, 2013). The odor of the fresh sewage is not as offensive but as it starts to get stale, it begins to give obnoxious and severely offensive odor. This bad smell is often produced due to the presence of gases like methane, hydrogen sulfide, and some other VOCs (volatile organic compounds) formed due to anaerobic decomposition of sewage and can pose a health risk to the surrounding inhabitants. The odor from the wastewater is removed during the treatment, using a variety of technologies such as chemical scrubbers and activated carbon towers which reduces the chemical odor up to the efficiency of 95–99% (Mohamed et al., 2014). In terms of odor, chemical scrubbers can reduce the overall odor by about 80–90%. The use of biotrickling filter technology has also been effective for odor and is perceived as an extremely cost-effective biological way of treating sewage odor (Wu et al., 2001; Lafita et al., 2012).

The temperature of the sewage is also an important parameter and has a direct bearing on the biological activity of native bacteria, solubility of gases and viscosity of the wastewater (McKinley and Vestal, 1985; Naka- saki et al., 1985). In general, the temperature of the raw sewage ranges between 15°C and 35°C at various places in different seasons. The devia- tion on either side of this range, coupled with the lower DO_2 adversely affects the efficiency of biological treatment for disposal of wastewater. The chemical characteristics of sewage water give an indirect indication of the extent of sewage decomposition that provides the basis for deciding the further treatment required for its safe disposal. The organic matter present in the sewage consists of carbohydrates (cellulose, starch, sugars), fats and fatty acids, nitrogenous compounds like urea and decomposed products of proteins (Shon et al., 2007). The measurement of the pH value of sewage remains a simple way to know the stage of the sewage decomposition. The pH of the fresh sewage is alkaline (slightly more than freshwater), which, in case of untreated standing sewage, tends to fall with passage of time (Nielsen et al., 2008). The fall in pH largely happens due to the produc- tion of acid by bacterial activity in anaerobic or nitrification processes. The pH of raw sewage ranges from 5.5 to 8.0. However, the pH of the sewage increases with each subsequent stage of the treatment and serves as an important parameter to evaluate the efficiency of certain treatment methods.

Proteins, amino acids, urea, albuminoid, nitrates, and other amines are the main nitrogenous compounds present in wastewater. Bacterial decomposition of these nitrogenous substances results in the produc- tion of ammonia. The presence of free ammonia in sewage indicates the very initial stage of decomposition of organic matter (hence, indicating recent pollution), whereas the presence of albuminoid nitrogen indicates the quantity of nitrogen in sewage before the decomposition of organic matter (Punmia et al., 1998). The nitrates and nitrites also indicate the progress of treatment as nitrates represent fully oxidized organic matter in sewage, whereas nitrites indicate the intermediate stage of conversion of organic matter. The presence of nitrites in any water source is extremely dangerous, but as oxidation of nitrites to nitrates is rapid, it is generally not found in water bodies. As nitrates represent fully oxidized matter, its presence in sewage is not dangerous. Sewage that contains higher nitrates, if disposed in a water body, will lead to increase in nitrate content in the water body. The phosphorus remains the essential nutrient for life, but, on the negative side, it is the decisive nutrient contributing to the increased eutrophication of lakes and natural waters (Carpenter, 2008). The sewage contains appreciable amounts of phosphorus, which largely come from human excreta, agricultural runoff containing fertilizers, food-processing industries. The increased use of synthetic detergents further adds to the phosphorus content of domestic sewage. The presence of phosphates in a water body indicates its pollution with sewage or other organic wastes. The concentration of phosphorus in raw sewage ranges between 5 and 10 mg L^{-1} that is generally adequate to sustain the aerobic biological wastewater treatment. The presence of phosphorus causes many water quality problems that include algal bloom in water bodies and the possible lethal effect of toxins produced by algae in drinking water.

7.7.2 MICROBIAL DIVERSITY, INTERACTION, AND FUNCTION IN WASTEWATER

Microbial community structure of the wastewater is highly complex, and a limited number of studies are undertaken to know their composition and dynamics. Microbial communities of the wastewater include diverse functional groups of microbes which broadly include populations of specific species of AOB, nitrite-oxidizing bacteria (NOB), denitrifying bacteria, and polyphosphate-accumulating organisms (PAOs). Some of these bacterial communities perform the key functions of reclamation wastewater like carbon and nutrient removal from sewage, while certain bacterial species (filamentous bacterial populations of *Microthrix parvicella*, *Thiothrix* spp., and *Nostocoida limicola*) can also be detrimental for sewage treatment by negatively influencing the settling properties of activated-sludge, contributing to the formation of foam and outcompeting microorganisms required for nutrient removal (Wagner and Loy, 2002)

$$
\begin{array}{cc}
\text{AOB} & \text{NOB} \\
\textit{Ammonia monooxygenase} & \textit{Nitrite oxidoreductase}
\end{array}
$$

$$NH_3 + O_2 + 2e^- + 2H^+ \rightarrow NH_2OH + H_2O \rightarrow NO^-_2 + 5H^+ + 4e^- \rightarrow NO^-_3$$

$$\textit{Hydroxylamine oxidoreductase}$$

Ammonia is the major nitrogen compound of sewage and is removed from wastewater treatment plants (WWTPs) by conversion to gaseous nitrogen via nitrification and denitrification. Ammonia oxidation is completed in two steps, that is, conversion of ammonia to nitrite and subsequently nitrite to nitrate. The first step of nitrification is carried out by AOB which have a key role in the WWTPs (Bellucci and Curtis, 2011). Other than AOB, some archaea, that is, ammonia-oxidizing archaea (AOA), can also oxidize ammonia. However, AOA remains the dominant nitrifiers in nitrogen removal from wastewater.

The quality of wastewater remains a key determinant of the distribution of AOB and AOA in aerobic reactors, but AOA populations have been shown to more sensitive to toxic compounds in the wastewater (Bai et al., 2012). The populations of these functional groups of slow-growing microbes in different WWTPs varies, as each reactor plant differs in configuration and operating conditions and receives compositionally different influence depending on the location and local conditions. The process of nitrification involves two-steps, namely, oxidation of ammonium to nitrite that is carried out by AOB. The second reaction involves the subsequent oxidation of nitrite to nitrate by NOB. Ammonia monooxygenase (AMO), a well-characterized transmembrane copper protein and hydroxylamine oxidoreductase, are the two enzymes which catalyze the conversion of ammonia to nitrite. The subsequent conversion of the nitrite to nitrate is catalyzed by a membrane-associated iron–sulfur molybdoprotein, nitrite oxidoreductase.

Traditional studies based on culture-dependent approaches have found the *Nitrosomonas europaea*, *Nitrosomonas eutropha*, *Nitrosococcus mobilis*, *N. mobilis*, and *Nitrobacter* spp. as the most abundant nitrifiers in samples from WWTPs (Henze et al., 1997). However, whole community DNA based studies have been used to reveal microbial diversity of particular functional groups of microbes in environmental samples on the basis of functional gene markers like soxB gene (unique gene to sulfur-oxidizing bacteria) and AMO, amoA (unique to ammonia oxidizing microbes) have been applied successfully, to ascertain the diversity of functional groups of microbes in environmental samples (Kumar et al., 2015). The metagenomics and metatrancriptomics sequence data are generated using NGS of activated sludge samples from WWTP in Hong Kong, reveals the abundance of *Proteobacteria*, *Actinobacteria*, *Bacteroidetes*, *Firmicutes*, and *Verrucomicrobia* in targeted sludge samples. Further, gene expression annotation studies on nitrogen removal and denitrification related genes sequences reveal the strong nitrification activity with most of the NOB belonging to *Nitrosomonas* and *Nitrosospira* species (Ye and Zhang, 2013). Further, some of the amoA gene based diversity studies on different samples from WWTPs have shown the genus *Nitrosospira*, is not an important ammonia oxidizer in sewage samples, though other ecosystem analysis reveals its dominant role in ammonia oxidation (Kowalchuk and Stephen, 2001). Likewise, culture-independent surveys have shown that instead of *Nitrobacter* spp., uncultured *Nitrospira* like microorganisms are the dominating nitrite oxidizers in most WWTPs (Dionisi et al., 2002). Besides, bacteria genera, namely, *Brocadia*, *Kuenenia*, *Anammoxoglobus*, *Jettenia*, and *Scalindua* belonging to *Planctomycetes*, are also detected in WWTPs (Jetten et al., 2009). The enhanced biological phosphorus removal (EBPR) is an economically beneficial and environment-friendly sewage treatment configuration that can be applied to activated sludge systems for phosphorus removal. Under EBPR conditions, PAOs are selectively augmented in the sludge bacterial community and accumulate large quantities of polyphosphate inside their cells resulting in improved removal of phosphorus. Understanding the identities and composition of PAOs can be of colossal help to stimulate conditions for phosphorus (P) removal. Molecular studies have demonstrated that *Acinetobacter*, the traditional model organism for EBPR, does not catalyze phosphorus removal in WWTP; however, this role is performed during the sewage treatment, by a novel, yet uncultured, *Betaproteobacteria* related to *Rhodocyclus* (Crocetti et al., 2000). The *polyphosphate kinase 1 gene* (*ppk1*) has been used as a genetic marker for the study of diversity and population structure of EBPR systems (He et al., 2007), and it is found that genetic diversity is essential for maintaining stable EBPR performance.

7.8 DRINKING WATER QUALITY AND MICROBES

Access to the adequate and safe drinking water is essential for the personal hygiene as well as the public health, but unfortunately, over billion of the people worldwide do not have access to this basic need. The adverse effects of unsafe drinking water and diseases caused by contamination of drinking water constitute a major burden on human health. The water quality standard parameters set for drinking water are territorial in nature (national or regional standards) (Shmueli, 1999). Many developed countries specify standards to be applied for an acceptable level of risk for constituents of water or indicators of water quality, in their own country. The European Drinking Water Directive, in the Europe, and the United States Environmental Protection Agency (EPA), in the USA, are such agencies which issue guidelines and keep surveillance for standards of safe drinking water. The World Health Organization (WHO) also publishes guidelines on the standards that should be achieved (WHO Report, 2011).

7.8.1 GEOCHEMICAL PROPERTIES OF DRINKING WATER

Drinking water have dissolved salts, minerals, essential elements, whose composition depends on the multiple factors like naturally occurring rocks, mineralogical profile of the area, soils, the effects of the geological setting and climate, depth of water table, and many other factors like mining and manufacturing activities in the area. For instance, the presence of high amounts of arsenic in groundwater in Bangladesh and West Bengal (India) is due to the naturally occurring high concentrations of arsenic in deeper levels of groundwater and also due to the dissolution of naturally occurring minerals and ores in the region. The WHO issues guidelines regarding the safe limits of various chemical pollutants found frequently present in drinking water (Table 7.2). These toxicants in drinking water often come from the untreated industrial effluents, weathering of parent rocks of the area, mixing of agricultural having high herbicide, pesticides, and fertilizer concentrations, domestic or municipal run off having high detergent concentration, seepage of these toxicants to shallow water tables.

7.8.2 MICROBIAL LOAD IN DRINKING WATER

Municipalities in urban areas often draw water from the surface water reservoirs, process it and supply to the households and domestic units. The remotely, located residential setup, which are not connected to such hydrologic supply, often depends on the nearest available water reservoir or the common community sources of groundwater like hand pumps and tube wells. Such water supply can often become contaminated with human pathogens (pathogenic bacteria, viruses, protozoa, and worm parasites), which may lead to the far-reaching health consequences. The infants and young children are at greatest risk of waterborne disease. The immunity of individuals, age, sex, state of general health also varies in a population, which determines the extent of the harmful effect of poor quality of drinking water. The water-disinfecting agents like chlorine reacts with organic content of the water and produce disinfection by-products like trihalomethanes, which are potently carcinogenic. Drinking water often get contaminated with fecal matter or by seepage or discharge from septic tanks, sewage treatment facilities, or during piped distribution and serves as vehicle of transmission of various bacterial, viral, and protozoal diseases. The another major concern in the supply of safe drinking water to households lies in the formation of undesirable biofilms (layers of bacteria) on the inner surfaces of water pipes hydrologic supply. Traditionally, biofilms are known to contain nonharmful bacteria and are used to pass wastewater through them, in order to remove organic matter (BOD) by overlaying bacteria, while protozoa and rotifers remove SS, including pathogens and other microorganisms. However, recent studies have shown that biofilms formed inside the pipes of water supply systems can be a safe hiding place for harmful bacteria such as *Escherichia coli* or *Legionella* (Ramalingam et al., 2013). In case of heavy microbial growth on these biofilms, microbial cells leak off into the water flow, which can make the water discolored or taste unpleasant and at worst can release more dangerous bacteria. The most common bacterial species with serious health consequences and of common prevalence in contaminated drinking water are *Salmonella typhi*, *Vibrio cholera*, *E. coli* (Serotype O148, O157, and O124), *Yersinia enterocolitica*, *Campylobacter jejuni*, *Legionella* spp. In addition to these bacteria, several protozoa like *Entamoeba histolytica*, *Giardia intestinalis*, *Cryptosporidium parvum*, *Toxoplasma gondii*, *Cryptosporidium parvum*, and some virus, namely, *Hepatitis A*, *Hepatitis E*, *Adenoviruses*, *Rotaviruses*, *Noroviruses*, and *Sapoviruses* are of common prevalence in drinking water and may cause diseases like intestinal infections, typhoid fever, cholera, dysentery, hepatitis, etc., if not kept under check with proper sanitary measures (WHO, 2011).

TABLE 7.2　A Brief Description of some Important Toxicants Often Present in Water is Presented.

Name of the toxicant	Source for water pollution	General occurrence	WHO safe limit
Acrylamide	Coagulants used in the treatment of drinking-water	A few $\mu g\ L^1$	0.0005 mg L^{-1}
Alachlor	Postemergence herbicide used to control annual grasses	<0.002 $\mu g\ L^{-1}$	0.02 mg L^{-1}
Aldicarb	Systemic pesticide used to control nematodes	500 $\mu g\ L^1$	0.01 mg L^{-1} frequently
Aldrin and dieldrin	Chlorinated pesticides, used against soil-dwelling pests	<0.01 $\mu g\ L^{-1}$	0.03 mg L^{-1}
Antimony	Antimony leached from antimony-containing materials	0.1–0.2 $\mu g\ L^{-1}$	0.02 mg L^{-1}
Arsenic	Present in earth's crust as arsenic sulfide, metal arsenates and arsenides	1–2 $\mu g\ L^{-1}$	0.01 mg L^{-1}
Atrazine	Pre- and early postemergence herbicide	<10 $\mu g\ L^1$	0.002 mg L^{-1}
Barium	Present in igneous and sedimentary rocks and comes in water naturally	100 $\mu g\ L^{-1}$	0.7 mg L^{-1}
Benzene	Introduced into water by industrial effluents and atmospheric pollution	<5 $\mu g\ L^{-1}$	0.01 mg L^{-1}
Boron	Boron is found naturally in ground-water, but its presence in surface water is frequently a consequence of the discharge of treated sewage effluent	0.1 and 0.3 μg L^{-1}	0.5 mg L^{-1}
Cadmium	Cadmium is released to the environment in wastewater	<1 $\mu g\ L^1$	0.003 mg L^{-1}
Chlorine	Widely used in the disinfection of swimming pools and is the most commonly used disinfectant and oxidant in drinking-water treatment	0.2–1 mg L^{-1}	5 mg L^{-1}
Chromium	Chromium is widely distributed in the Earth's crust	<2 $\mu g\ L^{-1}$	0.05 mg L^{-1}
Copper	Corrosion of interior copper plumbing	0.005–30 μg L^{-1}	2 mg L^{-1}

Source: WHO (2011).

7.8.3 PATHOGENIC BACTERIA DETECTED IN DRINKING WATER AND THE DETECTION TECHNIQUES

The contamination of water with human or animal feces containing pathogenic bacteria, viruses, protozoa, and helminths remains the greatest risk. This contamination of drinking water with pathogen cannot be detected by sight, smell, or taste, and hence, the water needs to be tested by employing various testing procedures. As there is wide range of pathogen and testing, the water for presence of all individual pathogens is impractical and expensive approach. The most pragmatic approach is the primary analysis, for the presence of indicator organisms like nonspecific *coliforms*, *Enterococci*, *E. coli*, and *Pseudomonas aeruginosa* rather than the pathogens themselves. The EPA needs that all public water suppliers routinely test for coliform bacteria and distribute water that meets the EPA standards. These indicator coliforms may not cause disease themselves but serves as a better indicator for the presence of contamination of pathogenic organisms. Further, the indicator organisms used are well known to be excreted in several fold higher amounts, in comparison to the pathogen, in the feces of the infected person and hence their level correlates well with pathogen levels in drinking water. Hence, initial water quality is judged by monitoring the level of indicator organisms, if their level exceeds beyond the prescribed set limits, specific analysis for pathogens (*S. typhi*, *Cryptosporidium*, *V. cholera*) may subsequently be undertaken using culture-based, biochemical, optical, or molecular methods. However, these indicator bacteria sometimes fail to give clear correlation with human viruses and protozoa cysts, and hence, its reliability has been challenged (Wéry et al., 2008). Further, there have been several reported cases of outbreaks of waterborne disease without being the indicator bacteria detected in the water source (Barrell et al., 2000). In such conditions, the correct identification of pathogen and understanding about its prevalence becomes of paramount importance in order to determine pathogen loads and concentrations in source waters, as it can serve as a basis for deciding treatment requirements to meet prescription of safe drinking water. Moreover, there are raised concern about higher sensitivity of indicator organism to the treatment process and their shorter survivability in comparison to pathogens, give a misleading picture of their correlation with pathogens (Girones et al., 2010).

In recent times, many sensitive, rapid, and quantitative analytical tools have been developed to detect the presence of pathogens in water. Classical culture-based methods of pathogen detection like multiple tube method, plate count, and pour plate method are tedious, time-consuming, insufficient, as they rely on the growth of specific bacteria in appropriate culture media (Köster et al., 2003). The growth on culture media is sometimes difficult to achieve due to various reasons encompassing the bacteria loosing viability, lack of proper culture conditions, bacteria may be embedded in biofilms, may undergo sporulation and several other reasons. Immunological methods that target detection of antigenic structures of pathogens are highly specific but often needs precultivation step that is also time consuming (Carey et al., 2004). These limitations of culture-based detection can be overcome by resorting to molecular approaches like molecular fingerprinting (ribotyping, RFLP, RAPD), polymerase chain reaction (PCR), quantitative PCR (qPCR), reverse-transcriptase PCR (RT-PCR), fluorescence in situ hybridization (FISH). These molecular techniques offer several advantages over conventional methods like high sensitivity and specificity, speed, ease of standardization and automation but do not differentiate between nucleic acid from viable or nonviable bacterial cells. The sensitivity of these techniques can be enhanced by performing an enrichment step prior to PCR reaction (Noble and Weisberg, 2005).

The qPCR-based analysis of specific DNA from water samples provides the estimate of number of genomes per volume of water of a specific pathogen. In qPCR, a specific-targeted DNA sequence is amplified and quantified simultaneously in real time, with the progress of amplification reaction. The value, so obtained, corresponds to the number of genomes in the water sample but does not give directly the number of cells. Several bacteria contain more than one copy of marker gene as rRNA genes and hence complicate the analysis. qPCR analysis has been extensively applied to detect and estimate the bacterial pathogens in a different water source, namely, *Salmonella* in surface water (Ahmed et al., 2009), *V. cholera, Shigella dysenteriae, Salmonella thyphimurium*, and *E. coli* in river water (Liu et al., 2009). Minogue et al. (2015) have developed a real-time PCR-based assay that can detect the common pathogenic bacteria like *Stenotrophomonas maltophilia, Burkholderia* spp., *P. aeruginosa*, and *Serratia marcescens* in extremely rapid (<5 h) manner.

FISH is based on the use of fluorescently labeled oligonucleotide probes often targeted against the ribosomal RNA (rRNA) and their subsequent visualization under a fluorescence microscope, which allow detection of a specific group of microbes within a mixed consortium. The fluorescence intensity of hybridized probes correlates well with the abundance of the target group. A typical FISH protocol includes the four basic steps involving fixation and permeabilization of the sample, hybridization with fluorescently labeled probe, washing steps to remove unbound probe and finally the detection of labeled cells by microscopy (Bottari et al., 2006). However, the technique is not so rapid and robust as of the PCR-based methods, and enrichment steps are often required in order to get desired sensitivity. The signal strength and sensitivity of the FISH can be increased using the multiple 16S rRNA gene targeted fluorescent probes (Lee et al., 1993). However, some ultra-expensive techniques like DNA chip array (Straub et al., 2002), fluorescence activated cell sorting, flow cytometry (Medema et al., 1998) based assay have also been employed to detect pathogens in drinking water, but the techniques are yet not widely available and are highly expensive.

7.9 SUMMARY

Microorganisms present in aquatic ecosystems (freshwater and saline water) are immensely diverse and play a crucial role in essentially all the biogeochemical cycles by contributing or mediating important geochemical functions like composition of organic molecules from simple elements (CO_2, N_2) and decomposition of dead organisms. Microorganisms present in deeper aquifers or intertidal/transition zone transform methane, sulfur, elemental nitrogen or nitrate by chemoautotrophic and/or chemolithotrophic metabolism under aerobic as well as anaerobic conditions. Due to the continuous increase in anthropogenic activities and other geogenic factors, diverse aquatic ecosystems are becoming severely affected; consequently, it may impact the microbial food web and diversity and in turn threaten freshwater and marine resources. In spite of the impressive advances of culture-independent studies like clone library, DGGE, NGS, etc. that have been made in assessing the diversity of aquatic microorganisms, the mechanisms that underlie the participation of microorganisms in aquatic food webs and biogeochemical cycles are inadequately understood. In this chapter, a detail description of the microbial community structure and dynamics occurring in various water environments, their survival, and role in the biogeochemical cycle is explained.

KEYWORDS

- **biogeochemical cycle**
- **geochemical factors**
- *Proteobacteria*

- *Actinobacteria*
- *Verrucomicrobia*

REFERENCES

Abreu, C.; Jurgens, G.; De Marco, P.; Saano, A.; Bordalo, A. A. Crenarchaeota and Euryar-chaeota in Temperate Estuarine Sediments. *J. Appl. Microbiol.* **2001,** *90,* 713–718.

Ahmed, W.; Sawant, S.; Huygens, F.; Goonetilleke, A.; Gardner, T. Prevalence and Occurrence of Zoonotic Bacterial Pathogens in Surface Waters Determined by Quantitative PCR. *Water Res.* **2009,** *43,* 4918–4928.

Amann, R. I.; Ludwig, W.; Schleifer, K. H. Phylogenetic Identification and In Situ Detection of Individual Microbial Cells Without Cultivation. *Microbiol. Rev.* **1995,** *59,* 143–169.

Antony, C. P.; Doronina, N. V.; Boden, R.; Trotsenko, Y. A.; Shouche, Y. S.; Murrell, J. C. *Methylophaga lonarensis* sp. nov., a Moderately Haloalkaliphilic Methylotroph Isolated from the Soda Lake Sediments of a Meteorite Impact Crater. *Int. J. Syst. Evol. Microbiol.* **2012,** *62,* 1613–1618.

Antony, C. P.; Kumaresan, D.; Ferrando, L.; Boden, R.; Moussard, H.; Scavino, A. F.; … and Murrell, J. C. Active Methylotrophs in the Sediments of Lonar Lake, a Saline and Alkaline Ecosystem Formed by Meteor Impact. *ISME J.* **2010,** *4,* 1470–1480.

Antony, C. P.; Kumaresan, D.; Hunger, S.; Drake, H. L.; Murrell, J. C.; Shouche, Y. S. Microbiology of Lonar Lake and Other Soda Lakes. *ISME J.* **2013,** *7*(3), 468–476.

Asano, T.; Burton, H.; Leverenz, H.; Tsuchihashi, R.; Tchobanoglous, G. *Water Reuse: Issues, Technologies, and Applications*; McGraw-Hill Professional: New York, USA, 2007; p 1570.

Auguet, J. C.; Barberan, A.; Casamayor, E. O. Global Ecological Patterns in Uncultured Archaea. *ISME J.* **2010,** *4,* 182–190.

Azam, F.; Fenchel, T.; Field, J. G.; Gray, J. S.; Meyer-Reil, L. A.; Thingstad, F. The Ecological Role of Water-column Microbes in the Sea. *Marine Ecology Progress Series,* **1983,** *10*(3), 257–263.

Bai, Y.; Sun, Q.; Wen, D.; Tang, X. Abundance of Ammonia-Oxidizing Bacteria and Archaea in Industrial and Domestic Wastewater Treatment Systems. *FEMS Microbiol. Ecol.* **2012,** *80,* 323–330.

Barrell, R. A.; Hunter, P. R.; Nichols, G. Microbiological Standards for Water and Their Relationship to Health Risk. *Commun Dis. Public Health* **2000**, *3*, 8–13.

Basak, P.; Majumder, N. S.; Nag, S.; Bhattacharyya, A.; Roy, D.; Chakraborty, A.; SenGupta, S; Roy, A; Mukherjee, A; Pattanayak, R; Ghosh, A; Chattopadhyay, D; Bhattacharyya, M. Spatiotemporal Analysis of Bacterial Diversity in Sediments of Sundarbans Using Parallel 16S rRNA Gene Tag Sequencing. *Microbial Ecol.* **2015a**, *69*, 500–511.Basak, P.; Pramanik, A.; Roy, R.; Chattopadhyay, D.; Bhattacharyya, M. Cataloguing the Bacterial Diversity of the Sundarbans Mangrove, India in the Light of Metagenomics. *Genomics Data* **2015b**, *4*, 90–92.

Behrenfeld, M. Biology: Uncertain Future for Ocean Algae. *Nat. Clim. Change* **2011**, *1*, 33–34.

Bellucci, M.; Curtis, T. P. Ammonia-Oxidizing Bacteria in Wastewater. *Methods Enzymol.* **2011**, *496*, 269–286.

Bernhard, A. E.; Donn, T.; Giblin, A. E.; Stahl, D. A. Loss of Diversity of Ammonia Oxidizing Bacteria Correlates with Increasing Salinity in an Estuary System. *Environ. Microbiol.* **2005**, *7*, 1289–1297.

Blum, J. S.; Bindi, A. B.; Buzzelli, J.; Stolz, J. F.; Oremland, R. S. *Bacillus arsenicoselenatis*, sp. nov., and *Bacillus selenitireducens*, sp. nov.: Two Haloalkaliphiles from Mono Lake, California That Respire Oxyanions of Selenium and Arsenic. *Arch. Microbiol.* **1998**, *171*, 19–30.

Boivin-Jahns, V.; Bianchi, A.; Ruimy, R.; Garcin, J.; Daumas, S.; Christen, R. Comparison of Phenotypical and Molecular Methods for the Identification of Bacterial Strains Isolated from a Deep Subsurface Environment. *Appl. Environ. Microbiol.* **1995**, *61*, 3400–3406.

Bottari, B.; Ercolini, D.; Gatti, M.; Neviani, E. Application of FISH Technology for Microbiological Analysis: Current State and Prospects. *Appl. Microbiol. Biotechnol.* **2006**, *73*, 485–494.

Branco, R.; Chung, A. P.; Veríssimo, A.; Morais, P. V. Impact of Chromium-Contaminated Wastewaters on the Microbial Community of a River. *FEMS Microbiol. Ecol.* **2005**, *54*, 35–46.

Carey, C. M.; Lee, H.; Trevors, J. T. Biology, Persistence and Detection of *Cryptosporidium parvum* and *Cryptosporidium hominis* Oocyst. *Water Res.* **2004**, *38*, 818–862.

Carpenter, S. R. Phosphorus Control Is Critical to Mitigating Eutrophication. *Proc. Natl. Acad. Sci.* **2008**, *105*, 11039–11040.Cerejeira, M. J.; Viana, P.; Batista, S.; Pereira, T.; Silva, E.; Valério, M. J.; Silva, A; Ferreira, M; Silva-Fernandes, A. M. Pesticides in Portuguese Surface and Ground Waters. *Water Res.* **2003**, *37*, 1055–1063.

Chen, L.; Wang, L. Y.; Liu, S. J.; Hu, J. Y.; He, Y.; Zhou, H. W.; Zhang, X. H. Profiling of Microbial Community During In Situ Remediation of Volatile Sulfide Compounds in River Sediment with Nitrate by High Throughput Sequencing. *Int. Biodeterioration and Biodegradation* **2013**, *85*, 429–437.

Conley, D. J.; Paerl, H. W.; Howarth, R. W.; Boesch, D. F.; Seitzinger, S. P.; Havens, K. E.; Lancelot, .; Likens, G. E. Controlling Eutrophication: Nitrogen and Phosphorus. *Science* **2009**, *323*, 1014–1015.

Costanza, R. The Value of Ecosystem Services. *Ecol. Econ.* **1998**, *25*, 1–2.

Cotner, J. B.; Biddanda, B. A. Small Players, Large Role: Microbial Influence on Biogeochemical Processes in Pelagic Aquatic Ecosystems. *Ecosystems* **2002**, *5*, 105–121.

Covich, A. P.; Austen, M. C.; Bärlocher, F.; Chauvet, E.; Cardinale, B. J.; Biles, C. L.; Inchausti, P.; Dangles, O.; Solan, M.; Gessner, M. O.; Statzner, B.; Moss, B. The Role of Biodiversity in the Functioning of Freshwater and Marine Benthic Ecosystems. *BioScience* **2004**, *54*, 767–775.

Cristobal, G.; Arbouet, L.; Sarrazin, F.; Talaga, D.; Bruneel, J. L.; Joanicot, M.; Servant, L. On-Line Laser Raman Spectroscopic Probing of Droplets Engineered in Microfluidic Devices. *Lab Chip* **2006**, *6*, 1140–1146.

Crocetti, G. R.; Hugenholtz, P.; Bond, P. L.; Schuler, A.; Keller, J.; Jenkins, D.; Blackall, L. L. Identification of Polyphosphate-Accumulating Organisms and Design of 16S rRNA-Directed Probes for Their Detection and Quantitation. *Appl. Environ. Microbiol.* **2000**, *66*, 1175–1182.

Crump, B. C.; Armbrust, E. V.; Baross, J. A. Phylogenetic Analysis of Particle-Attached and Free-Living Bacterial Communities in the Columbia River, Its Estuary, and the Adjacent Coastal Ocean. *Appl. Environ. Microbiol.* **1999**, *65*, 3192–3204.

Curtis, T. P.; Sloan, W. T.; Scannell, J. W. Estimating Prokaryotic Diversity and Its Limits. *Proc. Nat. Acad. Sci.* **2002**, *99*, 10494–10499.

Danielopol, D. L.; Griebler, C.; Gunatilaka, A.; Notenboom, J. Present State and Future Prospects for Groundwater Ecosystems. *Environ. Conserv.* **2003**, *30*, 104–130.

Danielopol, D. L.; Pospisil, P.; Rouch, R. Biodiversity in Groundwater: A Large-Scale View. *Trends Ecol. Evol.* **2000**, *15*, 223–224.

Dhar, R. K.; Zheng, Y.; Saltikov, C. W.; Radloff, K. A.; Mailloux, B. J.; Ahmed, K. M.; van Geen, A. Microbes Enhance Mobility of Arsenic in Pleistocene Aquifer Sand from Bangladesh. *Environ. Sci. Technol.* **2011**, *45*, 2648–2654.

Dionisi, H. M.; Layton, A. C.; Harms, G.; Gregory, I. R.; Robinson, K. G.; Sayler, G. S. Quantification of Nitrosomonas Oligotropha-Like Ammonia-Oxidizing Bacteria and *Nitrospira* spp. from Full-Scale Wastewater Treatment Plants by Competitive PCR. *Appl. Environ. Microbiol.* **2002**, *68*, 245–253.

Domenico, S.; Sinclair, L.; Ijaz, U. Z.; Parajka, J.; Reischer, G. H.; Stadler, P.; Blaschke, A. P.; Blöschl, G.; Mach, R. L.; Kirschner, A. K. T.; Farnleitner, A. H.; Eiler, A. Bacterial Diversity Along a 2600 km River Continuum. *Environ. Microbiol.* **2015**, *17*(12), 4994–5007.

Duckworth, A. W.; Grant, W. D.; Jones, B. E.; Steenbergen, R. V. Phylogenetic Diversity of Soda Lake Alkaliphiles. *FEMS Microbiol. Ecol.* **1996**, *19*, 181–191.

Dudgeon, D.; Arthington, A. H.; Gessner, M. O.; Kawabata, Z. I.; Knowler, D. J.; Lévêque, C.; Naiman, R. J.; Prieur-Richard, A. H.; Soto, D.; Stiassny, M. L.; Sullivan, C. A. Freshwater Biodiversity: Importance, Threats, Status and Conservation Challenges. *Biol. Rev.* **2006**, *81*, 163–182.

Duffy, J.; Stachowicz, J. J. Why Biodiversity is Important to Oceanography: Potential Roles of Genetic, Species, and Trophic Diversity in Pelagic Ecosystem Processes. *Mar. Ecol. Progr. Ser.* **2006**, *311*, 179–189.

Elderfield, H.; Upstill-Goddard, R.; Sholkovitz, E. R. The Rare Earth Elements in Rivers, Estuaries, and Coastal Seas and Their Significance to the Composition of Ocean Waters. *Geochim. Cosmochim. Acta* **1990**, *54*, 971–991.

Falkowski, P. G.; Fenchel, T.; Delong, E. F. The Microbial Engines That Drive Earth's Biogeochemical Cycles. *Science* **2008**, *320*, 1034–1039.

FAO. 2015. http://www.fao.org/docrep/t0551e/t0551e04.htm (accessed Nov 27, 2015).

Farkas, A.; Erratico, C.; Vigano, L. Assessment of the Environmental Significance of Heavy Metal Pollution in Surficial Sediments of the River Po. *Chemosphere* **2007**, *68*, 761–768.

Feris, K. P.; Ramsey, P. W.; Frazar, C.; Rillig, M.; Moore, J. N.; Gannon, J. E.; Holben, W. E. Seasonal Dynamics of Shallow-Hyporheic-Zone Microbial Community Structure Along a Heavy-Metal Contamination Gradient. *Appl. Environ. Microbiol.* **2004**, *70*, 2323–2331.

Field, C. B.; Behrenfeld, M. J.; Randerson, J. T.; Falkowski, P. Primary Production of the Biosphere: Integrating Terrestrial and Oceanic Components. *Science* **1998**, *281*, 237–240.

Gault, A. G.; Islam, F. S.; Polya, D. A.; Charnock, J. M.; Boothman, C.; Chatterjee, D.; Lloyd, J. R. Microcosm Depth Profiles of Arsenic Release in a Shallow Aquifer, West Bengal. *Mineral. Mag.* **2005**, *69*, 855–863.

Ghai, R.; Rodriguez-Valera, F.; McMahon, K. D.; Toyama, D.; Rinke, R.; de Oliveira, T. C. S.; … and Henrique-Silva, F. Metagenomics of the Water Column in the Pristine Upper Course of the Amazon River. *PLoS One* **2011**, *6*, e23785.

Girones, R.; Ferrús, M. A.; Alonso, J. L.; Rodriguez-Manzano, J.; Calgua, B.; Corrêa Ade, A.; Hundesa, A.; Carratala, A.; Bofill-Mas, S.Molecular Detection of Pathogens in Water—The Pros and Cons of Molecular Techniques. *Water Res.* **2010**, *44*, 4325–4339.

Glöckner, F. O.; Fuchs, B. M.; Amann, R. Bacterioplankton Compositions of Lakes and Oceans: A First Comparison Based on Fluorescence In Situ Hybridization. *Appl. Environ. Microbiol.* **1999**, *65*, 3721–3726.

Glöckner, F. O.; Zaichikov, E.; Belkova, N.; Denissova, L.; Pernthaler, J.; Pernthaler, A.; Amann, R. Comparative 16S rRNA Analysis of Lake Bacterioplankton Reveals Globally Distributed Phylogenetic Clusters Including an Abundant Group of Actinobacteria. *Appl. Environ. Microbiol.* **2000**, *66*(11), 5053–5065.

Goldstein, S. J.; Jacobsen, S. B. The Nd and Sr Isotopic Systematics of River-Water Dissolved Material: Implications for the Sources of Nd and Sr in Seawater. *Chem. Geol.: Isotope Geosci. Sect.* **1987**, *66*, 245–272.

Gough, H. L.; Stahl, D. A. Microbial Community Structures in Anoxic Freshwater Lake Sediment Along a Metal Contamination Gradient. *ISME J.* **2011**, *5*, 543–558.

Grant, W. D. *Alkaline Environments and Biodiversity. Extremophiles*; UNESCO, Eolss Publishers: Oxford, UK, 2006.

Grant, W. D.; Mwatha, W. E.; Jones, B. E. Alkaliphiles: Ecology, Diversity and Applications. *FEMS Microbiol. Rev.* **1990**, *75*, 255–270.

Ghiorse, W. C.; Wilson, J. T. Microbial Ecology of the Terrestrial Subsurface. *Adv. Appl. Microbiol.* **1988**, *33*, 107.

Hamonts, K.; Ryngaert, A.; Smidt, H.; Springael, D.; Dejonghe, W. Determinants of the Microbial Community Structure of Eutrophic, Hyporheic River Sediments Polluted with Chlorinated Aliphatic Hydrocarbons. *FEMS Microbiol. Ecol.* **2014**, *87*, 715–732.

He, S.; Gall, D. L.; McMahon, K. D. "Candidatus Accumulibacter" Population Structure in Enhanced Biological Phosphorus Removal Sludges as Revealed by Polyphosphate Kinase Genes. *Appl. Environ. Microbiol.* **2007**, *73*, 5865–5874.

Hemme, C. L.; Deng, Y.; Gentry, T. J.; Fields, M. W.; Wu, L.; Barua, S.; Barry, K.; Tringe, S. G.; Watson, D. B.; He, Z.; Hazen, T. C.; Tiedje, J. M.; Rubin, E. M.; Zhou,

J. Metagenomic Insights into Evolution of a Heavy Metal-Contaminated Groundwater Microbial Community. *ISME J.* **2010**, *4*, 660–672.

Henze, M.; Harremoës, P.; la Cour Jansen, J.; Arvin, E. Wastewater Treatment. In *Environmental Engineering*, 2nd ed; Förstner, U., Murphy, R. J., Rulkens, W. H., Eds.; Berlin: Springer, 1997.

Hoeft, S. E.; Blum, J. S.; Stolz, J. F.; Tabita, F. R.; Witte, B.; King, G. M.; Santini, J. M.; Oremland, R. S. *Alkalilimnicola ehrlichii* sp. nov., a Novel, Arsenite-Oxidizing Haloalkaliphilic Gammaproteobacterium Capable of Chemoautotrophic or Heterotrophic Growth with Nitrate or Oxygen as the Electron Acceptor. *Int. J. Syst. Evol. Microbiol.* **2007**, *57*, 504–512.

Hoeft, S. E.; Kulp, T. R.; Stolz, J. F.; Hollibaugh, J. T.; Oremland, R. S. Dissimilatory Arsenate Reduction with Sulfide as Electron Donor: Experiments with Mono Lake Water and Isolation of Strain MLMS-1, a Chemoautotrophic Arsenate Respirer. *Appl. Environ. Microbiol.* **2004**, *70*, 2741–2747.

Holmes, D. E.; Finneran, K. T.; O'neil, R. A.; Lovley, D. R. Enrichment of Members of the Family Geobacteraceae Associated with Stimulation of Dissimilatory Metal Reduction in Uranium-Contaminated Aquifer Sediments. *Appl. Environ. Microbiol.* **2002**, *68*, 2300–2306.

Humayoun, S. B.; Bano, N.; Hollibaugh, J. T. Depth Distribution of Microbial Diversity in Mono Lake, a Meromictic Soda Lake in California. *Appl. Environ. Microbiol.* **2003**, *69*, 1030–1042.

Jetten, M. S.; Niftrik, L. V.; Strous, M.; Kartal, B.; Keltjens, J. T.; Op den Camp, H. J. Biochemistry and Molecular Biology of Anammox Bacteria. *Crit. Rev. Biochem. Mol. Biol.* **2009**, *44*, 65–84.

Jiang, Z.; Li, P.; Wang, Y.; Li, B.; Deng, Y.; Wang, Y. Vertical Distribution of Bacterial Populations Associated with Arsenic Mobilization in Aquifer Sediments from the Hetao Plain, Inner Mongolia. *Environ. Earth Sci.* **2014**, *71*, 311–318.

Jones, B. E.; Grant, W. D.; Duckworth, A. W.; Owenson, G. G. Microbial Diversity of Soda Lakes. *Extremophiles* **1998**, *2*, 191–200.

Joshi, A. A.; Kanekar, P. P.; Kelkar, A. S.; Shouche, Y. S.; Vani, A. A.; Borgave, S. B.; Sarnaik, S. S. Cultivable Bacterial Diversity of Alkaline Lonar Lake, India. *Microbial Ecol.* **2008**, *55*, 163–172.

Kinegam, S.; Yingprasertchai, T.; Tanasupawat, S.; Leepipatpiboon, N.; Akaracharanya, A.; Kim, K. W. Isolation and Characterization of Arsenite-Oxidizing Bacteria from Arsenic-Contaminated Soils in Thailand. *World J. Microbiol. Biotechnol.* **2008**, *24*, 3091–3096.

Kolmakova, O. V.; Gladyshev, M. I.; Rozanov, A. S.; Peltek, S. E.; Trusova, M. Y. Spatial Biodiversity of Bacteria Along the Largest Arctic River Determined by Next-Generation Sequencing. *FEMS Microbiol. Ecol.* **2014**, *89*, 442–450.

Köster, W.; Egli, T.; Ashbolt, N.; Botzenhart, K.; Burlion, N.; Endo, T.; Grimont, P; Guillot, E; Mabilat, C; Newport, L; Niemi, M; Payment, P; Prescott, A; Renaud, P.; Rust, A. Analytical Methods for Microbiological Water Quality Testing. In *Assessing Microbial Safety of Drinking Water - Improving Approaches and Method*; Dufour, A., Snozzi, M., Köster, W., Bartram, J., Ronchi, E., Fewtrell, L., Eds.; OECD Publications: Paris, 2003; pp 237–295.

Kotelnikova, S.; Pedersen, K. Evidence for Methanogenic Archaea and Homoacetogenic Bacteria in Deep Granitic Rock Aquifers. *FEMS Microbiol. Rev.* **1997**, *20*, 339–349.

Kowalchuk, G. A.; Stephen, J. R. Ammonia-Oxidizing Bacteria: A Model for Molecular Microbial Ecology. *Annu. Rev. Microbiol.* **2001**, *55*, 485–529.

Kristiansson, E.; Fick, J.; Janzon, A.; Grabic, R.; Rutgersson, C.; Weijdegård, B.; Söderström, H.; Larsson, D. G. Pyrosequencing of Antibiotic-Contaminated River Sediments Reveals High Levels of Resistance and Gene Transfer Elements. *PLoS One* **2011**, *6*(2), e17038.

Kumar, S.; Krishnani, K. K.; Bhushan, B.; Brahmane, M. P. Metagenomics: Retrospect and Prospects in High Throughput Age. *Biotechnol. Res. Int.* **2015**, *2015*, 1–13.

Kundu, N.; Panigrahi, M.; Tripathy, S.; Munshi, S.; Powell, M. A.; Hart, B. Geochemical Appraisal of Fluoride Contamination of Groundwater in the Nayagarh District of Orissa, India. *Environ. Geol.* **2001**, *41*, 451–460.

Lafita, C.; Penya-Roja, J. M.; Sempere, F.; Waalkens, A.; Gabaldón, C. Hydrogen Sulfide and Odor Removal by Field-Scale Biotrickling Filters: Influence of Seasonal Variations of Load and Temperature. *J. Environ. Sci. Health A* **2012**, *47*, 970–978.

Lee, S.; Malone, C.; Kemp, P. F. Use of Multiple 16S Ribosomal-RNA-Targeted Fluorescent-Probes to Increase Signal Strength and Measure Cellular RNA from Natural Planktonic Bacteria. *Mar. Ecol. Progr. Ser.* **1993**, *101*, 193–201.

Lin, J. L.; Joye, S. B.; Scholten, J. C.; Schäfer, H.; McDonald, I. R.; Murrell, J. C. Analysis of Methane Monooxygenase Genes in Mono Lake Suggests That Increased Methane Oxidation Activity May Correlate with a Change in Methanotroph Community Structure. *Appl. Environ. Microbiol.* **2005**, *71*, 6458–6462.

Lin, J. L.; Radajewski, S.; Eshinimaev, B. T.; Trotsenko, Y. A.; McDonald, I. R.; Murrell, J. C. Molecular Diversity of Methanotrophs in Transbaikal Soda Lake Sediments and Identification of Potentially Active Populations by Stable Isotope Probing. *Environ. Microbiol.* **2004**, *6*, 1049–1060.

Linhoff, B. S.; Bennett, P. C.; Puntsag, T.; Gerel, O. Geochemical Evolution of Uraniferous Soda Lakes in Eastern Mongolia. *Environ. Earth Sci.* **2011**, *62*, 171–183.

Liu, Y. J.; Zhang, C. M.; Wang, X. C. Simultaneous Detection of Enteric Bacteria from Surface Waters by QPCR in Comparison with Conventional Bacterial Indicators. *Environ. Monit. Assess.* **2009**, *158*, 535–544.

Madsen, E. L.; Ghiorse, W. C. Groundwater Microbiology: Subsurface Ecosystem Processes. *Aquatic Microbiology: An Ecological Approach*; Blackwell, Cambridge, MA, 1993, 167–213.

Mason, A.; Korostynska, O.; Al-Shamma'a, A. I. Microwave Sensors for Real-Time Nutrients Detection in Water. In *Smart Sensors for Real-Time Water Quality Monitoring*; Springer Berlin Heidelberg, 2013, pp. 197–216.

McKinley, V. L.; Vestal, J. R. Effects of Different Temperature Regimes on Microbial Activity and Biomass in Composting Municipal Sewage Sludge. *Can. J. Microbiol.* **1985**, *31*, 919–925.

Medema, G. J.; Schets, F. M.; Ketelaars, H. A. M.; Boschman, G. Improved Detection and Vital Staining of Cryptosporidium and Giardia with Flow Cytometry. *Water Sci. Technol.* **1998**, *38*(12), 61–66.

Melack, J. M.; Kilham, P. Photosynthetic Rates of Phytoplankton in East African Alkaline, Saline Lakes1. *Limnol. Oceanogr.* **1974**, *19*(5), 743–755.

Methé, B. A.; Hiorns, W. D.; Zehr, J. P. Contrasts Between Marine and Freshwater Bacterial Community Composition: Analyses of Communities in Lake George and Six Other Adirondack Lakes. *Limnol. Oceanogr.* **1998**, *43*, 368–374.

Minogue, E.; Tuite, N. L.; Smith, C. J.; Reddington, K.; Barry, T. A Rapid Culture Independent Methodology to Quantitatively Detect and Identify Common Human Bacterial Pathogens Associated with Contaminated High Purity Water. *BMC Biotechnol.* **2015**, *15*, 6.

Miyoshi, T.; Iwatsuki, T.; Naganuma, T. Phylogenetic Characterization of 16S rRNA Gene Clones from Deep-Groundwater Microorganisms That Pass Through 0.2-Micrometer-Pore-Size Filters. *Appl. Environ. Microbiol.* **2005**, *71*, 1084–1088.

Mohamed, F.; Kim, J.; Huang, R.; Nu, H. T.; Lorenzo, V. Efficient Control of Odors and VOC Emissions via Activated Carbon Technology. *Water Environ. Res.* **2014**, *86*, 594–605.

Mueller-Spitz, S. R.; Goetz, G. W.; McLellan, S. L. Temporal and Spatial Variability in Nearshore Bacterioplankton Communities of Lake Michigan. *FEMS Microbiol. Ecol.* **2009**, *67*, 511–522.

Nakasaki, K.; Sasaki, M.; Shoda, M.; Kubota, H. Change in Microbial Numbers During Thermophilic Composting of Sewage Sludge with Reference to CO2 Evolution Rate. *Appl. Environ. Microbiol.* **1985**, *49*, 37–41.

Newton, R. J.; Jones, S. E.; Eiler, A.; McMahon, K. D.; Bertilsson, S. A Guide to the Natural History of Freshwater Lake Bacteria. *Microbiol. Mol. Biol. Rev.* **2011**, *75*, 14–49.

Nielsen, A. H.; Hvitved-Jacobsen, T.; Vollertsen, J. Effects of pH and Iron Concentrations on Sulfide Precipitation in Wastewater Collection Systems. *Water Environ. Res.* **2008**, *80*, 380–384.

Noble, R.; Weisberg, R. A Review of Technologies for Rapid Detection of Bacteria in Recreational Waters. *J. Water Health* **2005**, *3*, 381–392.

Nogales, B.; Lanfranconi, M. P.; Piña-Villalonga, J. M.; Bosch, R. Anthropogenic Perturbations in Marine Microbial Communities. *FEMS Microbiol. Rev.* **2011**, *35*, 275–298.

Nouri, J.; Mahvi, A. H.; Jahed, G. R.; Babaei, A. A. Regional Distribution Pattern of Groundwater Heavy Metals Resulting from Agricultural Activities. *Environ. Geol.* **2008**, *55*, 1337–1343.Orcutt, B. N.; Bach, W.; Becker, K.; Fisher, A. T.; Hentscher, M.; Toner, B. M.; Wheat, C. G.; Edwards, K. J. Colonization of Subsurface Microbial Observatories Deployed in Young Ocean Crust. *ISME J.* **2011**, *5*, 692–703.

Oremland, R. S.; Kulp, T. R.; Blum, J. S.; Hoeft, S. E.; Baesman, S.; Miller, L. G.; Stolz, J. F. A Microbial Arsenic Cycle in a Salt-Saturated, Extreme Environment. *Science* **2005**, *308*, 1305–1308.

Paul, D.; Kazy, S. K.; Gupta, A. K.; Pal, T.; Sar, P. Diversity, Metabolic Properties and Arsenic Mobilization Potential of Indigenous Bacteria in Arsenic Contaminated Groundwater of West Bengal, India. *PLoS One* **2015**, *10*(3), 1–40.

Paul, D.; Poddar, S.; Sar, P. Characterization of Arsenite-Oxidizing Bacteria Isolated from Arsenic-Contaminated Groundwater of West Bengal. *J. Environ. Sci. Health, A* **2014**, *49*, 1481–1492.

Pernthaler, J. Predation on Prokaryotes in the Water Column and Its Ecological Implications. *Nat. Rev. Microbiol.* **2005**, *3*, 537–546.

Pernthaler, J. Freshwater Microbial Communities. In *The Prokaryotes*; Springer Berlin Heidelberg, 2013, 97–112.

Petrie, L.; North, N. N.; Dollhopf, S. L.; Balkwill, D. L.; Kostka, J. E. Enumeration and Characterization of Iron (III)-Reducing Microbial Communities from Acidic Subsurface Sediments Contaminated with Uranium (VI). *Appl. Environ. Microbiol.* **2003**, *69*, 7467–7479.

Pomeroy, J. W.; Gray, D. M.; Brown, T.; Hedstrom, N. R.; Quinton, W. L.; Granger, R. J.; Carey, S. K. The Cold Regions Hydrological Model: A Platform for Basing Process Representation and Model Structure on Physical Evidence. *Hydrol. Process.* **2007**, *21*, 2650–2667.

Power, J.; Schepers, J. S. Nitrate Contamination of Groundwater in North America. *Agric. Ecosyst. Environ.* **1989**, *26*, 165–187.

Price, R. E.; Lesniewski, R.; Nitzsche, K. S.; Meyerdierks, A.; Saltikov, C.; Pichler, T.; Amend, J. P. Archaeal and Bacterial Diversity in an Arsenic-Rich Shallow-Sea Hydrothermal System Undergoing Phase Separation. *Front. Microbiol.* **2013**, *4*.

Proia, L.; Cassió, F.; Pascoal, C.; Tlili, A.; Romaní, A. M. The Use of Attached Microbial Communities to Assess Ecological Risks of Pollutants in River Ecosystems: The Role of Heterotrophs. In *Emerging and Priority Pollutants in Rivers*; Springer Berlin Heidelberg, 2012, pp. 55–83.

Punmia et al. Wastewater Characteristics. In: *Waste Water Engineering*; Punmia, B. C., Jain, A. K., Jain, A. K., Eds.; Firewall Media–Laxmi Publications: New Delhi, India, 1998; pp 180–217.

Ramalingam, B.; Sekar, R.; Boxall, J. B.; Biggs, C. Aggregation and Biofilm Formation of Bacteria Isolated From Domestic Drinking Water. *Water Sci. Technol. Water Supply* **2013**, *13*, 1016–1023.

Rascovan, N.; Maldonado, J.; Vazquez, M. P.; Farías, M. E. Metagenomic Study of Red Biofilms from Diamante Lake Reveals Ancient Arsenic Bioenergetics in *Haloarchaea*. *ISME J.* **2016**, *10*(2), 299–309.

Read, D. S.; Gweon, H. S.; Bowes, M. J.; Newbold, L. K.; Field, D.; Bailey, M. J.; Griffiths, R. I. Catchment-Scale Biogeography of Riverine Bacterioplankton. *ISME J.* **2015**, *9*, 516–526.

Reeder, S. W.; Hitchon, B.; Levinson, A. A. Hydrogeochemistry of the Surface Waters of the Mackenzie River Drainage Basin, Canada-I. Factors Controlling Inorganic Composition. *Geochim. Cosmochim. Acta* **1972**, *36*, 825–865.

Reza, R.; Singh, G. Heavy Metal Contamination and Its Indexing Approach for River Water. *Int. J. Environ. Sci. Technol.* **2010**, *7*, 785–792.

Rusch, D. B.; Halpern, A. L.; Sutton, G.; Heidelberg, K. B.; Williamson, S.; Yooseph, S.; Wu, D.; Eisen, J. A.; Hoffman, J. M.; Remington, K.; Beeson, K.; Tran, B.; Smith, H.; Baden-Tillson, H.; Stewart, C.; Thorpe, J.; Freeman, J.; Andrews-Pfannkoch, C.; Venter, J. E.; Li, K.; Kravitz, S.; Heidelberg, J. F.; Utterback, T.; Rogers, Y.-H.; Falcón, L. I.; Souza, V.; Bonilla-Rosso, G.; EguiarteL. E.; Karl, D. M.; Sathyendranath, S.; Platt, T.; Bermingham, E.; Gallardo, V.; Tamayo-Castillo, G.; Ferrari, M. R.; Strausberg, R. L.; Nealson, K.; Friedman, R.; Frazier, M.; Venter, J. C. The Sorcerer II Global Ocean Sampling Expedition: Northwest Atlantic Through Eastern Tropical Pacific. *PLoS Biol.* **2007**, *5*(3), 0398–0431.Sarkar, A.; Kazy, S. K.; Sar, P. Characterization of Arsenic Resistant Bacteria from Arsenic Rich Groundwater of West Bengal, India. *Ecotoxicology* **2013**, *22*, 363–376.

Sauvain, L.; Bueche, M.; Junier, T.; Masson, M.; Wunderlin, T.; Kohler-Milleret, R.; Diez, E. G.; Loizeau, J.-L.; Tercier-Waeber, M.-L.; Junier, P. Bacterial Communities in Trace Metal Contaminated Lake Sediments Are Dominated by Endospore-Forming Bacteria. *Aquat. Sci.* **2014**, *76*, 33–46.Savio, D.; Sinclair, L.; Ijaz, U. Z.; Parajka, J.; Reischer, G. H.; Stadler, P.; Blaschke, A. P.; Blöschl, G.; Mach, R. L.; Kirschner, A. K.; Farnleitner, A.

H.; Eiler, A. Bacterial Diversity Along a 2600 km River Continuum. *Environ. Microbiol.* **2015**, *17*(12), 4994–5007. Shmueli, D. F. Water Quality in International River Basins. *Polit. Geogr.* **1999**, *18(4), 437–476.*

Shon, H. K.; Vigneswaran, S.; Kandasamy, J.; Cho, J. *Characteristics of Effluent Organic Matter in Wastewater*; Faculty of Engineering, University of Technology, Sydney, Australia, 2007, pp 52–101.

Sigee, D. C. *Freshwater Microbiology: Biodiversity and Dynamic Interactions of Microorganisms in the Aquatic Environment*; John Wiley and Sons, Ltd., 2005. DOI: 10.1002/0470011254.fmatter.

Silveira, M. P.; Buss, D. F.; Nessimian, J. L.; Baptista, D. F. Spatial and Temporal Distribution of Benthic Macroinvertebrates in a Southeastern Brazilian River. *Braz. J. Biol.* **2006**, *66*, 623–632.

Simeonov, V.; Stratis, J. A.; Samara, C.; Zachariadis, G.; Voutsa, D.; Anthemidis, A.; Sofoniou, M.; Kouimtzis, T. Assessment of the Surface Water Quality in Northern Greece. *Water Res.* **2003**, *37*, 4119–4124.

Singh, A. K.; Mondal, G. C.; Kumar, S.; Singh, T. B.; Tewary, B. K.; Sinha, A. Major Ion Chemistry, Weathering Processes and Water Quality Assessment in Upper Catchment of Damodar River Basin, India. *Environ. Geol.* **2008**, *54*, 745–758.

Singh, D. D.; John, S. Study the Different Parameters of Sewage Treatment with UASB and SBR Technologies. *IOSR J. Mech. Civil Eng. (IOSR-JMCE)*, 6(1), 112–116.

Šlapeta, J.; Moreira, D.; López-García, P. The Extent of Protist Diversity: Insights from Molecular Ecology of Freshwater Eukaryotes. *Proc. R. Soc. Lond. B: Biol. Sci.* **2005**, *272*(1576), 2073–2081.

Sorokin, D. Y.; Foti, M.; Pinkart, H. C.; Muyzer, G. Sulfur-Oxidizing Bacteria in Soap Lake (Washington State), a Meromictic, Haloalkaline Lake with an Unprecedented High Sulfide Content. *Appl. Environ. Microbiol.* **2007**, *73*, 451–455.

Sorokin, D. Y.; Kuenen, J. G.; Muyzer, G. The Microbial Sulfur Cycle at Extremely Haloalkaline Conditions of Soda Lakes. *Front. Microbiol.* **2011**, *2*, 1–16.

Staley, Z. R.; Chase, E.; Mitraki, C.; Crisman, T. L.; Harwood, V. J. Microbial Water Quality in Freshwater Lakes with Different Land Use. *J. Appl. Microbiol.* **2013**, *115*, 1240–1250.

Stallard, R. F.; Edmond, J. M. Geochemistry of the Amazon: 2. The influence of Geology and Weathering Environment on the Dissolved Load. *J. Geophys. Res. Oceans (1978–2012)* **1983**, *88*, 9671–9688.

Stevens, T. O.; McKinley, J. P. Lithoautotrophic Microbial Ecosystems in Deep Basalt Aquifers. *Science* **1995**, *270*, 450.

Straub, T. M.; Daly, D. S.; Wunshel, S.; Rochelle, P. A.; DeLeon, R.; Chandler, D. P. Genotyping Cryptosporidium Parvum with an hsp70 Single-Nucleotide Polymorphism Microarray. *Appl. Environ. Microbiol.* **2002**, *68*, 1817–1826.

Sutton, N. B.; van der Kraan, G. M.; van Loosdrecht, M. C.; Muyzer, G.; Bruining, J.; Schotting, R. J. Characterization of Geochemical Constituents and Bacterial Populations Associated with As Mobilization in Deep and Shallow Tube Wells in Bangladesh. *Water Res.* **2009**, *43*, 1720–1730.

Ultee, A.; Souvatzi, N.; Maniadi, K.; König, H. Identification of the Culturable and Nonculturable Bacterial Population in Ground Water of a Municipal Water Supply in Germany. *J. Appl. Microbiol.* **2004**, *96*, 560–568.Van der Gucht, K.; Vandekerckhove, T.; Vloemans,

N.; Cousin, S.; Muylaert, K.; Sabbe, K.; Gillis, M; Declerk, S; De Meester, L; Vyverman, W. Characterization of Bacterial Communities in Four Freshwater Lakes Differing in Nutrient Load and Food Web Structure. *FEMS Microbiol. Ecol.* **2005**, *53*, 205–220.

Venter, J. C.; Remington, K.; Heidelberg, J. F.; Halpern, A. L.; Rusch, D.; Eisen, J. A.; Wu, D; Paulsen, I; Nelson, K. E; Nelson, W; Fouts, D. E; Levy, S; Knap, A. H; Lomas, M. W; Nealson, K; White, O; Peterson, J; Hoffman, J; Parsons, R; Baden-Tillson, H,' Pfannkoch, C; Rogers, Y. H; Smith, H. O. Environmental Genome Shotgun Sequencing of the Sargasso Sea. *Science* **2004**, *304*, 66–74.

Wagner, M.; Loy, A. Bacterial Community Composition and Function in Sewage Treatment Systems. *Curr. Opin. Biotechnol.* **2002**, *13*, 218–227.

Wang, X.; Jin, P.; Zhao, H.; Meng, L. Classification of Contaminants and Treatability Evaluation of Domestic Wastewater. *Front. Environ. Sci. Eng. China* **2007**, *1*, 57–62.

Watanabe, K.; Watanabe, K.; Kodama, Y.; Syutsubo, K.; Harayama, S. Molecular Characterization of Bacterial Populations in Petroleum-Contaminated Groundwater Discharged from Underground Crude Oil Storage Cavities. *Appl. Environ. Microbiol.* **2000,** *66*, 4803–4809.

Wéry, N.; Lhoutellier, C.; Ducray, F.; Delgenès, J. P.; Godon, J. J. Behaviour of Pathogenic and Indicator Bacteria During Urban Wastewater Treatment and Sludge Composting, As Revealed By Quantitative PCR. *Water Res.* **2008**, *42*, 53–62.

Whitman, W. B.; Coleman, D. C.; Wiebe, W. J. Prokaryotes: The Unseen Majority. *Proc. Natl. Acad. Sci.* **1998**, *95*, 6578–6583.

WHO (World Health Organization). *Guidelines for Drinking-water Quality*; 4th Ed., WHO Press, Geneva, Switzerland, 2011. URL: http://apps.who.int/iris/bitst ream/10665/44584/1/9789241548151_eng.pdf (accessed Jan 16, 2016).

Widmeyer, J. R.; Bendell-Young, L. I. Heavy Metal Levels in Suspended Sediments, *Crassostrea gigas*, and the Risk to Humans. *Arch. Environ. Contaminat. Toxicol.* **2008**, *55*, 442–450.

Wilhelm, S. W.; Matteson, A. R. Freshwater and Marine Virioplankton: A Brief Overview of Commonalities and Differences. *Freshw. Biol.* **2008**, *53*, 1076–1089.

Worm, B.; Barbier, E. B.; Beaumont, N.; Duffy, J. E.; Folke, C.; Halpern, B. S.; Jackson, J. B. C.; Lotze, H. K.; Micheli, F.; Palumbi, S. R.; Sala, E.; Selkoe, K. A.; Stachowicz, J. J.; Watson, R. Impacts of Biodiversity Loss on Ocean Ecosystem Services. *Science* **2006**, *314*, 787–790.

Wu, L.; Loo, Y.; Koe, L. A Pilot Study of a Biotrickling Filter for the Treatment of Odorous Sewage Air. *Water Sci. Technol.* **2001**, *44*, 295–299.

Ye, L.; Zhang, T. Bacterial Communities in Different Sections of a Municipal Wastewater Treatment Plant Revealed by 16S rDNA 454 Pyrosequencing. *Appl. Microbiol. Biotechnol.* **2013**, *97*, 2681–2690.

Zhang, Y.; Anninos, P.; Norman, M. L. A Multispecies Model for Hydrogen and Helium Absorbers in Lyman-Alpha Forest Clouds. *Astrophysical J. Lett.* **1995**, *453*, L57.

Zlatkin, I. V.; Schneider, M.; De Bruijn, F. J.; Forney, L. J. Diversity Among Bacteria Isolated from the Deep Subsurface. *J. Ind. Microbiol.* **1996**, *17*, 219–227.

Zwart, G.; Crump, B. C.; Kamst-van Agterveld, M. P.; Hagen, F.; Han, S. K. Typical Freshwater Bacteria: An Analysis of Available 16S rRNA Gene Sequences from Plankton of Lakes and Rivers. *Aquatic Microbial Ecol.* **2002**, *28*.

Zwart, G.; van Hannen, E. J.; Kamst-van Agterveld, M. P.; Van der Gucht, K.; Lindström, E. S.; Van Wichelen, J.; Lauridsen, T.; Crump, B. C.; Han, S.-K.; Declerck, S.. Rapid Screening for Freshwater Bacterial Groups by Using Reverse Line Blot Hybridization. *Appl. Environ. Microbiol.* **2003,** *69,* 5875–5883.

CHAPTER 8

BIOREMEDIATION OF PESTICIDES: AN ECO-FRIENDLY APPROACH FOR A CLEAN ENVIRONMENT

MANJULA P. PATIL[1], ARPANA H. JOBANPUTRA[1*],
DEEPAK KUMAR VERMA[2], SHIKHA SRIVASTAVA[3],
and ANIL KUMAR DWIVEDI[4]

[1]Department of Microbiology, PSGVPM's, SIP Arts, GBP Science & STSKVS Commerce College, District Nandurbar, Shahada 425409, Maharashtra, India

[2]Agricultural and Food Engineering Department, Indian Institute of Technology, Kharagpur 721302, West Bengal, India

[3]Department of Botany, Deen Dayal Upadhyay Gorakhpur University, Gorakhpur 273009, Uttar Pradesh, India

[4]Pollution and Environmental Assay Research Laboratory (PEARL), Department of Botany, Deen Dayal Upadhyay Gorakhpur University, Gorakhpur 273009, Uttar Pradesh, India

*Corresponding author. E-mail: arpana_j12@rediffmail.com

8.1 INTRODUCTION

A very hopeful statement had been written by one author while commencing a book write up "The history of man is the record of a hungry creature in search of food." India is an agriculture-based country and more than 60–70% of Indian population is dependent on agriculture as profession. By 2020, its population will cross 1.3 billion, and feeding and clothing this population is a very big challenge confronting us. Demand for agriculture-based food production and the progressive increase of pest problem necessitated the introduction of use of agrochemicals which ensure high yield and good quality crops (Graebing et al., 2002). Therefore, a tremendous pressure has been imposed on the annual food grain production and minimizing crop losses due to pest, plant diseases, and draught problems. Insects are responsible for heavy losses of commercial crops in both ways, quantitatively as well as qualitatively. In order to gain maximum yields and to fight against pest problems, the farmers resort to heavy use of pesticides. So in recent era, plant protection has gained due significance in crop production. Due to losses in crop production, it is a basic need to develop methods of crop production which include use of high yielding varieties (HYVs), changing cropping pattern, application of improved and high doses of fertilizers, water irrigation facilities, etc. Out of these, application of pesticides for pest control is more significant and leading to tremendous use of pesticides. Correlated to this context, it can be stated that pesticides are linked to pollutions affecting land, air, and water. Soil and water pollutions are more hazardous for health as persistent pesticides enter the food chain by these two main routes.

8.2   ENVIRONMENTAL POLLUTION

Day by day our life is becoming more and more comfortable, can we think how it could be possible? A simple and single line answer is there that it is only due to our inventions for getting such comforts. All the things created by us have two sides, upside has comfort and sophistication and downside has pollution and contamination due to incomplete knowledge and improper ways of application and disposal of such chemicals. This downside leads to adverse health effects and bad impacts on the human beings and environment as well as on the soil, sediments, ground water, and mainly food chain (Valentin et al., 2013).

Today, whole world is facing many problems; environmental problems are one among them in which contamination of soil, water, and air by all means of toxic chemicals is a major concern (Nyakundi et al., 2011). Such xenobiotic compounds like pesticides, solvents, fire retardants, pharmaceuticals, and lubricants are widely distributed in the environment due to their widespread use. Agricultural farms are producing about 80 billion pounds of hazardous organopollutants per year, and out of these, only 10% have been disposed of safely (Reddy and Mathew, 2001).

Although, several hazardous chemicals with their persistence and toxicity are present in our surrounding which cause very serious issues on environmental pollution and human health (Rossberg et al., 1986). Therefore, a contaminated site can be defined as "a site contaminated with harmful substances caused by man in or under the site to such extent, that pose the potential to cause a significant hazardous risk to the health or the environment in such a way that requires necessary action to manage the risk" (Van-Camp et al., 2004). Such harmful substances are presented in discarded materials so-called waste or wastes, generally comes from different sources after primary use by man. They are completely worthless, defective and of no longer use. Waste is grouped into different four categories depicted in Table 8.1 (Setyorini et al., 2002), they are (1) agricultural, (2) municipal, (3) industrial, and (4) nuclear. Some of the agricultural wastes are very valuable and beneficial for soil health if they are returned (plant residues, poultry wastes, slaughterhouses waste, piggery waste, dairy wastes, and livestock manure are examples of such waste) (Setyorini et al., 2002). However, improper handling and disposal of such waste may cause environmental pollution and health issues (Setyorini et al., 2002). The industrials waste produced by food-processing plants can

TABLE 8.1 Different Categories of Waste.

S. N.	Categories of waste	Remark	Example	References
1.	Agricultural	Result of agricultural practices	Organic materials (like farms wastes, fertilizer run-off, pesticides residue), animal wastes (like poultry wastes, slaughterhouses waste, piggery waste, dairy wastes, livestock manure), timber by-products (like harvest waste)	Setyorini et al. (2002)
2.	Municipal	Households and commercial establishments, and generally collected by local government	Food and kitchen waste, Green waste, paper, etc.	Sharholy et al. (2008)
3.	Industrial	Produced by industries which may be the products in gaseous, liquid or solid form	Carbon monoxide (CO), carbon dioxide (CO_2), nitrogen dioxide (NO_2), sulfur dioxide (SO_2)	Setyorini et al. (2002)
4.	Nuclear	Produced from nuclear reactors, generally known as radioactive garbage	Products of radioactive fission (uranium-234, neptunium-237, plutonium-238 and americium-241), transuranic elements	Robert (1999)

either be liquid or solid or both. Another is urban waste; they are municipal garbage made up of materials discarded by homes and industry which contains paper, plastic, and organic materials. Excess heavy metals in the soil originate from atmospheric deposition, improper stacking of the industrial solid waste, mining activities, sewage irrigation, and the use of pesticides and fertilizers (Zhang et al., 2011).

8.3 POLLUTION DUE TO PESTICIDES

Modern agricultural practices use various agrochemicals like pesticides, herbicides, insecticides, fungicides, etc. Such increased market demand for agricultural products leads to agricultural modernization which has been greatly facilitated by the industrial production and use of pesticides for vector control and pest management leading to alarming and diverse environmental contamination with pesticides (Ngowi et al., 2007). Pesticides are very large and varied group of substances produced for variety of purposes like agricultural, domestic, and industrial, but globally, they can be used extensively for controlling biological organisms like insects, rodents, and weeds. They are classified on the basis of their mode of action toward target pest which they kill. Distributions of pesticides on target and nontarget areas are depicted in Figure 8.1 (Gavrilescu, 2005). However, it is very necessary to monitor and control pesticide usage regularly as there will be high risk to environment, human beings, and animals as well as to plant health. Rajendran (2003) also stated that excessive and indiscriminate use of synthetic pesticides can damage the environment and agriculture and affect health and development by entering the food chain. Singh (2008) reported that in terrestrial ecosystem, there are three possible regions where pesticide contaminations can occur; surface soil, vadose zone (unsaturated zone) and ground water (saturated zone). Biodegradation process in these three regions is variable. In surface region, biodegradation takes place by aerobic mode and rapid due to presence of large number of aerobic microflora; in contrast, biodegradation is slow in another two regions vadose zone and ground water because microflora decreases with the depth. These are applied in various ways, directly on seed or by foliage which ultimately reaches to the soil and then to the water bodies. These pesticides accumulate in soil, water, and food, ultimately creating ecological stress. These applied pesticides persist in the environment through the

FIGURE 8.1 **(See color insert.)** Distribution of pesticides on target and nontarget areas. (*Source*: Adapted from Gavrilescu, 2005)

rainfall infiltration process which may be bioaccumulated or biomagnified by the biota causing air, water, and land contamination that may be toxic, mutagenic, or carcinogenic (Fan and Liao, 2009; Jilani, 2013).

The main source of environmental contamination is using of more pesticides for agricultural activities to increase crop yield, and the other one is agricultural industries that have no treatment facilities or have a grossly inadequate arrangement (Jilani, 2013), and this contamination becomes even more serious when persistent organic compounds are involved (Fan and Liao, 2009). The pesticides use is very serious in keeping the farmers' investment in fertilizer, labor, and seeds since they afford ensured cover from pest's damage. Therefore, their use is expected and the pollution due to the presence of pesticides and their residues in our environment will continue to be a bigger and serious challenge (Omolo et al., 2012). Due to all these aspects of terrestrial ecosystem, different bioremediation technologies need to evolve to remediate the different regions (Singh, 2008) and thereby removal of pesticides from environmental systems needs special focus and attention.

8.3.1 SOIL CONTAMINATION

We consume water and air on daily basis; that is why, quality of these natural resources is of immediate concern. Likewise, soil is a thin layer of unconsolidated material on bedrock. It is like the biogeochemical engine by which the Mother of earth gets support for their life, and in results, food, fodder, fiber, and fuel are received by us for survival of life (Valentin et al., 2013). It delivers ecosystem services like recycling of carbon and essential nutrients of all living materials, biological control of pests, regulation of the atmosphere, and filtering and storage of water which cannot be easily traded in markets (Haygarth and Ritz, 2009; Robinson et al., 2012). Soil is not an abiotic surface. It is compact, three-phase system composed of water, variety of solids, dissolved substances, and gases (Or et al., 2007). Many kinds of anthropogenic xenobiotics are produced at the beginning of industrialization for unlimited uses, and after their purpose has been served, they are often discarded in the environment. They often end up in the environment where soil is their final destination and where they persist for prolonged periods (Valentin et al., 2013). Natural soil capital is a large nonrenewable resource that has been threatened by continued urbanization, short-sighted agricultural practices, and desertification intensified by global change (Banwart, 2011) and loss due to erosion, sandstorm, and/or are severely degraded by persistent anthropogenic pollutants (Valentin et al., 2013). Currently, soil protection identifies the major threats to soils like salinization, erosion, landslides, organic matter decline, contamination, compaction, and sealing (Commission of the European Communities (CEC), 2006).

Soil contamination is always related to subsurface and ground water contamination which can be contaminated by point source and diffuse contamination. Point source includes improper waste disposal, industrial discharges, or accidental spills during handling or transportation of hazardous substances, and the latter is related to certain forestry and agricultural practices, transportations, improper waste and wastewater management, and atmospheric deposition of volatile compounds. Detrimental effects of pollutants on the soil microorganisms which may be directly related to loss of their biodiversity and functions like nutrients recycling (McDonald, 2016). However, this direct negative effect on soil microbes is still subjected to debate among microbiologist who study their ecology because the microorganism's communities may be surprisingly resilient from all those their activities in which recovery of contamination effects are involved (Valentin et al., 2013).

8.3.2 WATER CONTAMINATION

Major sources of water pollution due to pesticides include agricultural, domestic, and industrial waste. Nearly 98% of the pesticides imported had been classified as acutely toxic for fish and crustaceans and 73% for amphibians leads to eutrophication (Rani and Dhania, 2014). Mainly sewage disposal into rivers is the biggest pollution problem of fresh water; it can result in drop in amount of dissolved oxygen and reduction in biological oxygen demand which leads to reduction in flora and fauna of water ecosystem by suffocation, and in the developing countries, contamination of drinking water by untreated sewage also leads to approximately 14,000 death per day (Rani and Dhania, 2014). Discharge of industrial hot water from cooling engines is another source of water pollution which can decrease metabolic rate of organisms (Rani and Dhania, 2014). Another major source of water pollution is due to pesticides, while heavy rain fall excess fertilizers and pesticides are getting washed and introduced directly into water bodies which cause more serious threats to us.

8.4 BENEFITS AND HAZARDS OF PESTICIDES

8.4.1 BENEFITS OF PESTICIDES

There are several benefits of pesticides to human to protect from some environmental enemies and good lifestyle. The benefits of pesticides as reported by United States Environmental Protection Agency below are summarized as follows: (1) pesticides are the only effective means of controlling weeds, disease organisms or insect pests in many ways, (2) humans gain direct benefits from pesticides through wider selections and lower prices for food and clothing, (3) structural damage due to termite infestation of private, public, and commercial dwellings can be protected by pesticides, (4) human health can be protected due to great contribution of pesticide in preventing diseases outbreaks by controlling rodent and insect populations, (5) our drinking and recreational water has been sanitized by pesticides, (6) our indoor areas such as kitchens, nursing homes, operating rooms, etc. as well as dental and surgical instruments are disinfected by the use of pesticides, and (7) the pesticide industry also providing benefits to society by providing jobs to manufacturers, distributors, dealers, and farmers. In respect of the benefits of pesticides, refer Figure 8.2 for better understanding.

8.4.2 HAZARDS OF PESTICIDES

A report of Maksymiv (2015) stated that there are several hazards of pesticides which are summarized as follows:

1. Serious issue concerned with the environments including human and animal health hazards are raised because of the pesticides use.
2. Various diseases (such as neurological, psychological and behavioral dysfunctions; hormonal imbalances, leading to infertility, breast pain; immune system dysfunction; reproductive system defects; cancers; genotoxicity; blood disorders) are the result due to continue exposure of pesticides for a long period of time.
3. Pure air, soil, water, and other vegetation are contaminated by pesticides.
4. Pesticides can kill insects or weeds as well as may be toxic to a host of other organisms like birds, beneficial insects, fish, and nontarget plants, etc.
5. All kind of waters like surface and ground are contaminated by pesticides which have direct impact on aquatic fauna and flora as well as indirect impact on public health because of the consumption of water for survival of their life (Cerejeira et al., 2003).
6. Via surface runoff, aquatic organisms are directly exposed to pesticides.

8.5 NATIONAL AND INTERNATIONAL STATUS OF PESTICIDES

8.5.1 NATIONAL STATUS

Pesticides consumption is increased due to the promotion of HYVs that marked the Green Revolution (Rani and Dhania, 2014). In India, nearly 0.5 kg/ha of pesticides is applied (Rani and Dhania, 2014). This is because of the increased pest attack caused due to prevailing warm humid weather (Bhat and Padmaja, 2014). In 2007, Government of India predicted that in Asia, India was the largest pesticides producer and earned 12th rank in the world for pesticides use with 90,000 t annual production (Madhu et al., 2014). The total consumption of chemical pesticides, namely, herbicides, insecticides, fungicides, and bactericides in India is comparable in

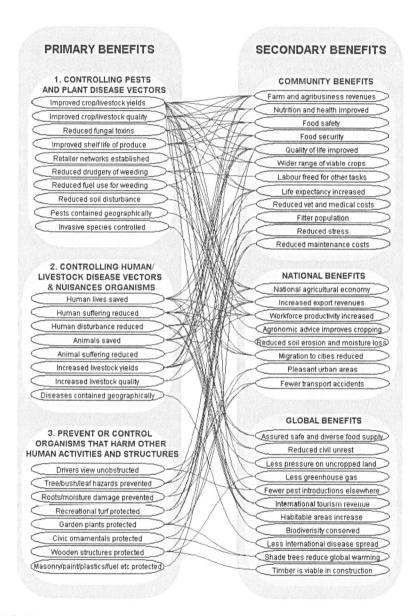

FIGURE 8.2 Benefits of pesticides. (*Note***:** For details on benefits of pesticides, reader should refer to report of Cooper, J. and Dobson, H. on *Pesticides and Humanity: The Benefits of Using Pesticides*; http://www.croplifeafrica.org/uploads/File/forms/resource_ center/fact_sheets/Pesticides%20and%20humanity,%20benefits.pdf, summited to Natural Resources Institute, University of Greenwich, Chatham Maritime, Kent, UK.)

the last 7 years, that is, from 2003 to 2010 (Table 8.2 and Fig. 8.3A), and it may also see the status of such technical grade pesticides consumption in different territories of India during 2005–2006 to 2012–2013 (Table 8.3). Continuous use of pesticides leads to their accumulation in soil and water (Rajendran, 2003), which pose significant effect on humans and animals and creates higher degree of lipopholicity and biological stability in food commodities (Tayade et al., 2013). In India, it is observed that 18% food production is lost due to insect pests, weeds, plant pathogen, birds, rodents, and in storage, for which the gain annual financial loss counts up to Rs. 70,000 crores. Currently, a total of 215 pesticides are registered for manufacturing and using in India (Central Insecticides Board & Registration Committee, Ministry of Agriculture (CIBRC), 2008). In India, annual crop loss due to pests is more than Rs. 6000 crores, of which 33% is due to weeds, 26% because of plant diseases, 20% by insects, 10% due to birds and rodents, and 11% by other miscellaneous factors (Rajendran, 2003).

Of total consumption of pesticides, insecticides are the predominant class in India (Indira Devi, 2016) and nearly account for 76% followed by fungicides (13%) and herbicides (10%) are comparable with the total consumption throughout the world (Table 8.4 and Fig. 8.3B) (Aktar et al., 2009). Furthermore, 54% in cotton, 17% in rice, and 13% in vegetable and fruits of the total quantity can be used in the country (Indira Devi, 2016). Related to pesticide residues in food, vegetables acquire highest score in the world due to unregulated use of pesticides (Navarro et al., 2007). Indira Devi (2016) stated in her lecture that in India about 72% of the food samples have shown the presence of pesticide residues within tolerance levels, while 28% samples showed above the tolerance level as compared to 1.25% globally. According to the report of Bhatnagar (2001) in Indian context, organic chlorine compounds were found comparatively at very high level in human blood, fat, and milk samples; due to this, about one-third of the total pesticides poisoning cases of the total world population are found in India (Puri, 1998).

8.5.2 INTERNATIONAL STATUS

The increasing difficulties experienced with pest attack have led to a concomitant increase in use of chemical pesticides. The crop losses amount to 20–30% with the use and 50–80% without the use of pesticides.

Pesticides are being widely used because they have been found to be convenient and provide quick control by reducing pest populations to extremely low levels. The uses of pesticides have increased since 1950, and now, it is used in the world each year with 2.5 million tons annual industrial production (Miller and Spoolman, 2012). It is estimated that for pest control annually 4 million tons of pesticides are applied to world crops, but of total applied pesticides less than 1% can reach to the target pest (Pimentel, 1983). According to the data obtained from World Health Organization (WHO), about 3×10^6 poisoning and 2.2×10^4 deaths cases

TABLE 8.2   Consumption of Chemical Pesticides in India from 2010 to 2003.

Year of consumption	Quantity in tons		
	Insecticides	**Fungicides and bactericides**	**Herbicides**
2010	20618.83	13055.44	6334.98
2009	14810.19	9626.71	4219.98
2008	3278.33	7491.31	3574.67
2007	14617.64	8297.14	4121.66
2006	16913.00	13367.00	6304.00
2005	21783.00	6566.00	6959.00
2004	21489.00	8435.00	5154.00
2003	22694.00	11028.08	7500.00

Source: FAOSTAT (2016).

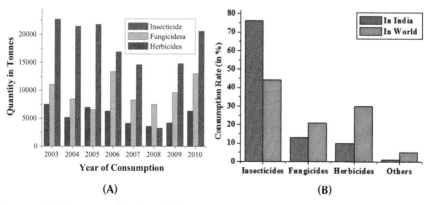

(A)   (B)

FIGURE 8.3   (See color insert.) General and comparative presentation of chemical consumption. (A) Pesticides consumption in India from 2003 to 2010 (adapted from FAOSTAT, 2016) and (B) chemical consumption pattern for agriculture India vs World. (Source: Adapted from Aktar et al., 2009)

TABLE 8.3 Consumption of Technical Grade Pesticides in Various States of India During 2005–2006 to 2012–2013.

States	Technical grade pesticides consumption (in tons)							
	2005–2006	2006–2007	2007–2008	2008–2009	2009–2010	2010–2011	2011–2012	2012–2013[a]
Andhra Pradesh	1997	1394	1541	1381	1015	8869	9289	6500
Arunachal Pradesh	2	17	16	10	10	10	17	–
Assam	165	165	158	150	19	150	160	183
Bihar	875	890	870	915	828	675	655	687
Chhattisgarh	450	550	570	270	205	570	600	675
Goa	05	09	2	9	10	9	8	9
Gujarat	2700	2670	2660	2650	2750	2600	2190	1210
Haryana	4560	4600	4391	4288	4070	4060	4050	4050
Himachal Pradesh	300	292	296	322	328	328	310	320
Jammu and Kashmir	1433	829	1248	2679	1640	1818	1711	NA
Jharkhand	70	82	81	85	89	84	151	151
Karnataka	1638	1362	1588	1675	1647	1858	1412	1225
Kerala	571	545	880	273	631	657	807	856
Madhya Pradesh	787	957	696	663	645	633	850	659
Maharashtra	3198	3193	3050	2400	4639	8317	6723	6617
Manipur	28	26	26	30	30	30	33	30
Meghalaya	06	09	06	NA	06	10	9	NA
Mizoram	25	40	44	44	39	4	4	4
Nagaland	05	05	05	18	14	NA	15	16
Odisha	963	778	NA	1156	1588	871	555	601
Punjab	5610	5975	6080	5760	5810	5730	5625	5725

TABLE 8.3 (Continued)

States	Technical grade pesticides consumption (in tons)							
	2005–2006	2006–2007	2007–2008	2008–2009	2009–2010	2010–2011	2011–2012	2012–2013[a]
Rajasthan	1008	3567	3804	3333	3527	3623	2802	1250
Sikkim	NA	02	06	3	4	NA	NA	NA
Tamil Nadu	2211	2048	3940	2317	2335	2361	1968	1919
Tripura	14	19	27	38	55	12	266	NA
Uttar Pradesh	6672	7414	7332	8968	9563	8460	8839	9035
Uttarakhand	141	207	270	221	222	199	206	220
West Bengal	4250	3830	3945	4100	NA	3515	3670	3390
Andaman and Nicobar Islands	03	NA	NA	6	14	NA	15	15
Chandigarh	0.78	NA	NA	NA	NA	NA	NA	NA
Delhi	39	NA	57	57	49	48	NA	NA
Dadra & Nagar Haveli	04	NA	NA	NA	NA	NA	NA	NA
Daman and Diu	01	NA	NA	NA	NA	NA	NA	NA
Lakshadweep	01	NA	NA	NA	NA	NA	NA	NA
Puducherry	41	40	41	39	39	39	38	40
Total	**39773**	**41515**	**43630**	**43860**	**41822**	**55540**	**52979**	**45386**

Source: IASRI (2016).
NA: not available.
[a]As on Feb 13, 2013.

TABLE 8.4   Chemical Consumption Pattern for Agriculture in India vs World.

Name of the chemical	Consumption rate in %	
	In India	**In world**
Insecticides	76	44
Fungicides	13	21
Herbicides	10	30
Others	1	5

Source: Aktar et al. (2009).

annually due to pesticides only across the globe of which majority of cases are observed from developing countries (Indira Devi, 2016).

According to a report, around one-third of the world's total agricultural production is lost in each year due to the insect-pests only. In the control of such insect and pests, about more than 2 million tons of pesticides are consumed at the annual rate (Rani and Dhania, 2014). Throughout the world, nearly 2 million tons of pesticides are consumed per year and among all 45% in consumed in Europe, 24% in USA, and 25% in the remaining part of worlds (Abhilash and Singh, 2009). Among the Asian countries, the highest consumption of pesticides is in China, Japan, India, and Korea (Rani and Dhania, 2014).

8.6 BIOREMEDIATION: AN ECO-FRIENDLY APPROACHE OF REMEDIATION

Biological fertilization is the traditional agro-ecological strategies are gaining special interest and possibly reverting, that includes the symbiosis association between the agricultural plants and mycorrhizal fungi (i.e., the rock phosphate-leaching fungi) or the bacteria (i.e., nitrogen-fixing symbiotic root nodule bacteria use leguminous plants for fixing atmospheric nitrogen) (Valentin et al., 2013), whereas in the case of conventional agriculture system process like tilling, deplete soil organic matter content leads to change in climatic conditions seems to intensify negative development (Kibblewhite et al., 2008; Banwart, 2011) and can only be sustainable if nutrients and carbon must be replenished and the soils are not over-exploited (Robinson et al., 2012); that is why, in future, recycling of nutrients from waste and by-products will be essential and more efficient way (Cordell et al., 2009).

8.6.1   BIOREMEDIATION

Traditional waste disposal methods like incineration, recycling, pyrolysis, and land filling can cost up to approximately \$1 trillion to decontaminate toxic waste sites in the agricultural farms (Reddy and Mathew, 2001), and remediation of contaminated sites using these methods is expensive and rather inefficient to clean contaminated sites (Sayler et al., 1990; Reddy and Mathew, 2001; Nyakundi et al., 2011) and may be responsible for formation of many different toxic intermediates (Sayler et al., 1990). The main problem lying behind using physicochemical methods for decontaminating polluted site is that these methods do not imply on large farms since only small soil samples are required and they are accomplished in the laboratories which require a lot of resources because the polluted soil has been excavated at a site and moved to storage area where it has been processed (Kearney and Wauchope, 1998; Nerud et al., 2003). In such a way, there is urgent need of a reasonable solution which is rapid, cost-effective, and eco-friendly that may help to solve this problem (Nyakundi et al., 2011). With the aim to solve this problem addition, many biological methods have been developed in which bacteria and fungi are involved for biodegradation of organic compounds (Schoefs et al., 2004; Levin et al., 2003). For treatment of hazardous waste, bioremediation technology has been found to be more effective and economical treatment technology among various technologies (Enrica, 1994; Kulkarni and Kaliwal, 2015). If we can consider the actual meaning of biodegradation and bioremediation both the processes are matching to each other in which organic compounds are metabolized with the help of microorganisms, but the only difference between both processes is that biodegradation is carried out by means of natural process and bioremediation is a technology-based process where microorganisms are used in situ to degrade pollutants (Singh, 2008). Bioremediation shall be considered up to the acceptable level only when an efficient bacterial strain can degrade large amount of pollutants to minimum amount with an adequate rate in limited time into its harmless or nontoxic degradation products. Bioremediation can be defined as the technology-based process, in which microorganisms or their enzymes are used to modify/detoxify or degrade organic contaminants in limited duration. General process of biodegradation of organic contaminant is shown on Figure 8.4. Nawaz et al. (2011) stated the term bioremediation for biodegradation of xenobiotic compounds as bioremediation which describe the use of living organisms to eliminate or minimize the hazardous problems of the environment that raised as the result of accumulated toxic chemicals and other such wastes. Bioremediation may be applied in order to degrade pesticides and to broken-down oil spills using microorganisms by its various technologies. It can be classified as ex situ in which contaminated material can be treated elsewhere and in situ bioremediation means treating contaminant at the site.

8.6.2   CONCEPT AND PRINCIPLES OF BIOREMEDIATION

Environmental biotransformation related to growth, survival, and development of organisms and microbes is very common process on earth. Growth in molecular biology and ecology gives various opportunities in development of biological processes like clean-up of polluted lands and water bodies. In this context development, in biotechnology, science produces recombinant organisms and proteins which are applicable for the decontamination of polluted sites. This can be applied for bioremediation of pollutants. The main principle of bioremediation includes transformation of organic compounds by living organisms through a series of metabolic reactions into nontoxic form (Fig. 8.5). In bioremediation technology

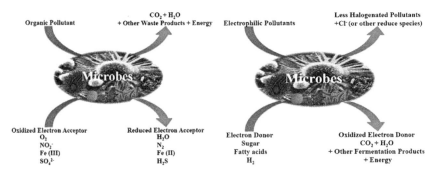

FIGURE 8.4   (See color insert.) General process of organic contaminant degradation (*Source*: Adapted from Rockne and Reddy, 2003). (A) Oxidative biodegradation and (B) reductive biodegradation.

naturally occurring microorganisms especially bacteria and fungi or plants are used to degrade pollutants into simple and nontoxic forms simply in this process; biological agents are used to reduce hazardous xenobiotics from the environment. The microorganisms used for bioremediation may be indigenous of the polluted habitat or may be isolated from different sites. When bioaugmentation is carried out for degradation of pollutants, capable microorganisms are carried out to the site of treatment; but it may be effective only when microorganisms can attack enzymatically on pollutants and convert them to harmless product. Since for effective results, it is necessary to maintain some environmental parameters very strictly for faster growth of microorganisms and degradation of contaminants. As

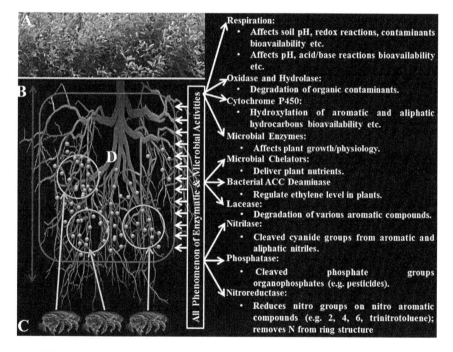

FIGURE 8.5   (See color insert.) Enzymatic and microbial activities responsible for the enhanced remediation in rhizospheric zone (*Source*: Adapted from Abhilash et al., 2009). (A) Standing plants in field indicated above the yellow line; (B) rhizospheric zone inside the ground surface indicated by blue arrow below the yellow line; (C) commonly found microorganisms in the rhizosphere, namely, *Pseudomonas* spp., *Flavobacterium* spp., and *Alcaligenes* spp. are involved in microbial activity of remediation (*Source*: Adapted from Barber and Martin, 1976; Barber, 1984); and (D) involved microorganisms in root of the plants inside the rhizosphere.

stated above in concept of bioremediation, it is very cost effective and easy way rather than other physicochemical methods.

8.6.3 STATUS OF BIOREMEDIATION

Due to low degradation rate and high cost, most of the approaches for organic compounds degradation have been trying to implement on large-scale basis. Nowadays, bioremediation shall gain more fame and interest for scientists toward biodegradation of organopollutants contaminated soil and water. Genetic engineering took special interest in this regard, but introduction of such genetically engineered recombinants for bioremediation process is a big problem; since there may be chances of release of genetically altered microorganisms in the environment concerning with unknown and potential environmental risks and public policy (Zhang and Qiao, 2002). They may be limited range of contaminants on which bioremediation is effective and the level achieved for residual contaminant may not be appropriate (Scott et al., 2008); as well as to design and implement a successful bioremediation there will be need of experienced expertise. As bioremediation seems to be good alternative to other conventional physicochemical methods of clean-up, there should be need to explore it with advanced innovations.

Bioremediation of ground water resources addresses very unique problem, because it is mobile; peoples and livestock can drink it frequently in untreated form; thus, there will be an urgency associated with groundwater treatment which can justify more drastic and expensive measures (Bonaventural and Johnson, 1997). Enzymes which may evolve in response to the presence of synthetic compounds are the foundation for pesticides bioremediation approaches which are currently available or under development (Scott et al., 2008).

8.6.4 APPLICATIONS OF BIOREMEDIATION

For the management or degradation of the waste, bioremediation is emerging as an alternative way which is considered as an eco-friendly and very case effective in the compression of remediation measures. A better understanding toward the application of bioremediation is given below which pays more attention.

1. Bioremediation techniques are used to degrade highly toxic organic pollutants, heavy metals, industrial effluents, chemicals, and pollutants from the environment.

2. Bioremediation of pesticides has been applied in surface soil treatment unit using microbial consortia and scale up in bioreactors was been reported by Geetha and Fulekar (2008). They concluded that using cow-dung microbial consortium pesticides like chlorpyrifos, cypermethrin, fenvalerate, and trichlopyr butoxyethyl ester had been bioremediated at varying concentrations under simulated environmental conditions and potential of *Pseudomonas aeruginosa* (NCIM 2074) had been assessed by exposing it under increasing concentration using scale up process.

3. Heavy metals released from tanneries produces toxic oxides which can cause lung cancer, paralysis, asthma, and mental disorders, or if accumulated in water, it causes harm to water fauna which can be remediated using bioremediation technology.

4. Oil spills in sea water can lead to death of water organisms and lead to ecological imbalance and disturbances in food chain; such oil spills have been treated using microorganisms by bioremediation technology.

5. Bioremediation can replace incineration which is an energy-dependent process that can ultimately save large amount of energy and protect us from global warming.

6. Genetically modified bacteria *Deinococcus radiodurans* has been used in bioremediation of mercury and aromatic hydrocarbons like toluene.

7. Soil structure has been greatly disturbed by organopollutants which can be remediated using indigenous or genetically modified microbes via bioremediation technology.

8. With the help of bioaugmentation, biostimulation or genetic engineering processes of bioremediation technology metabolic activities of microorganisms has been enhanced to increase degradation process of various pollutants.

9. Bioreactors used in bioremediation of polluted soil can enhance transport of nutrients and waste to degrading population (Bonaventural and Johnson, 1997).

10. Some naturally occurring or genetically modified microbes can produce high-affinity metal-binding protein for subsequent extraction, which are used for remediation of metal contaminated sites (Bonaventural and Johnson, 1997).
11. Environmentally undesirable components from air are another approach of bioremediation (Bonaventural and Johnson, 1997).

Earthworms, protozoans, nematodes, mites, isopods, collembola (springtails), enchytraeidae are also used in bioremediation, for cleaning of soil, water, air, nutrient mineralization, nitrification, enzyme activities, etc.; industrial effluents are also treated using them. They are also used in research purposes (Manzoor, 2011).

Another more recent approach of bioremediation is application of combine strategies, bioaugmentation and phytoremediation for pollutants. It is the most eco-friendly and efficient method of decontamination of pollutants. Plant exudates have been used by rhizospheric microbes, and they can degrade such pollutants very efficiently. This method can contribute toward restoration of polluted sites by vary safe means.

8.7 METHODS AND APPROACHES/STRATEGIES OF BIOREMEDIATION

8.7.1 METHODS OF BIOREMEDIATION

On the basis of removal and transportation of wastes for treatment, degree of saturation, and aeration of an area, the methods of bioremediation are basically grouped into two types; they are described in the subsections.

8.7.1.1 IN SITU BIOREMEDIATION

In situ technique is applied to the saturated soil and groundwater at the site with minimal level of disturbance (Kumar et al., 2011; Maheshwari et al., 2014). This technique has no need to excavate or remove water or soils to accomplish remediation (Kumar et al., 2011). In this method, oxygen and nutrients have been supplied by circulating aqueous solutions through contaminated soils for enhancing organic contaminant degrading ability in naturally occurring microorganisms (Kumar et al., 2011). While using in situ treatment method for ground water treatment, infiltration of water containing oxygen and nutrients or other electron acceptors conditions have been used (Vidali, 2001). Kumar et al. (2011) reported the principle of in situ bioremediation; it is based on the chemotactic abilities of motile microorganisms toward the pollutants as their nutrient source; so chemotactic abilities of such microorganisms have been enhanced and utilized in degradation of toxic compounds. In this regard, in situ bioremediation may be proved as superior and cheaper method for cleaning polluted environment (Kumar et al., 2011). The advantage of in situ bioremediation is of low cost and less disturbances in treatment as treatment can be accomplished in place which can avoid exhaustion and transportation of pollutants (Maheshwari et al., 2014). In situ bioremediation is further divided into two types, intrinsic bioremediation and engineered in situ bioremediation.

8.7.1.1.1   Intrinsic Bioremediation

The rate of utilization or degradation of substrate there have been need increase the metabolic rate of microorganism. This method utilizes stimulation of naturally occurring microorganisms by feeding nutrients and oxygen to them to enhance their metabolic activities (Kumar et al., 2011).

8.7.1.1.2   Engineered In Situ Bioremediation

In this method, physicochemical conditions have been enhanced when contaminated site conditions are not suitable for growth of certain microorganism with the help of engineered system; these parameters include nutrients (N and P), oxygen, and electron acceptors which can help to promote growth of microbes (Kumar et al., 2011).

8.7.1.2   EX SITU BIOREMEDIATION

Ex situ technique is applied to the soil and groundwater at the site which has been used to facilitate microbial degradation contaminant from the site via excavation of contaminated soil or pumping of groundwater (Kumar et al., 2011; Maheshwari et al., 2014). On the basis of state of contaminant, ex situ bioremediation is further divided into two classes—solid phase system and slurry phase system (Kumar et al., 2011).

8.7.1.2.1 Solid Phase Treatment

In solid phase treatment, organic waste such as plant litter, agricultural wastes and animal manures, and hazardous wastes like sewage sludge, industrial, domestic, and municipal waste can be treated by land farming, composting, and soil biopiles methods (Kumar et al., 2011).

8.7.1.2.2 Slurry Phase Bioremediation

It is more rapid than other treatment processes as contaminated soil is mixed with water and other additives in a bioreactor kept to mix microorganisms with contaminant present in soil; all the physicochemical parameters such as oxygen and nutrients are maintained up to optimum level for promoting growth of microorganisms and degradation of pollutants (Kumar et al., 2011). After completion of treatment, water has been removed from the solid part and disposed of or in case if it contains pollutants then further exposed for treatment (Kumar et al., 2011).

8.7.2 APPROACHES/STRATEGIES OF BIOREMEDIATION

During the last two decades, there were vast usages of pesticides worldwide, due to intensive agricultural practices and changes in farming. Such widespread use of pesticides for both agricultural and nonagricultural purposes has resulted in the contamination of various environmental matrices by residing of its residues. The applied pesticides have been transformed by microbes which have an ability to utilize it as an energy sources (Jaya and Rangaswamy, 2009). Otherwise, microorganisms develop new metabolic strategies and adapt to change the environment by mutation, induction, and/or through selective enrichment technique (Raymond et al., 2003). Bioaugmentation, biostimulation, composting, bioventing, bioreactor, and land farming are some of the important strategies of bioremediation.

8.7.2.1 IN SITU BIOREMEDIATION

8.7.2.1.1 Bioventing

Bioventing is the most common in situ treatment (Kumar et al., 2011; Maheshwari et al., 2014) used for soil and groundwater, contaminated with simple hydrocarbons and contamination to dip under the surface (Maheshwari et al., 2014). In this treatment, nutrients are supplied through wells to polluted sites for stimulation of indigenous bacteria, and air flow has been employed at low rate; due to it, only necessary amount of air required for biodegradation has been provided which minimizes volatilization and release of pollutants to the environment (Kumar et al., 2011; Maheshwari et al., 2014).

8.7.2.1.2 Biosparging

Biosparging treatment employs the injection of air under pressure below the water bodies to increase oxygen concentration of ground water which can help to enhance biodegradation of pollutants by indigenous bacteria (Kumar et al., 2011; Maheshwari et al., 2014). Contact between saturated zone of groundwater and soil can be increased through mixing via biosparging treatment. Flexible construction and design of this system is very simple and requires low cost, thereby installing small-diameter air injection points (Mendez and Maier, 2008; Kumar et al., 2011).

8.7.2.1.3 Bioaugmentation

Bioaugmentation is defined as the introduction of microorganisms exogenous or indigenous to the polluted site having ability to degrade pollutants (Scott et al., 2008; Kumar et al., 2011; Maheshwari et al., 2014). There are two factors reported by Kumar et al. (2011) and Maheshwari et al. (2014), mainly responsible to limit the application of added microbial cultures, they are (1) to sustain useful population levels non indigenous microbes can rarely compete with an indigenous population and (2) if the land treatment unit has been well managed, then indigenous population of microorganisms may be effective which are present in soil with long-term exposure to biodegradable waste. Similar to biostimulation it also requires suitable soil environment to achieve expected level of bioremediation (Scott et al., 2008).

8.7.2.2 EX SITU BIOREMEDIATION

8.7.2.2.1 Land Farming

Principle of land farming is stimulation of biodegradation capacity in indigenous microorganisms which is based on the aerobic degradation of contaminants where polluted soil has been excavated and spread over a prepared bed and tilled periodically until contaminants are degraded (Kumar et al., 2011; Maheshwari et al., 2014). This method gain much more attention as a disposal alternative because of clean-up liability, low maintenance cost, and no need to monitor the process (Kumar et al., 2011; Maheshwari et al., 2014).

8..2.2.2 Composting

In composting technique, contaminated soil has been amended with nonhazardous organic matter like agricultural waste, since organic manure can help to support increase in microbial population and enhance biodegradation (Kumar et al., 2011; Maheshwari et al., 2014).

8.7.2.2.3 Biopiles

Biopiles are also referred as hybrid of composting and land farming or refined version of land farming (Shannon and Unterman, 1993; Subramanian et al., 2006; Kumar et al., 2011) which can helps to control physical losses of pollutant by volatilization and leaching, typically used for petroleum hydrocarbons contaminated surfaces (Kumar et al., 2011; Maheshwari et al., 2014). It provides a favorable environment for growth of natural isolates as well as anaerobic microbes (United States Environmental Protection Agency (USEPA), 2016). In this method, engineered cells are made in the form of aerated composted piles (Kumar et al., 2011).

8.7.2.2.4   Bioreactors

Bioreactors used in this treatment are slurry reactor or aqueous reactor that is defined as a containment vessel used to create a condition for mixing three phases solid, liquid, or gas to enhance bioremediation rate of water soluble and soil bound contaminants in the form of water slurry of polluted soil and biomass capable to degrade desired pollutant (Kumar et al., 2011; Maheshwari et al., 2014). Bioreactors are used to treat polluted soil and water pumped up from a polluted plume in which contaminated solid materials like soil, sediment, sludge, or water can be processed via an engineered containment system (Kumar et al., 2011; Maheshwari et al., 2014). Bioreactor system proved more controlled and predictable as compared to in situ treatment methods because rate of biodegradation is more in bioreactor as compared to in situ treatment method due to presence of manageable environment in the container (Kumar et al., 2011; Maheshwari et al., 2014). As it gains above-said advantage, it also has a disadvantage that polluted soil needs pretreatment by excavation or pollutant that can be removed from soil via vacuum extraction before inoculation in bioreactor for treatment (Baker and Brooks, 1989; Sorek et al., 2008; United States Environmental Protection Agency (USEPA), 2016).

8.8   POTENTIAL MERITS AND DEMERITS OF BIOREMEDIATION

8.8.1   MERITS OF BIOREMEDIATION

Bioremediation should prove good alternative over physicochemical methods and have several advantages as follows:

1. Bioremediation is a natural process and have minimal environmental disruption; therefore, it gain greater public demand as well as acceptance for treatment of hazardous waste contaminating soil and water.
2. Very less expensive when bioremediation are used for cleaning-up of hazardous waste as compared to other such measures. Bioremediation also proven for permanent elimination of such waste.
3. End products or intermediates formed during biodegradation are usually harmless.

4. Bioremediation can use natural microorganisms.
5. It has reduced impact on natural ecosystem.
6. It is very effective in small quantities and often decompose quickly.
7. It lowers the exposures and largely avoiding the pollution problems.
8. It is less destructive and more cost effective.
9. It can eliminate the need of packaging, transport and disposal of contaminated material.
10. Bioremediation also reduces secondary risk of exposure and impact liability.

8.8.2 DEMERITS OF BIOREMEDIATION

1. Bioremediation is limited to only easily biodegradable compounds to achieve rapid and complete degradation; since many contaminants like polyaromatic hydrocarbons and polychlorinated organic compounds are resistance to microbial attack.

2. Sometimes biodegradation products are more toxic and persistence that the parent one.

3. As bioremediation is strictly biological process and to carry out any biological process, there should be a very strict need of some parameters such as capable microbial population, nutritional and physical parameters; since for success of bioremediation, these site factors are required to maintain.

4. It is very hard and tedious to put up from laboratory studies to field trials.

5. Contaminants may be present in any form solids, liquids and gases.

6. It takes longer time as compared to other physicochemical methods like incineration, land filling, or excavation.

7. Rate of clean-up of bioremediation should not be predicted because some microorganisms either degrade it very slowly or not at all.

8. Most of bioremediation processes may be carried out aerobically, but when it should be run under anaerobic conditions, there might be chances of other microorganisms' degradation rather than pollutants.

8.9   FACTORS INFLUENCING BIOREMEDIATION OF PESTICIDES

While studying bioremediation, it is essential to know about the population dynamics of pesticide degrading microbes in relation with other microorganisms present in that habitat and some variables like temperature, pH, nutrients, water availability, and pesticide or its metabolite concentration which also acts as limiting factors for pesticide degrading microbial population (Singh, 2008). The biodegradation of various pesticides under different physiological conditions using bacterial species has been reported by many researcher groups (Jaya and Rangaswamy, 2009; Naqvi, 2010).

Microbial growth and metabolic activities are greatly influenced by many physical parameters such as temperature, pH, aeration moisture, etc., and nutritional parameters like carbon, nitrogen, and phosphates. Such environmental parameters always influence the rate of biodegradation, so their optimization with respect to growth and degradation with physicochemical factors can help to enhance the rate of biodegradation and minimize clean-up time of pesticides. While conducting and monitoring bioremediation process, it is essential to consider above parameters along with redox potential in soil and ground water, oxygen content, electron donor or receptor concentrations and concentration of end product (Maheshwari et al., 2014). Pesticides applied in environment are transformed by biological and nonbiological processes by different mechanisms such as oxidation, reduction, hydrolysis and conjugation, etc., with the help of various enzymes and converts into biologically less active products (Michel et al., 1995).

8.9.1   ENVIRONMENTAL/PHYSICAL FACTORS

8.9.1.1   PESTICIDE STRUCTURE

Inherent biodegradability of pesticide has been depended upon the structure of pesticide molecule, that is, the substituent on phenyl ring can enhance degradation and addition of polar groups OH, NH_2, COOH, etc., decreases degradation and makes it more susceptible to microbial attack (Chowdhury et al., 2008); as well as substituent of alkyl or halogen can make pesticide more resistant to biodegradation (Cork and Krueger,

1991). There are very slight differences in the position of substituents in pesticides of the same class of pesticides that can affect the degradation rate (Vidali, 2001).

8.9.1.2   PESTICIDE CONCENTRATION

The pesticide concentration is considered as an important factor that determines the rate of biodegradation (Vidali, 2001). The kinetics behind the biodegradation of several pesticides approaches is first order in which the rate of degradation decreases is roughly proportionate to the residual pesticide concentration (Vidali, 2001; Chowdhury et al., 2008).

8.9.1.3   SOIL TYPES/SOIL STRUCTURE

Soil properties including pH, organic matter, clay content, etc. can affect the pesticides degradation in soil (Vidali, 2001; Chowdhury et al., 2008). Therefore, it is essential to study the effect of soil types in pesticide degradation (Chowdhury et al., 2008). Effective delivery of water, air, and nutrients has been controlled by soil structure, and for improvement of soil structure, organic material can be applied (Maheshwari et al., 2014). Soil with low permeability may not be appropriate for in situ bioremediation techniques because low soil permeability can alter in movement of nutrients, water, and oxygen in soil (Maheshwari et al., 2014).

8.9.1.4   SOIL MOISTURE

Available water is essential for all living organisms, and optimum moisture level can be achieved through irrigation only (Maheshwari et al., 2014). Pesticide movement and diffusion is based on water because it acts as solvent and essential for microorganisms functioning (Vidali, 2001; Chowdhury et al., 2008). Rate of degradation depends on the moisture content of the soil since degradation rate is slow in dry soil as compared to moist one (Vidali, 2001; Chowdhury et al., 2008).

8.9.1.5   TEMPERATURE

Temperature is another important physical factor which affects the degradation rate of the pesticides, but it depends upon the molecular structure of the pesticides and it also affects the adsorption of pesticide to microbes which are present in soil by altering the hydrolysis and solubility of pesticides in soil (Racke et al., 1997). Biochemical reaction rate can be greatly affected with rise or fall in environmental temperature. Increase in temperature can stimulate microbial activities, and some groups of microbes show their activities within certain range of temperature (Vidali, 2001; Chowdhury et al., 2008). In many biochemical reactions, each 10°C rise in temperature can doubles the rate of reaction, but above a certain limit, there have been chances of cell death (Maheshwari et al., 2014). When there will be rise in temperature in late spring, autumn and summer plastic covering can be used (Michel et al., 1995). Maximum growth and activity of bacteria in soil occurs at 25–35°C (Alexander, 1997), while optimum temperature range for pesticide degradation in soil is around 25–40°C (Jitender et al., 1993). Lower temperature and higher concentration of pesticide resulted in greater persistence (Vidali, 2001; Chowdhury et al., 2008).

8.9.1.6   SOIL pH

Several studies have shown that rate of degradation of pesticides can be greatly affected by pH (Vidali, 2001). Müller et al. (1998) reported that best degradation can be commenced in the range of pH 7 and below neutral range degradation rate can be slowed down (Andrea et al., 1994). Varies in pH greatly affects biodegradation rate, for example, if contaminated soil is acidic then before to proceeding for bioremediation it is necessary to balance pH by adding lime to the soil (Maheshwari et al., 2014). Susceptibility of pesticide to acid or alkaline catalyzed hydrolysis depends upon pH of soil (Racke et al., 1997). Adsorption, biotic, and abiotic degradation processes may be greatly affected by varying pH (Burns, 1975) and influences the sorptive behavior of pesticide molecule on organic surfaces and clay and make it bioavailable (Hicks et al., 1990).

8.9.1.7   OXYGEN

It is the most important parameter in bioremediation of pesticides. Increase in oxygen penetration in soil can decrease clean-up time of pesticides which increases the diffusion coefficient for oxygen (Huesemann and Trux, 1996).

8.9.1.8   SOIL SALINITY

There has been very limited information on pesticide degradation in saline soil although high salt concentration may be inhibitory to degradation process (Chowdhury et al., 2008).

8.9.2   BIOLOGICAL FACTORS

8.9.2.1   NUTRIENTS

Addition of nutrients like carbon, nitrogen, and phosphate—the building blocks of life to soil—can help in enhancing growth and activities of indigenous microbes for breakdown the pollutants like pesticides (Maheshwari et al., 2014).

8.9.2.2   ORGANIC CONTENT

Surface soil is nutritionally rich where organic matter contains diverse and high microbial population which can help to enhance plant growth. Organic matter contains carbon, nitrogen, phosphorus, and other micro and macronutrients, while subsurface soil and ground water sediments contains very less quantity of organic matter; therefore, there has been very less amount of microbial diversity and activities (Naqvi, 2010). Probably, greater than 1% of organic is crucial to confirm the presence of active indigenous flora and fauna of site which can degrade pesticide (Burns, 1975). On this basis, it has been predicted that biodegradation has been directly depends upon the organic matter contents which can increase biodegradation rate of pesticides (Adriaens and Hickey, 1993).

8.9.2.3   SOIL BIOTIC COMPONENTS

Varieties of biotic mechanisms are responsible for breakdown of pesticides in soil (Chowdhury et al., 2008). While studying the approach of rhizoremediation of pesticide contaminated soil, it has been indicated that pesticide degradation process is slow in case of sterilized soil, while the process is faster in unsterilized soil, which can prove directly that microorganisms play a vital role in pesticide bioremediation in soil. These microorganisms can use pesticide as their carbon, nitrogen, and energy source (Chowdhury et al., 2008).

8.9.3   ENZYMES

Enzymes are active in presence of microbial predators and antagonists even at low concentration (Rao et al., 2010), that is, the biological degradation may be very fast due to presence of enzymes (Gavrilescu, 2005). The ability of microbes to degrade pesticides has been depended on the ability to produce requisite enzymes and optimal environmental conditions to occur metabolic reactions (Gavrilescu, 2005). Bacteria, fungi (especially white-rot fungi), plants, and microbe–plant associations are main players to produce the enzymes (Rao et al., 2010), known as the main effectors of all kinds of transformations which occurs in biological system, also play a vital role as catalysts which can produce extensive transformations of toxicological and structural properties of pollutants along with their transformation into nontoxic products (Rao et al., 2010). Rate of biological degradation depends upon the presence of enzymes for specific chemicals; since in case of other chemicals process may be very slow (Gavrilescu, 2005).

8.10   BIOREMEDIATION OF PESTICIDES

Pesticides are greatly accumulated in food products and water supplies due to its unusual use, and regarding to environmental concerns of pesticides, there should be a need to develop convenient, safe, and cost-effective technologies for pesticide detoxification. There are many physical and chemical methods that can be applied to decontaminate toxic waste sites. The known methods for cleaning-up of pesticides residues from soil are chemical treatment, volatilization, and incineration which have met public opposition, due to the production of large volumes of acids and alkalis. Subsequently, they must be disposed-off because of their potential toxic emissions and lead to elevated economic costs (Richard et al., 1997; Zhang and Qiao, 2002; Lau et al., 2003). Bioremediation of xenobiotic compounds especially pesticides by microbes are an essential phenomenon in which these compounds are detoxified or removed from the environment that help in the prevention of environmental pollutions (Mervat, 2009; Kulkarni and Kaliwal, 2015). Utilizing microorganisms to degrade toxic organopollutants is an efficient, economical approach known as bioremediation that has been successful in laboratory studies (Boopathy, 2000; Mayer and Staples, 2002) and to achieve this screening of the potential degrading organism is one of the key steps (Kulkarni and Kaliwal, 2015). If the hazardous organic compounds have been provided in right environmental capabilities, bacterial culture is capable to degrade it (Smith-Greeier and Adkins, 1996; Hashmi, 2001). Thus, biological decontamination methods are preferable to conventional approaches; in general, microorganisms degrade numerous environmental pollutants without producing toxic intermediates (Pieper and Reineke, 2000; Furukawa, 2003). Bioremediation is a promising alternative to physicochemical methods of remediation, because it is less expensive and can selectively achieve complete destruction of organic pollutants (Alexander, 1999). Following are another four main variables on which the rate of biodegradation in soil depends (Singh, 2008), they are as follows: (1) availability of pesticide or metabolite to the microorganisms; (2) physiological status of the microorganisms; (3) survival and/or proliferation of pesticide degrading microorganisms at contaminated site; and (4) sustainable population of these microorganisms.

There are two bioremediation approaches used for xenobiotics treatment, one is directly based on microorganisms and the other is use of isolated enzymes (Zhang and Qiao, 2002). In the case of former one, naturally occurring microorganisms are used with genetic techniques and certain desirable biodegradation pathways are brought together from different organism in single host, but this technique has its own limitation associated with releasing of genetically altered microbes into environment; the second more feasible strategy based on enzymes can be preferred (Zhang and Qiao, 2002).

8.10.1   METHODS OF BIOREMEDIATION OF PESTICIDES

Recently, bioremediation of pesticides can be done by using the following three main approaches: (1) naturally occurring or genetically engineered microbes, (2) insect genes, and/or (3) isolated enzymes.

8.10.1.1 MICROORGANISM FOR PESTICIDES BIOREMEDIATION

8.10.1.1.1 Naturally Occurring Microorganisms

Bacteria and fungi are the major degraders of pesticides and their intermediate metabolites, which are isolated and characterized in the last decades because microbial metabolism ultimately degrades such compounds. Bacteria can easily degrade naturally occurring compounds but have less ability to directly degrade synthetic chemicals because of structural differentiation in between natural and synthetic one. In such cases, bacterial potential can be improved by exposing them with nutrient stress and under stressful conditions; they can utilize only synthetic chemicals as their nutrient (C and N) and energy source and can acquire an ability to degrade it.

8.10.1.1.2 Recombinant Microorganisms

In this era, there has been more development in recombinant DNA technology, and molecular probes like tools are used for the purpose of knowing biochemical mechanisms of microorganisms through which it can be easily isolated and identified for pesticide degradation potential from the environmental samples. Strain of interest for pesticide degradation can be detected using genes encoding enzymes for degradation (Zhang and Qiao, 2002). Among all the available methods to detect specific microbes in the environment gene probe and polymerase chain reaction have been extensively used (Picord et al., 1992; Tsai and Olson, 1992).

8.10.1.2 INSECT GENES FOR PESTICIDE BIOREMEDIATION

In insects, unique metabolic mechanisms are found for detoxification of a wide range of compounds and possess amazing adaptability for various physical and biological conditions so they can survive under a wide variety of physical and biological conditions (Zhang and Qiao, 2002). Nowadays, more than 500 insect species showed resistance against certain pesticides from various groups which include *Bt* biopesticides, organophosphates, organochlorines, carbamates, and pyrethroids (Sparks et al., 1993; Gould et al., 1995). Enzymes encoded with resistant genes of insects are very useful for degradation of pesticides like pollutants because in vivo some resistant insects can detoxify various kinds of pesticides (Zhang and Qiao, 2002). Zhang and Qiao (2002) reported three aspects for the molecular resistant mechanisms of insects to chemical insecticides summarized as follows in which latter the two mechanisms have been extensively studied:

1. Behavioristic resistance in which body exposure to insecticides have been reduced.
2. Enhanced detoxification in which hydrolases, cytochrome P450, glutathione S-transferase enzymes activities can sequestrate many kinds of pesticides.
3. Reduced target sensitivity can be changed in acetylcholinesterase (AChE), sodium channel, and gamma-aminobutyric acid receptor.

8.10.1.3   ISOLATED ENZYMES FOR PESTICIDE BIOREMEDIATION

Naturally occurring and genetically modified recombinant strains have potential impact on decontamination of pesticides but with respect to presence of pesticide residues in storage containers, spray tanks, and in fruits and vegetables as well as release of genetically altered microorganisms into environment; they do not seem to be effective and applicable (Zhang and Qiao, 2002). Therefore, science approached toward another option of use of crude or pure enzymes is convenient and reliable method to decontaminate pesticides from any site. Enzymes play a vital role in pesticide degradation. In the case of target organisms, it acts via mechanism of intrinsic detoxification and evolved metabolic resistance; while in open environment using metabolic assembly of soil and water microorganisms with the help of microbial enzymes (Scott et al., 2008).

8.10.2   ROLE OF MICROBIAL POPULATION IN BIOREMEDIATION OF PESTICIDES

Soil is called the most complex biomaterial on Earth (Young and Crawford, 2004). Soil biota confers the properties of soil and the biological, chemical, and physiological processes happening in it. In another way, quality and health of soil is evaluated on the basis of microflora present in the soil. Most of the properties of soil depend upon the activity of soil microflora which includes bacteria, fungi, and actinomycetes (Griffiths and Philippot, 2012; Prosser, 2012) and has been estimated that microbial decomposers are responsible for approximately 90% of the energy flow in soil (Nannipieri et al., 2003). Generally, contamination of soil with organic and inorganic pollutant decreases the biodiversity of soil (Griffiths and Philippot, 2012; Prosser, 2012). But many indigenous bacterial communities acts in such a way that they show resistance to the environment where these chemicals are applied repeatedly, and if specific, microbes show good ability to resist that environment; then, these microbes are modified genetically in such a way that they showed more tolerance to these chemicals or degrade them by using as a nutritional source for their growth and energy or by transferring their genes to other bacteria to improve the degradation capability of genetically modified bacteria. Many times a soil microbe shows obvious resistance to contaminants, it is due to nonbioavailability of the pollutants in the specific environment for the organisms, not just because of insensitivity of the exposed organism (Valentin et al., 2013). And if these pollutants in soil are adequately bioavailable to create toxic effects, it is also bioavailable to microbes which are able to degrade it (Valentin et al., 2013); but degradative bacteria always have specific catabolic mechanism to use and uptake organic contaminants (Stroud et al., 2007). Due to this catabolic mechanism, these organisms often have ability to use organic pollutants as C, N, and energy source (Leung et al., 2007). Degradation ability of bacteria is due to presence of catabolic pathways in microbes to degrade any organic compound which provides an advantage in the form of protection and cell building blocks (Robertson et al., 2007).

Bioremediation methodologies to treat xenobiotics such as pesticides in soil have gained considerable attention due to its eco-friendly nature and have been employed successfully in many countries (Enrica, 1994; Ritmann et al., 1988). To achieve this screening of the potential degrading organism is one of the key steps (Kulkarni and Kaliwal, 2015). El-Fakharany et al. (2011) reported that bioremediation using *Pseudomonas* sp. (EB20) has been regarded as a safe remediation technology of methomyl in drinking water. The biological effects of a pesticide in an aquatic environment depend on several factors like the concentration of the pesticide in water and the duration of time that the organism of interest is exposed. Information on the degradation rates of pesticides under ambient conditions is integral to estimating their aquatic environmental significance (Starner et al., 1999). In recent decade, the expansion and intensification of agricultural and industrial activities have led to pollution of groundwater and soil with pesticides, and many treatment processes have been developed to reduce the environmental impacts of this contamination (Nyakundi et al., 2011). In contaminated soils, microorganisms are more commonly found in mixtures.

Moreover, pesticide degradation in soil has been affected by a number of abiotic factors (Jaya and Rangaswamy, 2009). Shah and Thakur (2003) found that the pesticide degradation is under the impact of various physicochemical parameters and the major pathway governing degradation and ecotoxicity of these compounds has been microbial mediated. Microorganisms have been considered as the principal agents of xenobiotic degradation (Arbeli and Fuentes, 2007). It has thus become increasingly possible to isolate microorganisms which are capable of degrading recalcitrant and xenobiotic compounds from environments polluted with such toxic chemicals (Gibson and Harwood, 2002).

Carbamates are synthetic organic chemicals which are highly poisonous pesticides that have been widely used in agricultural farms as insecticides, fungicides, herbicides, nematicide, and acaricides (WHO, 1996). Its nontarget toxicity has been extended from human beings to both aquatic as well as terrestrial organisms with high sensitivity in fish and earthworms (WHO, 1996). Some of the carbamates used in horticultural farming include aldicarb, carbaryl, carbofuran (in form of Furadan), and methomyl (Omolo et al., 2012). The introduction of pesticides like methomyl, chlorpyrifos, dichlorvos, methyl parathion, and phorate into the soil environment has led to alteration in ecological balance that directly effect on fertility of soil (Graebing et al., 2002).

8.10.2.1 BIOREMEDIATION OF PESTICIDE CONTAMINATED SOIL

Waste generated from pesticide industries has been become a disposal problem (Geetha and Fulekar, 2008); this is a big environmental issue for today and tomorrow due to unavailability of proper treatment facility (Fulekar, 2005). Current situation for waste disposal and treatment has been accomplished by physicochemical methods, but these methods are quite inefficient and ineffective, and due to these, pesticides and its residues remain in the environment and causing toxicity by entering in the food chain. According to THE data obtained from WHO, only 2–3% of applied pesticides are used for controlling pests, while unused part remains in soil (United States Environmental Protection Agency (USEPA), 2016), which can cause contamination in soil.

Remediation of polluted soil by using plant–microbe interaction is based on the fact that plants can release some chemicals in like amino acids, sugars from its root called plant root exudates, which can help to enhance growth of microbial population and thereby stimulate microbes to degrade pollutants. Many scientists reported the mechanism of plant–microbe interaction involved in pesticide degradation. Possibly, there are three general mechanisms of rhizospheric enhancement of cometabolism of hazardous pollutants (Crowley et al., 2001).

1. Rhizospheric environment may be responsible for selective enrichment of degrader microbes where they can degrade pollutants at faster rate as compare to root free zone (Crowley et al., 1997).
2. When the pollutants' concentration has been low or unavailable to microbial population for their growth, at that time, rhizosphere may stimulate microbial growth by providing natural substrates excoriated by plants (Haby and Crowley, 1996; Alexander, 1999).
3. The rhizosphere is rich in natural compounds which may induce cometabolism of pollutants in certain microorganisms that carry degradative genes or plasmids; this may permit initial degradation of pollutants that would otherwise be unavailable as carbon sources.

8.10.3 RHIZOREMEDIATION: A BENEFICIAL PLANT–MICROBE INTERACTION

Conventional methods like incineration and landfilling are used for cleanup of polluted sites that can cost enormously (Kuiper et al., 2004) as well as these methods are not sufficient to clean the environment (Dixon, 1996). These methods have several drawbacks; incineration leads to the air pollution and leaches from landfills that can reach to groundwater and drinking wells in the form of water and gases, whereas excavation of soil can generate toxic air emissions (Kuiper et al., 2004). In the view of such hazardous and incomplete remediation methods, there has been need to search for alternative techniques to decontaminate polluted sites with environmental-friendly, safe, less-laborious, and cost-effective method of clean-up (Kuiper et al., 2004). Bioremediation is an alternative to physico-chemical decontamination method in which microbes or other biological systems are used to degrade the pollutants (Dua et al., 2002).

During phytoremediation plant enzymes are responsible for degradation of contaminants, while during bioaugmentation, degradation of pollutants can be carried out by indigenous microbial population (Kuiper et al., 2004). Microorganisms present in the rhizosphere of plants used during phytoremediation or of plants which are emerging as natural vegetation on a contaminated site contribute a lot to degradation of pollutants; this contribution of rhizomicrobial population is known as rhizoremediation (Anderson et al., 1993; Schwab and Banks 1994). Rhizoremediation is the combination of two approaches, phytoremediation and bioaugmentation, in which plant exudates can help to stimulate the survival and action of bacteria, which subsequently results in a more efficient degradation of pollutants (Kuiper et al., 2004). These plant root exudates contain sugars, organic acids, and amino acids, etc. (Vancura and Hovadik, 1965). The root system of plant can help to spread bacteria through soil and also penetrate through its impermeable layers (Kuiper et al., 2004). Rhizospheric microorganisms are main contributors of degradation process in many cases, but plants are function as solar-driven biological pump and treatment system which can attract water toward roots, accumulate water soluble contaminants in rhizosphere and help in degradation of contaminants (Erickson, 1997). The inoculation of pesticide degrading bacteria on plant seed can be an important additive to improve the efficiency of phytoremediation and bioaugmentation (Kuiper et al., 2004).

Better understanding of microbial processes can help in the development of bioremediation process (Kuiper et al., 2004). The main advantage behind using plants in combination with microorganisms is an increased microbial population and metabolic activities in rhizosphere (Kuiper et al., 2004) which can improve physicochemical properties of polluted soil, leads to make contact between microbes present on roots and pollutants present in soil (Schwab et al., 1995; Nichols et al., 1997; Kuiper et al., 2001). Most suitable plant species used in rhizoremediation are alfalfa, leguminous plants, and various grass varieties (Kuiper et al., 2001; Qiu et al., 1994) because these plants are highly branched root system and may have the ability to harbor large number of bacteria on their roots.

Another factor for success of rhizoremediation is primary and secondary metabolism, their establishment, survival, and interaction with other microorganisms (Kuiper et al., 2004). Shann and Boyle (1994) reported that existence of microbial diversity in rhizosphere depends upon the plants species and composition of root exudates; as well as plant age, root type, and soil type (Anderson et al., 1993). Bioremediation depends upon the nature of contaminant, structure of soil, and hydrogeology and nutritional availability along with microbial composition of site of clean-up (Dua et al., 2002). For effective results, combination of two technologies has been used which are under the development from many year; due to this problem, it is difficult to determine efficacy of bioremediation (Kuiper et al., 2004). *P. fluorescens* WCS365 was selected for its excellent competitive root colonizing abilities and was used as a model strain to study root colonization (Geels and Schippers, 1983).

The main principle of bioremediation is, when suitable rhizospheric bacteria inoculated together with a suitable plant by bacterial seed coating, these bacteria may adhere on roots together with natural flora of soil and spread through the soil which ultimately increases bioremediation of pollutants (Kuiper et al., 2004). The first studies of rhizospheric contaminants degradation was mainly highlighted on pesticides and herbicides (Hoagland et al., 1994; Zablotowicz et al., 1994; Jacobsen, 1997); which suggested that plants are protected against these pollutants with the help of efficient degrading bacteria (Kuiper et al., 2004). There are many reports on degradation of organopollutants such as polycyclic aromatic hydrocarbons, polychlorinated biphenyls, and pesticides too.

8.11 FUTURE OUTLOOK AND RESEARCH OPPORTUNITIES

Microorganisms play a very crucial role in bioremediation. Using microbes as tool for bioremediation, several in situ and ex situ strategies have been developed, bioaugmentation, bioventing, biosparging, composting, land-filling, etc. All the methods are successful up to some extent. However, there has been need to develop these methods for more efficient and complete clean-up of pollutants. Bioremediation can help to clean polluted sites by inducing natural biodegradation process. But it can gain success when we can understand the process exactly by knowing microbes, their metabolism, and survival in environment and degradation mechanism, although it is essential to exploit the bioremediation up to field trials after its pilot scale success. Bioremediation of pesticides is most focused strategy as tremendous application of pesticides for improved crop yield is the primary need to serve food for such huge population. Several microbial strains have been genetically modified for efficient degradation, but insertion of genetically modified microbes into soil can increase in genetically altered microflora in soil. To overtake this problem, isolated enzyme technology has been adopted which can give good results and serve as efficient technology of cleanup.

8.12   CONCLUSION

Bioremediation is an innovative and preferred technology used for clean-up of pesticide contaminated sites because of its easy operation and low cost and specially can give complete destruction of pesticides and its residues present in soil. This process can enhance biodegradation of pollutants by supplying indigenous microflora with additional nutrients carbon sources. The whole process of bioremediation for pesticide contaminated soil has been accomplished by stimulating growth of indigenous microbes or by inoculating efficient degrading strain of bacteria or fungi to the soil which results into complete mineralization of pesticides into CO_2 and H_2O. Bioremediation can be accomplished using in situ methods such as bioventing, biosparging, biostimulation, and bioaugmentation or ex situ technologies including bioreactors, biofilters, land farming, and composting. All the methods are included above in strategies of bioremediation.

Environmental contamination due to several toxic/xenobiotic compounds simply called pollutants/contaminants is a very big issue in front of each environmental protection agency as well as us. To save earth from such anthropogenic pollutants is the prime need of our society, and for that, several physical, chemical, and biological treatment technologies are adopted. Each technology can serve for degradation purpose but up to very less extent they are successful. While now a day out of all these technologies biological treatment methods are preferred due to low cost, no side effects, complete destruction of pollutants from site of contamination and specially requires less time. Bioremediation is serving as suitable method to restore polluted sites. Regarding advantages and disadvantages of bioremediation technology, it can prove most acceptable and suitable methods as compared to other physicochemical methods of decontamination.

8.13 SUMMARY

The overall summary of this book chapter includes the issues related to environmental pollution and alternative approaches for its remediation. Regarding overall the best way to decrease pesticide pollution is to use biological, safer, cost-effective technologies like bioremediation.

KEYWORDS

- **soil contamination**
- **bioremediation**
- **pesticides**
- **biopiles**
- **composting**

REFERENCES

Abhilash, P. C.; Jamil, S.; Singh, N. Transgenic Plants for Enhanced Biodegradation and Phytoremediation of Organic Xenobiotics. *Biotechnol. Adv.* **2009,** *27*, 474–488.

Abhilash, P. C.; Singh, N. Pesticide Use and Application: An Indian Scenario. *J. Hazard. Mater.* **2009,** *165*(1 3), 1–12.

Adriaens, P.; Hickey, W. J. Physiology of Biodegradative Microorganisms. In *Biotechnology for the Treatment of Hazardous Waste;* Stoner, D. L., Ed.; Lewis Publishers, CRC Press: USA, 1993; pp 97–136.

Aktar, W.; Sengupta, D.; Chowdhury, A. Impact of Pesticides Use in Agriculture and Phytoremediation of Organic Xenobiotics. *Interdiscipl. Toxicol.* **2009,** *27*, 474–488.

Alexander, M. *Biodegradation and Bioremediation.* 2nd Ed., Academic Press: San Diego, USA, 1999, pp 453.

Alcxander, M. *Introduction to Soil Microbiology.* 2nd Ed.; Wiley Eastern Limited: New Delhi, India, 1977.

Anderson, T. A.; Guthrie, E. A.; Walton, B. T. Bioremediation in the Rhizosphere. *Environ. Sci. Technol.* **1993,** *27*, 2630–2636.

Andrea, M. M.; Tomita, R. Y.; Luchini, L. C.; Musumeci, M. R. Laboratory Studies on Vola-tilization and Mineralization of 14c-p,p'-DDT in Soil, Release of Bound Residues and Dissipation from Solid Surfaces. *J. Environ. Sci. Health, Part B* **1994,** *29*(1), 133–139.

Arbeli, Z; Fuentes, C. L. Accelerated Biodegradation of Pesticides: An Overview of the Phenomenon, Its Basis and Possible Solutions; and a Discussion on the Tropical Dimen-sion. *Crop Protect.* **2007,** *26,* 1733–1746.

Baker, A. J. M.; Brooks, R. R. Terrestrial Higher Plants Which Hyperaccumulate Metallic Elements—A Review of their Distribution, Ecology and Phytochemistry. *Biorecovery* **1989,** *1*(2), 81–126.

Banwart, S. Save our Soils. *Nature* **2011,** *474,* 151–152.

Barber, D. A.; Martin, J. K. The Release of Organic Substances by Cereal Roots into Soil. *New Phytol.* **1976,** *76*: 68.

Barber, S. A. *Soil Nutrient Bioavailability*; Wiley-Interscience: New York, 1984.

Bhat, D.; Padmaja, P. Estimation of Pesticides in Soil Samples in Ghaziabad (UP) India. *Int. J. Adv. Technol. Eng. Sci.* **2014,** *2*(6), 2348–7550.

Bhatnagar, V. K. Pesticide Pollution Trends and Perspectives. *ICMR Bull.* **2001,** *31*(9), 87–88.

Bonaventural, C.; Johnson, F. M. Healthy Environments for Healthy People: Bioremedia-tion Today and Tomorrow. *Environ. Health Perspect.* **1997,** *105*(Supplement 1), 5–20.

Boopathy, R. Bioremediation of Explosives Contaminated Soil. *Int. Biodeterior. Biode-grad.* **2000,** *46*(1), 29–36.

Burns, R. G. Factors Affecting Pesticides Loss From Soil. In *Soil Biochemistry;* Paul, E. A., McLaren, A. D. Eds.; Marcel Dekker, Inc.: New York, USA, 1975; Vol. 4, pp 103–141.

Central Insecticides Board & Registration Committee, Ministry of Agriculture (CIBRC), 2008. www.cibrc.nic.in (accessed Nov 15, 2015).

Cerejeira, M. J.; Viana, P.; Batista, S.; Pereira, T.; Silva, E.; Valerio, M. J.; Silva, A, Ferreira, M.; Silva-Fernandes, A. M. Pesticides in Portuguese surface and ground waters. *Water Res.* **2003,** *37*(5), 1055–1063.

Chowdhury, A.; Pradhan, S.; Saha, M.; Sanyal, M. Impact of Pesticides on Soil Microbio-logical Parameters and Possible Bioremediation Strategies. *Indian J. Microbiol.* **2008,** *48,* 114–127.

Commission of the European Communities (CEC). Communication from the Commission to the Council, the European Parliament, the European Economic and Social Committee and the Committee of the Regions: Thematic Strategy for Soil Protection, 2006. http://ec.europa.eu/environment/archives/soil/pdf/SEC_2006_620.pdf (accessed Feb 10, 2016).

Cordell, D.; Drangert, J.-O.; White, S.; The Story of Phosphorus: Global Food Security and Food for Thought. *Global Environ. Change* **2009,** *19*(2), 292–305.

Cork, D. J.; Krueger, J. P. Microbial Transformation of Herbicides and Pesticides. *Adv. Appl. Microbiol.* **1991,** *36,* 1–66.

Crowley, D. E.; Luepromechai, E.; Singer, A. Metabolism of Xenobiotics in the Rhizo-sphere. In *Pesticide Biotransformation in Plants and Microorganisms: Similarities and Divergences;* Hall, J. C., Hoagland, R. E., Zablotowicz, R. M. Eds.; ACS Symposium Series 777; American Chemical Society: Washington, DC, 2001; pp 333–352.

Crowley, D. E.; S. Alvey; E. S. Gilbert. Rhizosphere Ecology of Xenobiotic-Degrading Microorganisms. In *Phytoremediation of Soil and Water Contaminants;* Kruger, E. L., Anderson, T. A., Coats, J. R. Eds.; ACS Symposium Series 777. American Chemical Society: Washington, DC, 1997; pp 20–36.

Dixon, B. Bioremediation Is here to Stay. *ASM News* **1996,** *62,* 527–528.

Dua, M.; Sethunathan, N.; Johri, A. K. Biotechnology and Bioremediation: Successes and Limitations. *Appl. Microbiol. Biotechnol.* **2002,** *59,* 143–152.

El-Fakharany, I. I.; Massoud, A. H.; Derbalah, A. S.; Saad Allah, M. S. Toxicological Effects of Methomyl and Remediation Technologies of Its Residues in an Aquatic System. *J. Environ. Chem. Ecotoxicol.* **2011,** *3*(13), 332–339.

Enrica, G. The Role of Microorganism in Environmental Decontamination. In *Contaminants in the Environment—A Multidisciplinary Assessment of Risk to Man and Other Organisms;* Renzoni, A., Mattei, N., Lari, L., Fossi, C. Eds. Lewis Publishers, CRC Press: USA, 1994; pp 235–246.

Erickson, L. E. An Overview of Research on the Beneficial Effects of Vegetation in Contaminated Soil. *Ann. N.Y. Acad. Sci.* **1997,** *829*(1), 30–35.

Fan, C.; Liao, M.-C. The Mechanistic and Oxidative Study of Methomyl and Parathion Degradation by Fenton Process. *World Acad. Sci. Eng. Technol. Int. J. Chem. Mol. Nucl. Mater. Metall. Eng.* **2009,** *3*(11), 611–615.

FAOSTAT. *Pesticides (Use)*; Food and Agriculture Organization of the United Nations Statistics Division, 2016. http://faostat3.fao.org/download/R/RP/E (accessed July 5, 2016).

Fulekar, M. H. Bioremediation Technologies for Environment. Indian J. *Environ. Protect.* **2005,** *25*(4), 358–364.

Furukawa, K. Super Bugs for Bioremediation. *Trends Biotechnol.* **2003,** *21,* 187–190.

Gavrilescu, M. Fate of Pesticides in the Environment and Its Bioremediation. *Eng. Life Sci.* **2005,** *5*(6), 497–526.

Geels, F. P.; Schippers, B. Selection of Antagonistic Fluorescent *Pseudomonas* spp. and Their Root Colonization and Persistence Following Treatment of Seed Potatoes. *J. Phytopathol.* **1983,** *108,* 193–206.

Geetha, M.; Fulekar, M. H. Bioremediation of Pesticides in Surface Soil Treatment Unit Using Microbial Consortia. *Afr. J. Environ. Sci. Technol.* **2008,** *2*(2), 036–045.

Gibson, J.; Harwood, C. S. Metabolic Diversity in Aromatic Compounds Utilization by Anaerobic Microbes. *Annu. Rev. Microbiol.* **2002,** *56,* 345–369.

Gould, F.; Anderson, A.; Reynolds, A.; Bumgarner, L.; Moar, W. Selection and Genetic Analysis of a *Heliothis virescens* (Lepidoptera: Noctuidae) Strain with High Levels of Resistance to *Bacillus thuringiensis* Toxins. *J. Econ. Entomol.* **1995,** *88,* 1545–1559.

Graebing, P.; Frank, M.; Chib, J. S. Effects of Fertilizers and Soil Components on Pesticide Phyotolysis. *J. Agric. Food Chem.* **2002,** *50,* 7332–7339.

Griffiths, B. S.; Philippot, L. Insights into the Resistance and Resilience of the Soil Microbial Community. *FEMS Microbiol Rev.* **2012,** *37*(2), 112–129.

Haby, P. A.; Crowley, D. E. Biodegradation of 3-Chlorobenzoate as Affected by Rhizodeposition and Selected Carbon Substrates. *J. Environ. Qual.* **1996,** *25,* 304–310.

Hashmi, I. Microbiological Transformation of Hazardous Waste During Biological Treatment. PhD Thesis, Institute of Environmental Studies, University of Karachi, Pakistan, 2001.

Haygarth, P. M.; Ritz, K. The Future of Soils and Land Use in the UK: Soil Systems for the Provision of Land-Based Ecosystem Services. *Land Use Policy* **2009**, *26S*, S187–S197.

Hicks, R. J.; Stotzky, G.; Voris, P. V. Review and Evaluation of the Effects of Xenobiotic Chemicals on Microorganisms in Soil. *Adv. Appl. Microbiol.* **1990**, *35*, 195–253.

Hoagland, R. E.; Zablotowicz, R. M.; Locke, M. A. Propanil Metabolism by Rhizosphere Microflora. In *Bioremediation Through Rhizosphere Technology;* Anderson, T. A., Coats, J. R., Eds.; American Chemical Society: Washington, DC, 1994; pp 160–183.

Huesemann, M. H.; Trux, M. J. The Role of Oxygen Diffusion in Passive Bioremediation of Petroleum Contaminated Soils. *J. Hazard. Mater.* **1996**, *51*, 93–113.

IASRI. *State-wise Consumption of Pesticides (Technical Grade) in India*; Indian Agricultural Statistics Research Institute: New Delhi, India, 2016. .res.in/agridata/13data%5Cchapter2%5Cdb2013tb2_16.pdf (accessed Sept 23, 2016).

Indira Devi, P. Pesticides or "Healthicides"? An Attempt at Estimating the Health Costs of Pesticide Applicators, 2016. http://www.webmeets.com/files/papers/ERE/WC3/1084/Pesticide_Health_Cost_June%20Indira.pdf (accessed Feb 10, 2016).

Jacobsen, C. S. Plant Protection and Rhizosphere Colonization of Barley by Seed Inoculated Herbicide Degrading *Burkholderia (Pseudomonas) cepacia* DBO1 (pRO101) in 2,4-D Contaminated Soil. *Plant Soil* **1997**, *189*, 139–144.

Jaya, M. .; Rangaswamy, V. Biodegradation of Selected Insecticides by *Bacillus* and *Pseudomonas* sps in Ground Nut Fields. *Toxicol. Int.* **2009**, *16*(2), 127–132.

Jilani, S. Comparative Assessment of Growth and Biodegradation Potential of Soil Isolate in the Presence of Pesticides. *Saudi J. Biol. Sci.* **2013**, *20*, 257–264.

Jitender, K.; Kumar, J.; Prakash, J. Persistence of Thiobencarb and Butachlor in Soil Incubated at Different Temperatures. In *Integrated Weed Management for Sustainable Agriculture*, Proc. Indian Soc. Weed. Sci. Int. Seminar, Hisar, India, 1993; pp 123–124.

Kearney, P.; Wauchope, R. Disposal Options Based on Properties of Pesticides in Soil and Water. In *Pesticide Remediation in Soils and Water;* Kearney, P., Roberts, T. Eds.; Wiley Series in Agrochemicals and Plant Protection; John Wiley & Sons, Inc.: India, 1998.

Kibblewhite, M. G.; Ritz, K.; Swift, M. J. Abscessing the Impact of Agricultural Intensification on Biodiversity: A British Prospective. *Philos. Trans. R. Soc. London, Ser. B: Biol. Sci.* **2008**, *363*(1492), 777–787.

Kuiper, I.; Bloemberg, G. V.; Lugtenberg, B. J. J. Selection of a Plant–Bacterium Pair as a Novel Tool for Rhizostimulation of Polycyclic Aromatic Hydrocarbon-Degrading Bacteria. *Mol. Plant-Microbe Interact.* **2001**, *14*, 1197–1205.

Kuiper, I.; Lagendijk, E. L.; Bloemberg, G. V.; Lugtenberg, B. J. J. *Review:* Rhizoremediation: A Beneficial Plant–Microbe Interaction. *Mol. Plant-Microbe Interact.* **2004**, *17*(1), 6–15.

Kulkarni, A. G.; Kaliwal, B. B. Bioremediation of Methomyl by Soil Isolate—*Pseudomonas aeruginosa. IOSR J. Environ. Sci. Toxicol. Food Technol. (IOSR-JESTFT)* **2015**, *8*(12), 1–10.

Kumar, A.; Bisht, B. S.; Joshi, V. D.; Dhewa, T. Review on Bioremediation of Polluted Environment: A Management Tool. *Int. J. Environ. Sci.* **2011**, *1*(6), 1079–1096.

Lau, K. L. Tsang, Y. Y.; Chiu. S. W. Use of Spent Mushroom Compost to Bioremediate PAH Contaminated Samples. *Chemosphere* **2003**, *52*, 1539–1546.

Leung, K. T.; Nandakumar, K.; Sreekumari, K.; Lee, H.; Trevors, J. T. Biodegradation and Bioremediation of Organic Pollutants in Soil. In *Modern Soil Microbiology,* 2nd Ed.; van Elsas, J. D., Trevors, J. T., Jansson, J. K., Nannipieri, P. Eds.; CRC Press: USA, 2007.

Levin, L.; Viale, A.; Forchiassin, A. Degradation of Organic Pollutants by the White Rot Basidiomycete *Trametes trogii. Int. Biodeterior. Biodegrad.* **2003,** *52,* 1–5.

Madhu, K. B.; Wankhede, P. P.; Mankar, D. M.; Debbarma, K. Studies on Farmers Knowledge About Pesticide Usage. *Karnataka J. Agric. Sci.* **2014,** *27*(4), 545–547.

Maheshwari, R.; Singh, U.; Singh, P.; Singh, N.; Jat, B. L.; Rani, B. To Decontaminate Wastewater Employing Bioremediation Techniques. *J. Adv. Sci. Res.* **2014,** *5*(2), 7–15.

Maksymiv, I. Pesticides: Benefits and Hazards. *J. Vasyl Stefanyk Precarpathian Nat. Univ.* **2015,** *2*(1), 70–76.

Manzoor, S. A. Introduction and Applications of Bioremediation, 2011. http://www.biotecharticles.com/Environmental-Biotechnology-Article/Introduction-and-Applications-of-Bioremediation-1078.html (accessed Jan 12, 2016).

Mayer, A. M.; Staples, R. C. Laccase: New Functions for an Old Enzyme. *Phytochemistry* **2002,** *60,* 551–565.

McDonald, G. V. Soil Microorganisms, 2016. http://statebystategardening.com/state.php/newsletters/stories/soil_microorganisms/ (accessed Sept 23, 2016).

Mendez, M. O.; Maier, R. M. Phytostabilization of Mine Tailings in Arid and Semiarid Environments—An Emerging Remediation Technology. *Environ. Health Perspect.* **2008,** *116*(3), 278–283.

Mervat, S. M. Degradation of Methomyl by the Novel Bacterial Strain *Stenotrophomonas maltophilia* M1. *Electron. J. Biotechnol.* **2009,** *12*(4), 1–6.

Michel, F. C.; Reddy, C. A.; Forney, L. J. Microbial Degradation and Humification of the Lawn Care Pesticide 2,4-Dichlorophenoxyacetic Acid During the Composting of Yard Trimmings. *Appl. Environ. Microbiol.* **1995,** *61*(7), 2566–2571.

Miller, G. T. J.; Spoolman, S. E. Living in the Environment. 17th Ed.; Brooks/Cole, Cengage Learning: Belmont, CA, USA, 2012; p 800.

Müller, R.; Antranikian, G.; Maloney, S.; Sharp, R. Thermophilic Degradation of Environmental Pollutants, In: *Biotechnology of Extremophiles.* Antranikian G., Ed.; (Advances in Biochemical Engineering/Bio-technology) Ed.; Springer: Berlin, 1998; Vol. 61, pp 155–169.

Nannipieri, P.; Ascher, P. J.; Ceccherini, M. T.; Landi, L.; Pietramellara, G.; Renella, G. Microbial Diversity and Soil Functions. *Eur. J. Soil Sci.* **2003,** *54*(4), 655–670.

Naqvi, S. T. A. Biodegradation of Carbamates by soil Bacteria and Characterization of the Methyl Carbamates Degradation Hydrolase "Med" Heterologously Expressed in *Escherichia coli.* Ph.D. Thesis, Quaid-e-Azam University, Islamabad, Pakistan, 2010.

Navarro, S.; Vela, N.; Navarro, G. Review. An Overview on the Environmental Behaviour of Pesticide Residues in Soils. *Spanish J. Agric. Res.* **2007,** *5*(3), 357–375.

Nawaz, K.; Hussain, K.; Choudary, N.; Majeed, A.; Ilyas, U.; Ghani, A.; Lin, F.; Ali, K.; Afghan, S.; Raza, G.; Ismail Lashari, M. Eco-friendly Role of Biodegradation Against Agricultural Pesticides Hazards. *Afr. J. Microbiol. Res.* **2011,** *5*(3), 177–183.

Nerud, F.; Baldrian, J.; Gabriel, J.; Ogbeifun, D. Onenzymic Degradation and Decolorization of Recalcitrant Compounds. In *The Utilization of Bioremediation to Reduce Soil Contamination: Problems and Solutions;* Václav, S., Glaser, J. A., Baveye, P. Eds.;

NATO Science Series; Kluwer Academic Publisher, Springer: Netherlands, 2003; Vol. 19, pp 127–133.

Ngowi, A. V. F.; Mbise, T. J.; Ijani, A. S. M.; London, L.; Ajayi, O. C. Pesticides Use by Smallholder Farmers in Vegetable Production in Northern Tanzania. *Crop Protestation* **2007,** *26*(11), 1617–1624.

Nichols, T. D.; Wolf, D. C.; Rogers, H. B.; Beyrouty, C. A.; Reynolds, C. M. Rhizosphere Microbial Populations in Contaminated Soils. *Air Water Soil Pollut.* **1997,** *95*, 165–178.

Nyakundi, W. O.; Magoma, G.; Ochora, J.; Nyende, A. B. (2011). Biodegradation of Diazinon and Methomyl by White Rot Fungi From Selected Horticultural Farms in Rift Valley and Central Kenya. *J. Appl. Technol. Environ. Sanitation* **2007,** *1*(2), 107–124.

Omolo, K. M.; Magoma, G.; Ngamau, K.; Muniru, T. Characterization of Methomyl and Carbofuran Degrading-Bacteria from Soils of Horticultural Farms in Rift Valley and Central Kenya. *Afr. J. Environ. Sci. Technol.* **2012,** *6*(2), 104–114.

Or, D.; Smets, B. F.; Wraith, J. M.; Dechesne, A.; Friedman, S. P. Physical Constraints Affecting Bacterial Habitats and Activity in Unsaturated Porous Media—A Review. *Adv. Water Res.* **2007,** *30*(6–7), 1505–1527.

Picord, C.; Ponsonnet, C.; Paget, E.; Nesmex, T.; Simonet, P. Detection and Enumeration of Bacteria in Soil by Direct DNA Extraction and Polymerase Chain Reaction. *Appl. Environ. Microbiol.* **1992,** *58*, 2717–2719.

Pieper, D. H.; Reineke, W. Engineering Bacteria for Bioremediation. *Curr. Opin. Biotechnol.* **2000,** *11*(3), 262–270.

Pimentel, D. In *Effects of Pesticides on the Environment*. 10th International Congress on Plant Protection, Croydon, UK, 1983; Vol. 2, pp 685–691.

Prosser, J. I. Ecosystem Processes and Interactions in a Morass of Diversity. *FEMS Microbiol. Ecol.* **2012,** *81*(3), 507–519.

Puri, S. N. In *Present Status of Integrated Pest Management in India*. Paper Presented At Seminar on Integrated Pest Management (IPM), Asian Productivity Organization at Thailand Productivity Institute, Bangkok, 1998.

Qiu, X.; Shah, S. I.; Kendall, E. W.; Sorensen, D. L.; Sims, R. C.; Engelke, M. C. Grass-Enhanced Bioremediation for Clay Soils Contaminated with Polynuclear Aromatic Hydrocarbons. In *Bioremediation Through Rhizosphere Technology*; Anderson, T. A., Coats, J. R. Eds.; American Chemical Society: Washington, DC, 1994; pp 142–157.

Racke, K. D.; Skidmore, M. W.; Hamilton, D. J.; Unsworth, J. B.; Miyamoto, J.; Cohen, S. Z. Pesticide Fate in Tropical Soils. *Pure Appl. Chem.* **1997,** *69*, 1349–1371.

Rajendran, S. In *Environment and Health Aspects of Pesticides Use in Indian Agriculture*, Proceedings of the 3rd International Conference on Environment and Health, Chennai, India; Bunch, M. J., Suresh, V. M., Kumaran, T. V. Eds., 2003; pp 353–373.

Rani, K.; Dhania, G. Bioremediation and Biodegradation of Pesticide from Contaminated Soil and Water—A Novel Approach. *Int. J. Curr. Microbiol. Appl. Sci.* **2014,** *3*(10), 23–33.

Rao, M. A.; Scelza, R.; Scotti, R.; Gianfreda, L. Role of Enzymes in the Remediation of Polluted Environments. *J. Soil Sci. Plant Nutr.* **2010,** *10*(3), 333–353.

Raymond, D.; Jeffrey, K.; Alan, S. Detection of a Methyl Carbamate Degrading Gene in Agricultural Soils. *Commun. Soil Sci. Plant Anal.* **2003,** *34* (3, 4), 393–406.

Reddy, C. A.; Mathew, Z. Bioremediation Potential of White Rot Fungi. In *Fungi in Bioremediation;* Gadd, G. M. Ed.; British Mycological Society Symposia (No. 23); Cambridge University Press, Cambridge, UK, 2001; pp 52–78.

Richard, D. R.; Irinak, K.; Ashok, M.; WiHred, C. Biodegradation of Organophosphorus Pesticides by Surface-Expressed Organophosphorus Hydrolase. *Nat. Biotechnol.* **1997,** *15*, 984–987.

Ritmann, B. E.; Jacson, D. E.; Storck, S. L. Potential for Treatment of Hazardous Organic Chemicals With Biological Process. *Biotreatment Syst.* **1988,** *3*, 15–64.

Robert, C. *The Nuclear Fuel Cycle: Analysis and Management*; American Nuclear Society: La Grange Park, IL, 1999; pp 52–57. ISBN 0-89448-451-6.

Robertson, S. J.; McGill, W. B.; Massicotte, H. B.; Rutherford, P. M. Petroleum Hydrocarbon Contamination in Boreal Forest Soils: A Mycorrhizal Ecosystems Perspective. *Biol. Rev.* **2007,** *82*, 213–240.

Robinson, D. A.; Hockley, N.; Dominati, E.; Lebron, I.; Scow, K. M.; Reynolds, B.; Emmett, B. A.; Keith, A. M.; De Jonge, L. W.; Schjonning, P.; Moldrup, P.; Jones, S. B.; Tuller, M. Natural Capital, Ecosystem Services, and Soil Change: Why Soil Science Must Embrace an Ecosystems Approach. *Vadose Zone J.* **2012,** *11*(1). DOI: 10.2136/vzj2011.0051.

Rockne, K.; Reddy, K. R. *Bioremediation of Contaminated Sites*; University of Illinois at Chicago, 2003. http://tigger.uic.edu/~krockne/proceeding9.pdf#search=%22bioremediation%20of%20pesticides%20and%20herbicides%22 (accessed April 23, 2016).

Rossberg, M.; Lendle, W.; Togel, A. E. L.; Dreher, L. E.; Rassaeerts, H.; Kleinschmid, P.; Strack, H.; Beck, U.; Lipper, K. A.; Trokelson, T. R.; Loser, E.; Beutel, K. K. Chlorinated hydrocarbon. In *Ullmann's Encyclopedia of Industrial Chemistry;* Gerhartz, W. Ed.; VCH: Weinheim, Germany, 1986; pp 233–398.

Sayler, G. S.; Hooper, S. W.; Layton, A. C.; King, J. M. H. Catabolic Plasmids of Environmental and Ecological Significance. *Microb. Ecol.* **1990,** *19*, 1–20.

Schoefs, O.; Perrier, M.; Samson, R. Estimation of Contaminant Depletion in Unsaturated Soils Using a Reduced-Order Biodegradation Model and Carbon Dioxide Measurement. *Appl. Microbiol. Biotechnol.* **2004,** *64*, 256–261.

Schwab, A. P.; Banks, M. K. Biologically Mediated Dissipation of Polyaromatic Hydrocarbons in the Root Zone. In *Bioremediation Through Rhizosphere Technology*; Anderson, T. A., Coats, J. R. Eds.; American Chemical Society: Washington, DC, 1994; pp 132–141.

Schwab, A. P.; Banks, M. K.; Arunachalam, M. Biodegradation of Polycyclic Aromatic Hydrocarbons in Rhizosphere Soil. In *Bioremediation of Recalcitrant Organics*; Hinchee, R. E., Anderson, D. B., Hoeppel, R. E. Eds.; Battelle Memorial Institute: Columbus, OH, USA, 1995; pp 23–29.

Scott, C.; Pandey, G.; Hartley, C. L.; Jackson, C. J.; Cheesman, M. J.; Taylor, M. C.; Pandey, R.; Khurana, J. L.; Teese, M.; Coppin, C. W.; Weir, K. M.; Jain, R. K.; Lal, R.; Russell, R. J.; Oakeshott, R. G. The Enzymatic Basis for Pesticide Bioremediation. *Indian J. Microbiol.* **2008;** 48: 65–79.

Setyorini, D.; Prihatini, T.; Kurnia, U. (2002). Pollution of Soil by Agricultural and Industrial Waste. Extension Bulletins, FFTC Publication Database, Centre for Soil and Agroclimate Research and Development, Jalan Ir. Juanda, Indonesia. .http://www.agnet.org/library.php?func=view&style=type&type_id=&id=20110804163924&print=1 (accessed Dec 10, 2015).

Shah, S.; Thakur, I. S. Enzymatic Dehalogenation of Insecticide by *Pseudomonas fluorescens* of the Microbial Community from Tannery Effluent. *Curr. Microbiol.* **2003,** *47*, 65–70.

Shann, J. R.; Boyle, J. J. Influence of Plant Species on In Situ Rhizosphere Degradation. In *Bioremediation Through Rhizosphere Technology*; Anderson, T. A., Coats, J. R. Eds.; American Chemical Society: Washington, DC, 1994; pp 70–81.

Shannon, M. J.; Unterman, R. Evaluating Bioremediation: Distinguishing Fact From Fiction. *Annu. Rev. Microbiol.* **1993**, *47*, 715–736

Sharholy, M.; Ahmad, K.; Mahmood, G.; Trivedi, R. C. Municipal Solid Waste Management in Indian Cities—A Review. *Waste Manage.* **2008**, *28*, 459–467.

Singh, D. K. Biodegradation and Bioremediation of pesticide in Soil: Concept, Method and Recent Developments. *Indian J. Microbiol.* **2008**, *48*, 35–40.

Smith-Greeier, L. L.; Adkins, A. Isolation and Characterization of Soil Microorganisms Capable of Utilizing the Herbicide Dichloro-*p*-methyl as a Sole Source of Carbon and Energy. *Can. J. Microbiol.* **1996**, *42*, 221–226.

Sorek, A.; Atzmon, N.; Dahan, O.; Gerstl, Z.; Kushisin, L.; Laor, Y.; Mingelgrin, U.; Nasser, A.; Ronen, D.; Tsechansky, L.; Weisbrod, N.; Graber, E. R. "Phytoscreening": The Use of Trees for Discovering Subsurface Contamination by VOCs. *Environ. Sci. Technol.* **2008**, *42*(2), 536–542.

Sparks, T.C.; Graves, J.B.; Leonard, B.R. Insecticide Resistance and the Tobacco Budworm: Past, Present, and Future. *Rev. Pestic. Toxicol.* **1993**, *2*, 149–183.

Starner, K.; Kuivila, K. M.; Jennings, B.; Moon, G. E. Degradation Rates of Six *Pesticides in Water from the Sacramento River, California. Published by U.S. Geological Survey Toxic Substances Hydrology Program—Proceedings of the Technical Meeting, Charleston, South Carolina, March 8–12, 1999, Vol. 2. Contamination of hydrologic Systems and Related Ecosystems, 1999*; U.S. Geological Survey Water-Resources Investigations Report 99-4018 B, 1999. http://ca.water.usgs.gov/archive/reports/wrir994018/CA-0216.pdf (accessed Nov 26, 2015).

Stroud, J. L.; Paton, G. I.; Semple, K. T. Microbe–Aliphatic Hydrocarbon Interactions in Soil: Implications for Biodegradation and Bioremediation. *J. Appl. Microbiol.* **2007**, *102*(5), 1239–1253.

Subramanian, M. O.; David, J.; Shanks, J. V. TNT Phytotransformation Pathway Characteristics in *Arabidopsis*: Role of Aromatic Hydroxylamines. *Biotechnol. Prog.* **2006**, *22*(1), 208–216.

Tayade, S.; Patel, Z. P.; Mutkule, D. S.; Kakde, A. M. Pesticide Contamination in Food: A Review. *IOSR J. Agric. Vet. Sci.* **2013**, *6*(1), 7–11.

Tsai, L.; Olson, B. H. Detection of Low Number of Bacterial Cells in Soils and Sediments by Polymerase Chain Reaction. *Appl. Environ. Microbiol.* **1992**, *58*, 754–758.

United States Environmental Protection Agency (USEPA). *Databases Related to Pesticide Risk Assessment*; Washington, DC, 2016. https://www.epa.gov/pesticide-science-and-assessing-pesticide-risks/databases-related-to-pesticide-risk-assessment (accessed April 23, 2016).

Valentin, L.; Nousiainen, A.; Mikkonen, A. Introduction to Organic Contaminants in Soil: Concepts and Risks. In *Emerging Organic Contaminants in Sludges: Analysis, Fate and Biological Treatment;* Vicent, T., Caminal, G., Eljarrat E., Barceló, D. Eds.; *The Handbook of Environmental Chemistry*; Springer-Verlag: Berlin: Heidelberg, 2013; pp 1–29.

Van-Camp, L.; Bujarrabal, B.; Gentile, A. R.; Jones, R. J. A.; Montanarella, L.; Olazabal, C.; Selvaradjou, S. K. Reports of the Technical Working Groups Established Under the

Thematic Strategy for Soil Protection. Vol.-4. Contamination and Land Management. EUR 21319 EN/4, Office for Official Publications of the European Communities: Luxembourg, 2004. http://eusoils.jrc.ec.europa.eu/esdb_archive/Policies/STSWeb/vol4.pdf (accessed April 23, 2016).

Vancura, V.; Hovadik, A. Root Exudates of Plants II. Composition of Root Exudates of Some Vegetables. *Plant Soil* **1965,** *22,* 21–32.

Vidali, M. Bioremediation—An Overview. *Pure Appl. Chem.* **2001,** *73*(7), 1163–1172.

WHO. *Environmental Health Criteria for Methomyl; 178.* World Health Organization: Geneva, 1996. http://www.inchem.org/documents/ehc/ehc/ehc178.htm (accessed April 23, 2016).

Young, I. M.; Crawford, J. W. A Review: Interactions and Self-Organization in the Soil–Microbe Complex. *Science* **2004,** *304,* 1634–1637.

Zablotowicz, R. M.; Hoagland, R. E.; Locke, M. A. Glutathione *S*-Transferase Activity in Rhizosphere Bacteria and the Potential for Herbicide Detoxification. In *Bioremediation Through Rhizosphere Technology;* Anderson, T. A., Coats, J. R. Eds.; American Chemical Society: Washington, DC, 1994; pp 184–198.

Zhang, J. L.; Qiao, C. L. Novel Approaches of Remediation for Pesticide Pollutants. *Int. J. Environ. Pollut.* **2002,** *18*(5), 423–433.

Zhang, W.; Jiang, F.; Ou, J. *Global Pesticide Consumption and Pollution: With China as a Focus,* 2011. http://agris.fao.org/agris-search/search.do?recordID=CN2011200030 (accessed April 23, 2016).

INDEX

...d by CPI Group (UK) Ltd, Croydon, CR0 4YY
23/10/2024
01777703-0013